Books are to be returned
the la

LIVERPOOL JMU LIBRARY

3 1111 01161 4987

PHYTOREMEDIATION OF TOXIC METALS

PHYTOREMEDIATION OF TOXIC METALS

Using Plants to Clean Up the Environment

ILYA RASKIN
Rutgers University
New Brunswick, New Jersey

BURT D. ENSLEY
Phytotech, Inc.
Monmouth Junction, New Jersey

A WILEY-INTERSCIENCE PUBLICATION

JOHN WILEY & SONS, INC.
New York • Chichester • Weinheim • Brisbane • Singapore • Toronto

This book is printed on acid-free paper. ∞

Copyright © 2000 by John Wiley & Sons, Inc. All rights reserved.

Published simultaneously in Canada.

No part of this publication may be reproduced, stored in a retrieval system or transmitted in any form or by any means, electronic, mechanical, photocopying, recording, scanning or otherwise, except as permitted under Sections 107 or 108 of the 1976 United States Copyright Act, without either the prior written permission of the Publisher, or authorization through payment of the appropriate per-copy fee to the Copyright Clearance Center, 222 Rosewood Drive, Danvers, MA 01923, (978) 750-8400, fax (978) 750-4744. Requests to the Publisher for permission should be addressed to the Permissions Department, John Wiley & Sons, Inc., 605 Third Avenue, New York, NY 10158-0012, (212) 850-6011, fax (212) 850-6008, E-Mail: PERMREQ@WILEY.COM.

For ordering and customer service, call 1-800-CALL-WILEY.

Library of Congress Cataloging-in-Publication Data:

Raskin, Ilya.
    Phytoremediation of toxic metals : using plants to clean up the environment / Ilya
Raskin, Burt D. Ensley.
      p. cm.
    Includes index.
    ISBN 0-471-19254-6 (cloth : alk. paper)
      1.Phytoremediation. 2. Metals--Environmental aspects. I. Title. II. Ensley, Burt D.
    (Burt DeWitt), 1948-

    TD192.75 .R37 2000
    628.5'2 21--dc21   99-043427

Printed in the United States of America.

10 9 8 7 6 5 4 3 2 1

CONTENTS

CONTRIBUTORS

Judith Auer Shaw, Department of Environmental Protection, Trenton, New Jersey

Alan J. M. Baker, School of Biological Sciences, Department of Animal and Plant Sciences, The University of Sheffield, United Kingdom

W. R. Berti, Principal Researcher, DuPont Co., Newark, Delaware

Scott P. Bizily, Genetics Department, University of Georgia, Athens, Georgia

Michael J. Blaylock, Phytotech Inc., Monmouth Junction, New Jersey

Christopher S. Cobbett, Department of Genetics, University of Melbourne, Parkville, Australia

Scott D. Cunningham, Principal Researcher, DuPont Co., Newark, Delaware

M. P. de Souza, Department of Plant and Microbial Biology, University of California, Berkeley, California

Slavik Dushenkov, Phytotech Inc., Monmouth Junction, New Jersey

Burt D. Ensley, Phytotech Inc., Monmouth Junction, New Jersey

David J. Glass, President, D. Glass Associates, Inc., Needham, Massachusetts

Peter B. Goldsbrough, Purdue University, Department of Horticulture, West Lafayette, Indiana

Mary Lou Guerinot, Dartmouth College, Biological Sciences Department, Hanover, New Hampshire

Ray Hinchman, Energy Systems Division, Argonne National Laboratory, Argonne, Illinois

Jianwei W. Huang, Phytotech Inc., Monmouth Junction, New Jersey

Yoram Kapulnik, Institute of Field and Garden Crops, ARO, The Volcani Center, Bet Dagan, Israel

Ute Krämer, Universität Bielefeld, Bio IV-Plant Physiology and Biochemistry, Fakultät fuer Biologie-W5, Bielefeld, Germany

Richard B. Meagher, Genetics Department, University of Georgia, Athens, Georgia

M. Cristina Negri, Energy Systems Division, Argonne National Laboratory, Argonne, Illinois

E. A. H. Pilon-Smits,[*] Department of Plant and Microbial Biology, University of California, Berkeley, California

Roger D. Reeves, Department of Chemistry and Biochemistry, Massey University, Palmerston North, New Zealand

Steve Rock, U.S. EPA National Risk Management, Research Laboratory, Cincinnati, Ohio

Clayton L. Rugh, Genetics Department, University of Georgia, Athens, Georgia

David E. Salt, Chemistry Department, Northern Arizona University, Flagstaff, Arizona

Philip G. Sayre, U.S. EPA National Risk Management, Research Laboratory, Washington, DC

Jerald Schnoor, Department of Civil and Environmental Engineering, University of Iowa, Iowa City, Iowa

Norman Terry, Department of Plant and Microbial Biology, University of California, Berkeley, California

Robert K. Tucker, Ecopolicy Center, Cook College, Rutgers University, New Brunswick, New Jersey

[*]E. A. H. Pilon-Smits is currently at Colorado State University, Department of Biology, Fort Collins, Colorado 80523.

PREFACE

Soils and waters contaminated with toxic metals pose a major environmental and human health problem that needs an effective and affordable technological solution. The partial success of microbial bioremediation has been limited to the degradation of some organic contaminants and has been ineffective at addressing the challenge of toxic metal contamination, particularly in soils. Current methodologies for remediating toxic metal-polluted soils rely mainly on excavation and burial at a hazardous waste site at an average cost of $1 million per acre. Many sites remain contaminated with no remediation in sight simply because it is too expensive to clean them up with the available technologies.

Phytoremediation may provide an economically viable solution for remediating some of these sites. Specifically, several subsets of metal phytoremediation have been developed and targeted for commercialization.

1. Phytoextraction, in which high-biomass, metal-accumulating plants and appropriate soil amendments are used to transport and concentrate metals from the soil into the above-ground shoots, which are harvested with conventional agricultural methods

2. Rhizofiltration, in which plant roots grown in aerated water, precipitate and concentrate toxic metals from polluted effluents

3. Phytostabilization, in which plants stabilize the pollutants in soils, thus rendering them harmless

4. Phytovolatilization, in which plants extract volatile metals (e.g., mercury and selenium) from soil and volatilize them from the foliage

Clearly, phytoextraction of metals represents one of the largest economic opportunities because of the size and scope of environmental problems associated with metal-contaminated soils and the competitive advantage offered by a plant-based technology. Plants are particularly well suited for phytoextraction, since they not only take up toxic metals from soil but can also deposit them into tissues that can be easily harvested. Soils contaminated with lead, cadmium, chromium, and radionuclides can be found in almost every country. Phytoextraction has a potential to develop into a cost-effective "green" remediation technology based on the use of specially selected or engineered plants to remove toxic metals, including radionuclides, from soils and water. This emerging technology takes advantage of the fact that a living plant can be compared to a solar-driven pump, which can extract and concentrate particular elements from the environment. Throughout millions of years of evolution plants have perfected this ability, which is essential to plant growth and reproduction. Elemental metals cannot be degraded by living organisms. Therefore, they can only be remediated by concentration and, in the special case of mercury and selenium, by volatilization. Bacteria and fungi can degrade many organic pollutants and accumulate many metals probably better than many plants. However, these microbes cannot be used for the effective phytoextraction of metals since they cannot be separated from soil. Only plants can do this job well by using their roots as giant underground collecting systems and their shoots as harvestable metal concentrators. Phytoextraction tricks the plant into accumulating environmentally toxic metals in addition to or instead of the essential nutrients it needs.

The successful transfer of phytoremediation from the laboratory to the field is a crucial step in the development of this technology. Also a receptive regulatory environment will foster its use. Plants have an enviable position of being the only cheap, renewable resource available to our civilization. Gradual depletion of natural resources pushes scientists to explore the use of plants and their products as substitutes for oil- and mining-based commodities and as alternatives to costly and energy-intensive engineering processes. It may be the right time to utilize the plant's potential for the metal remediation and to develop engineering and biological approaches that further improve this process.

This book is dedicated to discussing the progress that occurred in the area of metal phytoremediation. The basic idea that plants can be used for environmental remediation is certainly very old and cannot be traced to any particular source. However, a series of fascinating scientific discoveries combined with interdisciplinary research approaches allowed the development of this idea into a promising environmental technology. Phytoremediation research is rapidly gaining momentum in companies and universities. The number of phytoremediation-related publications is growing yearly, as is funding for this research. By the mid-1990s, the efforts of a handful of groups led by Scott Cunningham, Rufus Chaney, Alan Baker, Milton Gordon, Gary Banuelos, Norman Terry, Richard Meagher, Roger Reeves, and others led to the creation of a new scientific discipline of metal phytoremediation with its own annual conferences, meeting sessions, workshops, information networks, and calls for proposals.

Ilya Raskin
Rutgers University

Burt D. Ensley
Phytotech, Inc.

PHYTOREMEDIATION OF TOXIC METALS

SECTION I

INTRODUCTION

1

RATIONALE FOR USE OF PHYTOREMEDIATION

BURT D. ENSLEY, PH.D.

INTRODUCTION

The discovery of metal hyperaccumulating properties in certain plants has suggested that it may be feasible to use these plants for the cleanup of heavy-metal contamination in soil and water. This approach to environmental restoration has been termed phytoremediation. The relatively low potential cost of phytoremediation allows the treatment of many sites that cannot be addressed with currently (at the time of writing) available methods, and phytoremediation preserves the topsoil and reduces the amount of hazardous materials generated during cleanup. The economic and environmental advantages provide an excellent reason for the use of this approach in the treatment of contaminated sites.

Heavy-metal and radionuclide contamination in soils, surface water, and groundwater poses major environmental problems worldwide with substantial economic consequences. Since most current technologies cannot selectively remove heavy metals, many contaminated sites can be remediated only by using labor-intensive and costly excavation and landfilling technology. In contrast, phytoremediation uses plants to accumulate heavy metals from contaminated soil and water. These plants can be grown and harvested economically, leaving only residual levels of pollutants.

The goal of current phytoremediation efforts is to develop innovative, economical, and environmentally compatible approaches to remove heavy metals from the

Phytoremediation of Toxic Metals: Using Plants to Clean Up the Environment, edited by Ilya Raskin and Burt D. Ensley.
ISBN: 0-471-19254-6 ©2000 John Wiley & Sons, Inc.

environment. The exploration of this use of plants has led to the development of two active phytoremediation methods, termed *phytoextraction* and *rhizofiltration* (Salt et al., 1995). The important features of these approaches include lower costs for treatment, generation of a potentially recyclable metal-rich plant residue, applicability to a range of metals and radionuclides, minimal environmental disturbance, reduced secondary airborne or waterborne wastes, and increased public acceptance.

DEFINITION OF PHYTOREMEDIATION

The terms *bioremediation* and *phytoremediation* are occasionally used as if they were interchangeable. They actually designate quite different concepts and potential applications.

Bioremediation

Bioremediation companies often employ microbe-based technology for the degradation of organic compounds such as polycyclic aromatic hydrocarbons (PAHs). Such approaches have no effect on most heavy metals. Phytoremediation of metals and bioremediation of organics may soon be used together on sites contaminated with both types of material.

Phytoremediation for Nonmetals

Several organizations and individuals are currently developing plant-based remediation technology. In the context of this technology, the term *phytoremediation* refers to the use of green or vascular plants for removal of organic contaminants. Plants have been used to treat organic materials found in petroleum, nitrotoluenes, a basic constituent of a class of explosives, and chlorocarbons such as poly(chlorinated biphenyls) (PCBs). It may also be possible to apply phytoextraction and organic phytoremediation to clean sites containing both hydrocarbons and heavy metals.

Non-Metal-Removal Use of Plants

A specialized application is the use of trees such as poplars to control the movement of water underground. Rows of fast-growing poplars with deep roots are able to absorb and transpire large quantities of water from the roots through shoots into the atmosphere. This can be an effective tool in immobilizing water-soluble contamination and preventing migration, but it does not remove contaminants such as heavy metals from the site.

PHYTOEXTRACTION

Phytoextraction is the use of metal-accumulating plants that can transport and concentrate metals from the soil to the roots and aboveground shoots. By screening and selection procedures scientists have identified promising metal-accumulating lines of crop plants. These plants can remove heavy metals from soils by concentrating the metals in their aboveground stems and leaves (Kumar et al., 1995).

In developing phytoextraction, researchers have focused on plant species that have such desirable characteristics as high biomass production, handling ease, genetic characterization, and established cultivation practices. Many scientists are now investigating approaches to increasing the phytoextraction capacity of plants. Screening isolates the more desirable plants, and efforts have been directed at identifying soil amendments and metal-uptake-enhancing chemicals to increase the efficiency of phytoextraction. The observation that the addition of the chelators to contaminated soil during phytoextraction increases the uptake and transport of lead into the shoots of plants provided a major advance in phytoextraction technology (Huang et al., 1997; Blaylock et al., 1997).

Phytoextraction may potentially be used in large-scale, cost-effective, on-site treatment because of its relatively low cost. Topsoil would be preserved and the amount of hazardous materials generated reduced significantly. This environmentally compatible approach is gaining acceptance by the public and regulatory agencies, and may eventually emerge as the preferred method of treating sites contaminated with heavy metals or radionuclides.

RHIZOFILTRATION

Rhizofiltration is the use of plant roots to absorb, concentrate, and precipitate heavy metals from water. The ability to absorb different metal ions is a common property of all plant roots but varies between plant species. The roots of sunflowers have been used to treat water containing lead, uranium, strontium, cesium, cobalt, and zinc to concentrations below the accepted water standards (Dushenkov et al., 1995).

A rhizofiltration approach has been used to remove uranium from groundwater on sites at Ashtabula, OH (Dushenkov et al., 1997a), and Oak Ridge, TN, and to remove cesium and strontium from a pond near the Chernobyl reactor (Dushenkov et al., 1997b). The presence of other metal ions does not substantially interfere with rhizofiltration of lead, cesium, or strontium. This suggests that rhizofiltration can be used to remove target metals from mixtures containing different metal ions that are likely to occur in surface runoff and groundwater.

SITES TREATABLE BY PHYTOREMEDIATION

The U.S. Environmental Protection Agency (EPA) Comprehensive Environmental Response Compensation Liability Information System (CERCLIS) estimates that there are 30,000 candidate sites for hazardous waste treatment services in the United States alone, including industrial sites containing water and solid waste contaminated with heavy metals. Potential sites would include those owned or contaminated by battery manufacturers; electroplating, metal finishing, and mining companies; producers of solvents, coated glass, paints, leather, and chemicals; as well as municipal, state, and federal agencies responsible for the reclamation of previously contaminated sites such as landfills, railroad yards, and abandoned manufacturing facilities. The EPA, which administers, supervises, or approves the cleanup of the so-called Superfund sites, has a unique role as both the largest customer and the agency generating the regulations defining this market. The costs estimated for the treatment of sites contaminated with heavy metals, and heavy metals mixed with organics over a 5-year period are shown in Table 1-1 (U.S. EPA, 1995).

The types of site that can be treated by phytoremediation include the following:

Superfund.   National Priority (Superfund) sites ranked on the National Priorities List (NPL) are some of the most contaminated and hazardous waste sites in the United States. Current cleanup costs for all Superfund sites are estimated at approximately $16.5 billion. Approximately 15% of these sites are contaminated only by heavy metals, whereas 64% of the sites contain heavy metals mixed with organic waste.

RCRA.   Hazardous waste treatment, storage, and disposal facilities are regulated by the Resource Conservation Recovery Act (RCRA). Between 1500 and 3000 of these sites are expected to require corrective action under RCRA to remove contaminants. The wastes have characteristics similar to those of Superfund wastes; 64% of the sites contain metals mixed with organics and 15% are contaminated solely by heavy metals.

DOD.   The Department of Defense (DOD) has identified 7313 sites covering 26,000 acres that are contaminated with hazardous wastes. The total cleanup

TABLE 1-1.   Projected 5-Year Costs for Remediation of Metal-Containing Sites ($ U.S. Millions)

Sector	Metals and Organics	Metals Only
Superfund	$10,400	$2,400
RCRA	12,800	3,000
DOD	2,400	400
DOE	6,500	900
State	800	200
Private Sector	2,500	200
Total	$35,400	$7,100

Source: U.S. EPA 542-R-92-012, 1993 and Hazardous Waste Remediation Project, 1992, *Hazardous Waste Remediation: The Task Ahead*. Waste Management and Education Institute, The University of Tennessee, Knoxville.

costs for DOD sites are estimated at approximately $25 billion, with approximately $14 billion in remedial action. DOD sites contaminated with heavy metals represent 11% of the total, with 15% of that number contaminated with metals alone.

DOE. The Department of Energy (DOE) has identified 4000 contaminated sites, with 23 sites on the Superfund list. Some of these sites contain radioactive materials, making handling and treatment particularly difficult and expensive. Of those DOE sites that have been evaluated, 53% contain heavy-metal or radionuclide contamination, with 7% of the sites contaminated by metals alone.

State Sites. It has been estimated that 19,000 state-funded contaminated sites require remediation. Expenditures currently total approximately $427 million for all the states. The EPA estimates that 38% of state-funded sites contain heavy metals mixed with organic waste and 9% contain metals alone.

Private-Party Sites. Eventually an estimated 24,000 sites will require remediation with funding from private parties, at an estimated total cost of approximately $24 billion.

Radioactive Contamination

Remediation of soil and water contaminated with radioactive elements represents up to 50% of the estimated $230 billion that the DOE expects (at the time of writing) to spend over the next 75 years (U.S. DOE, 1995). Up to $100 million per year may eventually be spent on the domestic and international application of phytoremediation to the treatment of such contamination, because alternative solutions are so expensive. For example, the excavation, packaging, transport, analysis, and disposal at a low-level radioactive waste landfill of uranium-contaminated soil costs at least $1600 per cubic yard (U.S. DOE, 1996).

It may also be possible to use phytoextraction and rhizofiltration in the treatment of areas contaminated by the Chernobyl reactor accident and in the treatment of certain mixed wastes containing radionuclides that pose a serious problem at many sites. *Mixed waste* is defined as waste that displays hazardous characteristics, must be disposed of as hazardous waste, and contains a radioactive element in sufficient quantities to require disposal as a low-level radioactive waste.

FIELD TRIALS

The efficacy of phytoremediation has been demonstrated by scientists and engineers at public and private institutions through a series of field trials of heavy-metal and radionuclide removal from soil and water. These trials establish the field performance characteristics of phytoremediation methods developed in the laboratory and greenhouse, and present opportunities to

- Validate the technology applications and assumptions
- Educate the public about phytoremediation
- Accrue on-site experience
- Evaluate methods and approaches
- Identify development and research needs and priorities
- Provide actual field results
- Train personnel
- Evaluate performance
- Generate data for models and estimates of costs and performance

ELEMENTS OF A PHYTOEXTRACTION PROJECT

The specialized elements of a phytoremediation project include treatability testing, site layout and design, supply and application of proprietary materials and seeds, field engineering, and metals analysis. A phytoextraction project often begins with a treatability study. Soil samples from the contaminated site are analyzed in a greenhouse to determine the potential effectiveness of phytoextraction with that particular soil–contaminant mixture. Although treatability studies can be time-consuming and unattractive to site owners because of the extra expense, two seemingly similar sites can produce markedly different growth and metal uptake performance by the plants. A disappointing outcome on a full-scale project can damage the reputation of the technology and the provider, but can be prevented by adequate testing prior to field work.

The site layout and remediation design includes the design of monitoring and irrigation systems based on the results of the treatability study. Monitoring and modeling ensures the contamination does not migrate out of the root zone, which typically corresponds to the depth of contamination. Field engineering includes periodic visits by staff responsible for inspection of the site prior to cultivation, seeding, examination of plants, application of soil amendments, supervision of harvest, and technical guidance on biomass disposal.

Nonspecialized site activities such as routine construction, installation, monitoring, and operation can be conducted by either a phytoremediation provider or a remediation contractor. Collaboration between a specialized technical provider and a remediation contractor can reduce project costs, since remediation contractors bring extensive site remediation experience, properly trained personnel, and aggressive pricing for their part of a project.

Once on-site activities begin, project management requires close cooperation between a phytoremediation technology provider and a remediation contractor performing the basic services and maintenance essential to complete the project. The primary functions performed by a remediation contractor on a phytoremediation project are site preparation, regulatory compliance, cultivation of crops, harvest and disposal of biomass, metal analysis, and site closure.

Site preparation involves preparing the area for initial cultivation. Cultivation encompasses tilling, installation of irrigation and monitoring systems, power and water hookups, inspection of plants during the growing season, and the application of fertilizers and insecticides. Mature plants are harvested and prepared for shipment and disposal.

The administrative and regulatory responsibilities include all work related to filing the remediation program (if necessary), quality assurance and control (QA/QC), health and safety, certification of employees regarding environmental and OSHA training, documentation regarding contaminant levels, and compliance with regulatory issues. These activities commence prior to actual site work and continue after the project is completed.

Analytical work covers the sampling and analysis of soil in an EPA-certified analytical laboratory. To confirm that the remediated soil achieves the regulatory standards, the final responsibility of the contractor is site closure, which includes the transmission of data and reports to the appropriate regulatory agency to obtain certification that the site has been successfully treated.

THE ECONOMICS OF PHYTOREMEDIATION

Phytoextraction

Phytoremediation is proposed to be a cost-effective alternative for the treatment of contaminated soil. The most widely utilized alternative to phytoextraction is excavation and disposal of hazardous soil. A direct comparison of the costs associated with landfill–excavation and phytoremediation reveals what is believed to be the clear economic advantage of phytoextraction. The estimated expenses incurred in the remediation of a site contaminated with lead using the conventional excavation–landfill approach most commonly practiced in the United States are approximately $150–$350 per ton. The four primary variables that determine the cost for this technique are the total acreage of the contaminated site, the depth of the excavation required to remove the contamination, the distance to the landfill, and the facility cost of stabilizing and landfilling per ton or cubic yard of soil.

The primary variables in the cost of phytoremediation are the total acreage of the contaminated site, depth of the contamination, effectiveness of the plants in removing the contamination, and the total amount of contaminant to be removed. These factors determine the total number of crops per project. The estimated costs for the phytoremediation treatment of a lead-contaminated site, including off-site disposal of the biomass as hazardous waste, are $20–$80 per ton.

Rhizofiltration

Heavy metals in industrial process water and groundwater are most commonly removed by precipitation or flocculation, followed by sedimentation and disposal of the resulting sludge. For many of these streams, treatment is relatively simple and inexpensive, costing a few cents up to $1.50 per thousand gallons of water processed.

The special characteristics of some heavy metals in water make these streams much more difficult and expensive to treat. The precipitation of some metals requires large amounts of flocculating agents, resulting in the formation of heavy sludges and high disposal costs. The treatment of chelated copper in certain rinse waters can cost up to $40 per pound of copper removed ($350 per thousand gallons at 1000 ppm Cu). Over 80% of that cost is for disposal of the hazardous sludge. An ion-exchange process for treating chelated copper has recently been developed by Bio-recovery Systems. This method, which assumes that the copper can be electrochemically recovered from the residue to eliminate disposal, reduces treatment costs to a still-expensive $15 per pound of copper removed. Rhizofiltration has targeted costs of $1–$5 per pound of metal removed for treatment of chelated copper in water.

Because of their highly toxic nature, certain metals must be "polished" after conventional treatment to meet discharge standards. An important example of this class of contaminant is radionuclide waste. Precipitation and filtration to recover radionuclides from water has been projected to cost over $1 million per year to treat a relatively small flow (2 gal/min); this estimate was based primarily on the cost of radioactive sludge disposal. Rhizofiltration offers a cost advantage in water treatment because plants can remove up to 60% of their dry weight as heavy metals, thus markedly reducing the disposal cost of the hazardous or radioactive residue.

Rhizofiltration is also expected to be a cost-competitive technology in the treatment of surface water or groundwater containing low, but significant, concentrations of heavy metals such as chromium, lead, and zinc.

REFERENCES

Blaylock, M. J. et al., 1997. Enhanced accumulation of Pb in Indian mustard by soil applied chelating agents. *Environ. Sci. Technol.* **31**:860–865.

Dushenkov, S. et al., 1997a. Removal of uranium from water using terrestrial plants. *Environ. Sci. Technol.* **31**:3468–3474.

Dushenkov, S. et al., 1997b. Phytoremediation: A novel approach to an old problem. In *Global Environmental Biotechnology*, D. L. Wise, ed., Elsevier Science B.V., Amsterdam, pp. 563–572.

Dushenkov, S. et al., 1995. Rhizofiltration: The use of plants to remove heavy metals from aqueous streams. *Environ. Sci. Technol.* **29**:1239–1245.

Huang, J. W. et al., 1997. Phytoremediation of lead-contaminated soils: Role of synthetic chelates in lead phytoextraction. *Environ. Sci. Technol.* **31**:800–805.

Kumar, P. et al., 1995. Phytoextraction: The use of plants to remove heavy metals from soils. *Environ. Sci. Technol.* **29**:1232–1238.

Salt, D. et al., 1995. Phytoextraction: A novel strategy for the removal of toxic metals from the environment using plants. *Biotechnology* **13**:468–474.

U.S. DOE, 1995. *Estimating the Cold War Mortgage. The 1995 Baseline Environmental Management Report.* United States Department of Energy Center for Environmental Management Information, DOE/EM-0232.

U.S. EPA, 1993. *Cleaning up the Nation's Waste Sites: Markets and Technology Trends.* Office of Solid Waste and Emergency Response, Technology Innovation Office (OS-110W) Washington, DC, EPA 542-R-92-012.

AVRIL ROBARTS LRC

SECTION II

ENVIRONMENTAL POLLUTION AND GREEN PLANTS

2

ECONOMIC POTENTIAL OF PHYTOREMEDIATION

DAVID J. GLASS, PH.D.

INTRODUCTION

In the past several years (at the time of writing), phytoremediation has attracted the interest of private and industrial site owners, regulators, and the environmental engineering community as a potentially useful remediation tool, but it remains at a very early stage of commercial development. This chapter assesses how well phytoremediation competes against the large number of conventional and innovative treatment technologies that are available or being developed today, and will assess the extent of the remediation market that might be available to phytoremediation in the coming years. The chapter analyzes potential markets for phytoremediation for several specific remediation scenarios. These include: remediation or immobilization of metals or radionuclides from soils, using phytoextraction (Salt et al., 1995), phytostabilization (Cunningham et al., 1996), or phytovolatilization (Rugh et al., 1996); rhizofiltration of metals or radionuclides from aqueous solutions (Salt et al., 1995); and the use of deep-rooted trees to remove large volumes of water from aquifers to treat organic contaminants (Schnoor et al., 1995).

GENERAL FEATURES OF PHYTOREMEDIATION

Table 2-1 lists some of the general advantages and disadvantages associated with phytoremediation. The pros and cons of phytoremediation for specific applications are examined in the case studies below.

Phytoremediation of Toxic Metals: Using Plants to Clean Up the Environment, edited by Ilya Raskin and Burt D. Ensley.
ISBN: 0-471-19254-6 ©2000 John Wiley & Sons, Inc.

TABLE 2-1.   General Advantages and Disadvantages of Phytoremediation

Advantages	Disadvantages
Cost	Time
Low capital and operating costs	Slower than some alternatives; seasonally
Metal recycling provides further economic advantages	dependent
	Many natural hyperaccumulators are slow growers
Performance	Performance
Permanent treatment solution	Biological methods are not capable of
In situ application avoids excavation	100% reduction
Capable of remediating bioavailable fraction	May not be applicable to all mixed wastes
	High metal concentrations or other
Capable of mineralizing organics	contaminants may be toxic
Applicable to a variety of contaminants, including some recalcitrants	Soil phytoremediation applicable only to surface soils
Trees are capable of high hydraulic pumping pressures	Space
	Groundwater, wastewater applications may require large available surface area
Other	
Public acceptance; aesthetically pleasing	Other
Compatible with risk-based remediation, brownfields	Need to displace existing facilities (e.g., wastewater treatment)
Can be used during site investigation or after closure	Regulators may be unfamiliar with the technology and its capabilities
	Lack of recognized economic performance data

General Advantages

The most widely touted advantage of phytoremediation is its cost; phytoremediation is generally believed to entail capital and operating costs far lower than those of many competing technologies. Discounting the costs of monitoring the site, the major costs of phytoremediation are the tilling and preparation of the soil, planting the seeds, weed and pest control, and harvesting and disposal of the biomass. However, at the present time there are few documented estimates of the cost of phytoremediation, and this is one factor that has hindered its more widespread acceptance. Table 2-2 lists some of the published cost estimates for phytoremediation of soils and water. These estimates derive from several different economic assumptions and involve several different contaminants (e.g., metals vs. organics) and have been left in their original units, but some comparisons are possible. The consensus estimate for the cost of phytoremediation of soils is about $25–$100 per ton; the cost of phytoremediation of

TABLE 2-2.   Cost Estimates of Phytoremediation

	Soils
$1–$10/m³	($2500–15,000/ha; S. Cunningham, DuPont)
$15–$20/ton	(E. Drake, Exxon)
$25–$50/ton	(Phytotech)
$29–$48/m³	($60,000–$100,000/acre at depth of 50 cm; Salt et al.)
$80/yd³	(R. Levine, DOE)
$100–$150/m³	($200,000–$300,000/acre; R. Chaney, USDA)
	Water
$0.64 per 1000 gal treated	(V. Medina, EPA)
$2.00–$6.00 per 1000 gal treated	(Phytotech)

waters is likely in the range of $0.60–$6.00 per 1000 gal treated; the low end of these ranges are organic compounds, and the high end are metals and radionuclides. As will be seen in the text that follows, these figures indeed compete well with many alternative technologies in specific markets.

In addition to cost, phytoremediation offers a number of performance-based advantages. It can be a permanent treatment solution (i.e., permanently removing the contaminant from the media), and as an in situ method it holds several advantages, including the fact that remediation can take place without unduly disturbing the site. Because phytoremediation can generally remediate the fraction of the contamination that is most bioavailable, it is compatible with risk-based remediation approaches that allow the unleachable fraction to remain in the soil or aquifer. For this reason, phytoremediation is generally compatible with Brownfields remediations, where target cleanup levels may be more lenient. Finally, remediating a hazardous waste site with plants is believed to be more aesthetically pleasing and possibly more acceptable to the public than other remediation methods.

General Disadvantages

If cost is the greatest potential advantage of phytoremediation, time is the greatest disadvantage. Use of plants for remediation at many sites will not be fast—one estimate is that 18–60 months may be needed for site closure in some cases (Neidorf, 1996). In addition to the general reliance on the life cycles of plants and the need to coordinate plantings with ordinary growing seasons, many naturally occurring hyperaccumulator plants are extremely slow growers—it has been estimated that natural hyperaccumulators might take 13–16 years to clean a typical site (Boyd, 1996; Salt et al., 1995).

The use of plants, like all biological methods, does not allow 100% removal or reduction of contaminants, because reaction rates decrease as concentrations of contaminant decrease. In addition, high levels of heavy-metal contaminants may be toxic to plants. Finally, soil phytoremediation is generally believed to be limited in applicability to surface soils (i.e., perhaps the top 1 m of the soil), and would be limited by the solubility or availability of the contaminant (which is especially a problem with metals).

Among the strategies being used to overcome these disadvantages are the use of chelators to enhance metal solubility in soils (Salt et al., 1995; Huang et al., 1997), combining phytoremediation with other in situ technologies, and the selection or breeding of improved plant varieties, using classical genetics or advanced molecular biology tools (Raskin, 1996; Cunningham and Ow, 1996; Rugh et al., 1996). Because the regulatory and public acceptance barriers are not as severe, it is reasonable to expect that transgenic plants will find quicker and wider use in phytoremediation than have engineered microbes in bioremediation (Glass, 1997a).

COMPANIES DEVELOPING PHYTOREMEDIATION

Overview

In just a few years, a commercial industry has begun to develop, primarily in the United States, for the application of phytoremediation. This industry is developing along the lines previously seen for the microbial bioremediation industry in the United States (Glass et al., 1995). Much of the innovation and research is being driven by smaller companies, whereas certain diversified environmental engineering firms have become involved in field or full-scale projects. Also involved are a number of industrial site owners, as well as universities, nonprofit laboratories, and government agencies.

Dedicated Phytoremediation Companies

We can define "dedicated" phytoremediation companies as those whose sole, or primary, treatment technology is phytoremediation, although we also include some small companies working more broadly in plant-related activities such as constructed wetlands. Table 2-3 lists dedicated phytoremediation firms in the United States and Europe.

Phytotech, Inc. (Monmouth Junction, NJ), founded on work from Ilya Raskin's laboratory at Rutgers University, is developing plants for remediation of metals and radionuclides from soil and water, and has performed field demonstrations of the use of *Brassica* species to remove lead from soils, and the use of sunflowers to remove uranium and cesium from aqueous wastestreams. PhytoWorks, Inc. (Gladwyne, PA) is focusing on both the remediation of organics and phytoremediation of mercury; the latter is based on the work of Richard Meagher (Rugh et al., 1996), with the goal of

TABLE 2-3. For-Profit Companies Developing Phytoremediation

Dedicated Phytoremediation Companies	Diversified Consulting/Engineering Companies	Industrial Companies
United States	ARCADIS Geraghty & Miller	Alcoa
Applied Natural Sciences	CH2M Hill	Arco Chemical
EarthCare	Geomatrix Consultants	BP Amoco
Ecolotree	Geosyntec Consultants	Chevron
Phytokinetics	MSE Technology	DuPont
Phytotech	Applications	Exxon
PhytoWorks	Parsons Engineering	Kaiser Aluminum
Thomas Consultants	Roy F. Weston	Monsanto
Verdant Technologies		Occidental Petroleum
Viridian Resources		Rhone-Poulenc
Overseas		Union Carbide
Aquaphyte Remediation (Canada)		
BioPlanta (Germany)		
Consulagri (Italy)		
OEEL (U.K.)		
Piccoplant (Germany)		
Plantechno (Italy)		
Slater (U.K.) Ltd.		

creating transgenic plants capable of extracting and volatilizing mercury in reduced form. EarthCare, Inc. (Venison, TX) is also attempting to address both metals and organic contaminants, using a variety of plants and grasses. Viridian Resources, LLC (Houston, TX) has recently acquired rights to metal phytoremediation technology developed at the U.S. Department of Agriculture. Phytoremediation of organic contaminants in groundwater is being pursued by Applied Natural Sciences (Hamilton, OH), Ecolotree (Iowa City, IA), and Phytokinetics, Inc. (Logan, UT). Phytokinetics is also using grasses and other plants to stimulate rhizosphere biodegradation of organics. As shown in Table 2-3, at least seven companies outside the United States are developing phytoremediation, although most maintain expertise in other commercial uses of plants such as constructed wetlands, reed beds, or agriculture. One company, Slater (U.K.) Ltd., was founded to develop technologies for metal phytoremediation.

Although dedicated phytoremediation companies can be the drivers of innovation, these companies face substantial barriers to success. Many of these companies are collaborating with consulting/engineering (C/E) firms. Although the C/E firms offer access to national markets and the ability to compete for large sites, these relationships carry the risk that the C/E firms will eventually be able to commercialize the technology on their own, without the dedicated company, as has happened in other segments of the site remediation market. Thus, dedicated firms need to develop proprietary technologies that can be protected by patents or trade secrets, to avoid losing market share to the C/E firms or to companies with access to the right type of plant or tree (e.g., nurseries) once the skills to practice the technology become more widespread.

It is, however, difficult to develop proprietary technologies in environmental markets. In general, the small profit margins of the remediation business make it very difficult to support expensive research and development (R&D) activities, and the marketplace is still somewhat hostile to innovative technologies. Furthermore, companies offering one or a very small number of remedial options are disadvantaged because they cannot compete for remediation jobs for which their technology is inappropriate. The phytoremediation industry may evolve, as has the microbial bioremediation industry before it, to the point where the market is dominated by the diversified C/E firms, as it is today; in fact, dedicated microbial bioremediation firms have shown a very high failure rate (Glass et al., 1995), as a result of many of the same factors discussed here.

Diversified Consulting/Engineering Companies

Table 2-3 also shows several diversified consulting/engineering firms that are known to have phytoremediation capabilities or that have worked with dedicated phytoremediation firms. For example, Geosyntec Consultants and Geraghty & Miller have in-house expertise from prior work with managed or constructed wetlands. One joint effort relating to metals is the collaboration between MSE Technology Applications, Inc., Phytotech, Argonne National Labs, and Cornell University on a treatability study of phytoremediation for radionuclides.

As noted in the text above, C/E firms are generally better positioned to commercialize treatment technologies for hazardous wastes, because they are capable of applying or investigating a variety of remedial techniques at any given site and have the size and marketing muscle to address sites anywhere in the country. Today, relatively few C/E firms have credible phytoremediation expertise, but that is already beginning to change as the parameters to practice the technology are better understood and more widely disseminated. It is reasonable to predict that C/E firms may someday dominate the phytoremediation market.

Industrial Companies

A number of large industrial companies, principally in the oil or chemicals industry, are also active in conducting or supporting phytoremediation research (Table 2-3), with several—such as Kaiser Aluminum and Alcoa—involved in metals mining or manufacture. In addition, DuPont has been one of the pioneers in phytoremediation of metals, through the work of Scott Cunningham and colleagues on lead removal.

These companies are interested primarily in phytoremediation for its possible use on their own contaminated sites, and are either funding internal research or are collaborating with dedicated phytoremediation companies on such research projects. Several of these industrial companies have entered consortia to support phytoremediation research; for example, a Petroleum Environmental Research Forum working group conducts research on phytoremediation of hydrocarbon contaminated soils and sludges, and the Phytoremediation of Organics Action Team was organized by the Environmental Protection Agency (EPA) under its Remediation Technologies Development Forum in 1997.

Universities, Nonprofit Labs, and the Government

Almost all the pioneering work in phytoremediation took place in academic laboratories, and today many university research groups produce valuable basic and applied research in this field (Glass, 1997b). Several government agencies and laboratories are conducting or funding research on phytoremediation, notably EPA, the U.S. Department of Agriculture, the Department of Energy (DOE) and its National Laboratories, and several branches of the Defense Department (DOD). In addition, EPA is evaluating phytoremediation at four sites in its Superfund Innovative Technology Evaluation (SITE) program (Rock and Beckman, 1997), one of which involves the use of phytoextraction to treat lead, cadmium, and hexavalent chromium at a former metal plating facility. This level of government support is similar to that that has been devoted to other emerging technologies such as microbial bioremediation, and may ultimately help phytoremediation's market acceptance.

ASSESSING THE MARKET POTENTIAL OF PHYTOREMEDIATION

Case Studies

The remainder of this chapter examines five potential markets for phytoremediation, as case studies of its commercial potential. First, we attempt to quantify the total market opportunity (i.e., 100% market share) for each potential market sector. Next, we analyze how phytoremediation measures up against available alternative treatment technologies, using several factors, including cost, efficiency, time, and overall performance criteria. Many of these factors are fairly subjective, and are reviewed on a pro-and-con basis for each case study market. This leads to predictions of the market for phytoremediation over the coming years, in each of the sectors and for the overall market. These case study markets are

- Remediation of organics from groundwater
- Remediation of metals from soils
- Remediation of metals from groundwater

AVRIL ROBARTS LRC

- Treatment of metals in industrial wastewater
- Remediation of radionuclides (soils and water)

No treatment technology is capable of garnering anywhere near 100% market share within any given market sector. Each sector includes existing well-entrenched methodologies that, for the most part, are effective in meeting regulatory endpoints (e.g., chemical treatment for industrial wastewater, landfilling and containment for soil remediation, and pump-and-treat methods for groundwater remediation). These entrenched technologies make up a large percentage of the market for each category, leaving little room for the numerous alternative or innovative treatment technologies vying for this remaining market share. No innovative technology will be able to muster more than a few percent of the overall market in any sector [e.g., after nearly 10 years' serious consideration as a viable treatment alternative, microbial bioremediation makes up perhaps 2% of the overall site remediation market (Glass, et al., 1995)].

Overall Hazardous Waste Remediation Market

The total U.S. market for hazardous site remediation in 1997 was in the vicinity of $8–$9 billion per year, after several years of flat or even negative growth (Basta and Veasey, 1996; The Hazardous Waste Consultant, 1996). The U.S. EPA (1997) contains data on hazardous waste sites subject to the Superfund law, the Resource Conservation and Recovery Act (RCRA), and DOE and DOD remediation programs, from which one can estimate how this market breaks down into discrete segments. Although many sites include more than one type of media, the relative proportions at Superfund, RCRA, DOE and DOD sites are approximately 50% groundwater, 45% soils, and 5% sludges and sediments. However, in considering the sites nationwide that are actually being remediated, a greater percentage is likely attributable to groundwater, so we can use an estimated ratio of 60% groundwater to 30% soils.

Similarly, the EPA (U.S. EPA, 1997) presents reasonably detailed figures on the presence of metals at sites in these programs and in federal and state underground storage tank programs, and a relative ratio of 90% organics to 10% metals can be calculated on the basis of the number of sites.[1] However, the bias in favor of organics is due to the many (> 160,000) underground storage tank sites, which have little, if any, metal contamination, and because these sites are generally the least expensive to clean up, they contribute proportionally less to the dollar value of the overall market. A calculation weighing the relative ratio of metals to organics against the estimated cost to remediate these sites leads to an estimate that 35% of the dollar value of the overall remediation market is attributable to metals and 65% to organics.

---

[1]Many sites include both organics and metals. We have therefore used relative ratios, calculating the ratio $M/(M+O)$, where M = the number of sites with metal contamination and O = the number of sites with organic contamination (where $M + O$ is greater than the total number of sites as a result of the overlap), treating these sites separately in the calculation, to create a fraction by which to divide total dollars spent on remediation.

These calculations lead to the following estimates of the total market sizes for the sectors of concern to this analysis:

- $3.0–$3.5 billion per year for organic contaminants from groundwater
- $0.8–$1.0 billion per year for metals from soils
- $1.6–$2.0 billion per year for metals from groundwater

Remediation of Organics from Groundwater

The total U.S. market of as much as $3.5 billion per year makes this sector the most attractive of the case studies (although it does not relate to metals), and indeed, a substantial amount of remediation activity in the United States is directed at organic contaminants in groundwater. The most common conventional technologies are various pump-and-treat techniques, where methods such as activated carbon, ion exchange, or UV treatment are employed to treat contaminated water above ground. The pump-and-treat method is used alone or in combination with other methods at 99% of Superfund sites with groundwater contamination (U.S. EPA, 1997). In situ technologies such as soil vapor extraction, bioventing, and conventional biostimulation are also being increasingly used, either alone or with pump-and-treat techniques (U.S. EPA, 1997).

As discussed in Glass (1997b), phytoremediation, at an estimated cost for organics of $0.64 per 1000 gal, measures up well against several alternatives described in two published cost comparisons (Medina and McCutcheon, 1996; Quinton et al., 1997). Furthermore, phytoremediation can be used in combination with other in situ techniques such as natural attenuation or biostimulation. However, the numerous methods currently used for groundwater remediation are better studied and accepted than phytoremediation, creating a substantial barrier to market entry [see Glass (1997b) for a fuller description].

Phytoremediation of organics in groundwater nevertheless remains the best established application of the technology today, and numerous full-scale cleanups have already been conducted by companies such as Ecolotree and Applied Natural Sciences. We estimate the 1997 market in this sector to be $2–$3 million. Considering its positive track record, this market share should grow steadily in years to come, perhaps to $35–$70 million by 2005.

Remediation of Metals from Soils

Use of phytoremediation for metals-contaminated soils presents a U.S. market opportunity of just below $1 billion per year. Phytoremediation is a viable technology in this marketplace because there are relatively few available alternatives, and those that are practiced tend to be quite expensive (see Table 2-4). The most common methods of remediation of metals-contaminated soils are landfilling and containment (i.e., fixation or capping); innovative techniques such as acid extraction, soil washing, and in situ flushing also being used (U.S. EPA, 1997; Salt et al., 1996). Other

TABLE 2-4. Available Technologies for Remediation of Metals and Radionuclides in Soil and Water

Soils[a]	Cost	Water[b]	Cost
Chemical treatment	$100–$500	Ion exchange	$0.13–$1.75
Soil washing	$75–$200	Electrochemical reduction	$0.60
Soil flushing (in situ)	$40–$190	Adsorption	$1.00–$20.00
Vitrification (reagent)	$75–$90	Membrane	$1.20–$5.80
Vitrification (thermal)	$250–$425	Separation–	
Thermal desorption	$150–$500	filtration	
Thermal treatment	$170–$300	Chemical	$0.80–$1.20
Electrokinetics	$20–$200	precipitation	
Incineration	$200–$1500	Reverse osmosis	$3.00–$6.20
Landfilling	$100–$500	Activated carbon	$119–$210
		Biosorption	$9–$3400
Phytoremediation	$25–$100		
		Phytoremediation	$0.60–$6.00

[a]All figures in dollars per ton.

[b]All figures in dollars per 1000 gal treated.

techniques shown in Table 2-4 are being practiced or developed for in situ or ex situ treatment of metals-bearing soils. Although cost estimates vary considerably (primarily because of the differences in chemistry and physical properties among different metals), the technologies requiring excavation are generally greater than $100/ton (except soil washing), whereas in situ technologies can cost less than $100/ton. Phytoremediation, at the consensus estimate of $25–$100 per ton, is quite competitive with these techniques.

Phytoremediation is attractive for metal remediation for other reasons as well. As with all in situ methods, phytoremediation can be used at sites where excavation is not practical or possible. The ability to recover and recycle metals from the plant biomass is a potential added advantage for those metals for which a market exists [one estimate, cited in Watanabe (1997), is that zinc and cadmium removed from a typical site could have a resale value of $1069 per hectare]. However, the need to harvest the biomass, and in some cases to dispose it as hazardous waste subject to RCRA standards, creates an added cost that can be a potential drawback to the technology.

Among other potential disadvantages, phytoremediation can be slower than other methods and more seasonally dependent because plants may not grow well at lower soil temperatures that are nevertheless sufficiently high to allow the practice of other techniques. Other factors are the slow growth of natural hyperaccumulators, the limitation of phytoremediation to the upper soil layers, and the potentially limited applicability of phytoremediation to mixed wastes.

Phytoremediation of metals from soils has been successfully demonstrated in the field, and is being aggressively promoted. We estimate that the 1997 market was low, probably $1–$2 million per year, but the potential for growth is great because of these favorable factors and because revenue per site is likely to be substantial (i.e., a few successful jobs can lead to a substantial increase in the overall market size). We estimate that the market may grow to $15–$25 million by the year 2000 and to $70–$100 million by 2005.

Remediation of Metals from Groundwater

The U.S. market for remediation of metals from groundwater may be as high as $2.0 billion per year. Much of this results from common industrial sources, but much metals-contaminated groundwater arises from acid mine drainage (polluted water from mining that contains sulfur, iron, and other metals) and acid rock drainage (acidic runoff from mining that often leaches minerals from rock). In the United States, it has been estimated that $1 million is spent per day to clean up 12,000 mi of rivers and 180,000 acres of lakes that are contaminated by mining wastes, and the total cleanup costs of acid mine drainage have been estimated to be as much as $70 billion. The U.S. Bureau of Mines estimates that 5–10% of the nation's one million mining-related sites may pose environmental problems (Thomson and Turney, 1995). In Canada, there are estimated to be over 12,500 ha of tailings and 740 million tonnes (metric tons) of waste mine rock in nine provinces, with an estimated total cleanup cost of CA $2–$5 billion (MEND, 1997). A reasonable estimate would be that the annual U.S. market for AMD cleanup alone is $500–$700 million, representing a large portion of the overall market for metals in groundwater.

The most common method for cleanup of metals-contaminated water, particularly AMD, is chemical treatment, in which metals are precipitated from solution using basic reagents such as hydroxides (e.g., lime), to form sludge, which is then disposed as a hazardous waste, or in some cases reused or recycled. AMD is commonly treated using constructed wetlands, in which plant and microbial systems are used to remove or precipitate metals from solution, sometimes combined with limestone or other precipitation systems. Other treatment methods for aqueous wastestreams are also being used or investigated (Table 2-4). Chemical treatment remains the favored approach because it is reliable, efficient, and reasonably inexpensive ($0.80–$1.20 per 1000 gal). Phytoremediation, at $6.00 or less per 1000 gal, is potentially cost-competitive. However, one drawback is that it cannot be used with highly concentrated wastestreams such as AMD, since high metal concentrations can be toxic to plants, and can generally be used only with shallow aqueous wastestreams, and thus would not be favored for deeper aquifers, except via an aboveground flow-through reactor in a pump-and-treat system. Although the use of constructed wetlands for AMD creates a favorable precedent for the applicability of phytoremediation, it remains to be seen how well metal hyperaccumulators can be incorporated into a constructed wetland, or grow under the conditions where mining wastes are found.

AVRIL ROBARTS LRC

For these reasons, we believe that phytoremediation will be slow to be adopted for removal of metals from groundwater. The 1997 market was probably less than $0.5 million, and growth only as high as $1–$3 million may be possible by 2005.

Treatment of Metals in Industrial Wastewater

In its regulations under the Clean Water Act (40 CFR 403-471), EPA divides the industrial wastewater market into categories grouped by industry. Those whose wastestreams include significant amounts of heavy metals include electroplating and metal finishing; metal machinery and manufacturing (a regulatory category still under development); various types of mining, battery manufacture, metal molding, and casting; and several types of chemical manufacturing. EPA published a series of regulatory development documents in the 1970s and 1980s surveying all the regulated industries and characterizing the size of the industry and the amount and composition of the wastestreams (e.g., U.S. EPA, 1984). The data in these documents can be used to estimate the total amount of effluent produced in each segment in a year, and by applying an arbitrary cost of treatment (e.g., $2 per 1000 gal of water), one can estimate the total treatment costs for that subsector. This calculation places the larger market sectors at several hundred million dollars per year each and the smaller ones at several tens of millions of dollars per year. Altogether, the totals of all the 40 or 50 discrete industry segments total $5 billion, a figure consistent with several published estimates of the total industrial wastewater treatment market. Applying this methodology to the major metals-bearing industrial wastestreams, the U.S. market for treatment of metals is likely in the range of $1–$2 billion per year.

The water treatment technologies shown in Table 2-4 are applicable to metals-bearing industrial wastestreams. As noted above, chemical treatment is the most common method by far, and most existing waste treatment plants use this method. Some of the other methods shown in the table, such as filtration, are being used, and are generally cost-effective, but their usefulness may be limited by problems such as membrane fouling, requiring frequent replacement of the filtration unit, and by other factors.

Phytoremediation is cost-competitive with these techniques, including chemical treatment, and often involves much lower capital expenses. Phytoremediation stacks up well on a performance basis, since the efficiencies of many of these techniques may vary considerably according to the chemistry of the metal being treated (Glass, 1997b). Phytoremediation eliminates sludge creation, and plant biomass is generally easier or cheaper to handle or dispose of than is sludge. The technology can also be used in combination with other treatment technologies, for example, as a polishing step to remove relatively low metal concentrations remaining in treated wastestreams.

In spite of these advantages, it is difficult to see phytoremediation gaining a significant share of this market, at least not in the United States and other developed countries where industrial facilities have already made substantial capital investments in conventional technologies to comply with government regulations; displacing such investments will be difficult. Other difficulties that can be foreseen include the possible need for large surface areas for the treatment unit (depending on the uptake rate of the plants), and the fact that plants, as living biological systems, are more easily damaged or stressed than are physicomechanical systems. We therefore estimate the 1997 market to be essentially zero (i.e., less than $100,000 per year), with only very modest potential for growth in the future. It is difficult to envision a U.S. market of greater than $2 million by 2005.

Remediation of Radionuclides (Soils and Water)

One sector of the market for metals remediation is the remediation of radioactive metals (i.e., radionuclides). This market is dominated by the large potential market for cleanup of federal facilities operated by the U.S. Department of Energy (DOE), although there is also a smaller market of privately owned sites. The DOE is committed to an aggressive 75-year program to clean all its sites (U.S. DOE, 1995), but the continuance of this program in the coming years and decades at current funding levels is clearly dependent on budgetary and political pressures (in fact, such pressures have led DOE more recently to revise its plan so that most clean-up activities are completed by 2006).

According to several published estimates, DOE might need to spend a total of $10–$22 billion on actual remediation costs at these facilities, but these estimates do not necessarily distinguish between the radioactive contamination at these facilities and that contamination that is solely hazardous. According to a recent DOE publication (U.S. DOE, 1997), there are 1.8 billion m^3 of contaminated water and 79 million m^3 of contaminated solid media at these DOE sites. Considering only those fractions containing radioactive waste (either alone or in combination with hazardous wastes), there are 400 million gallons of contaminated water and 67 million m^3 of solid media. Estimating the cost of the needed cleanup, using the conservative figures of $5 per 1000 gal to treat the water and $200 per cubic meter to treat the soils and solid media, gives an overall cost of about $15 billion, which is in line with the estimates cited above. Given that DOE originally expected most of the actual remediation activities to be completed in 20–25 years (U.S. DOE, 1995), this leads to an average of $700–$750 million per year (now higher in view of the expedited schedule). We can expect the early years to be far below this average but costs at the peak years to exceed this, reaching as high as $1 billion per year. Including the private sites, we feel that an estimate of $300–$500 million per year is a reasonable estimate of today's market.

The relative economics of phytoremediation for radionuclide remediation have been discussed above: Phytoremediation in both soils and water should be cost-competitive with the few other technologies that are available (although radionuclide cleanup costs will be at the high side of the various cited ranges). In fact, there are few available technologies for radionuclide remediation, and those that are available are not entirely satisfactory. Phytoremediation, as an in situ technology, avoids increased exposure that would be caused by disruption of soils by excavation, and the concentration of radioactivity in plant biomass should reduce the ultimate cost of disposal.

Phytoremediation nevertheless suffers from many of the drawbacks cited above for metals, as well as some other drawbacks. The radioactivity of the biomass will require worker protection during harvest and will likely require some way of limiting access of animals to the plants during remediation. In addition, the highly hazardous nature of radioactive material means that the residual presence of even extremely low concentrations of radionuclide can still present a substantial public health risk, and phytoremediation, as a biological method, will never be able to approach 100% removal of the contaminant.

Nevertheless, this sector of the market shows the same potential for dramatic growth as do metals in general, for the simple reason that the choice of phytoremediation for a small number of sites is likely to have a large impact on the dollar value of the market. However, the technology has not yet been as well demonstrated in the field as have other uses of phytoremediation, and the current market is probably fairly small (less than $500,000 per year in 1997). We predict rapid growth once the technology is used at radioactive waste sites, and the market could reach $4–$10 million by 2000 and $40–$80 million by 2005.

Other Markets

Phytoremediation is being used for a variety of groundwater-related activities (e.g., vegetative caps, treatment of landfill leachate) not considered in the first case study, and the technology may also be used as a temporary or interim solution at small, privately owned sites where other remediation methods are not practical because of cost. Plants are also used for removal or degradation of organics from soil, and for the stimulation of microbial biodegradation. Finally, the phytoremediation market overlaps to some extent with existing uses of constructed wetlands, which are not considered in this study. Several of these applications are among the major current uses for phytoremediation and represent some of the largest potential markets (Table 2-5; Glass, 1998a; and 1998b); however, as they do not relate to metal remediation, they are outside the scope of this chapter.

Overall Market Estimates

Table 2-5 shows the cumulative estimates for these case studies and the other uses. The 1997 U.S. phytoremediation market was about $7–$14 million, most of which is attributable to remediation of organics from groundwater and the other market sectors discussed above. Growth to $55–$103 million may be seen by 2000, and the market in 2005 may reach $214–$370 million. The most rapid growth should be seen in metals and radionuclide remediation, given the higher costs of these cleanups, but steady, substantial growth should be seen in the better established sectors relating to remediation of organics.

TABLE 2-5.   U.S. Phytoremediation Market Estimates for 1997–2005 (in millions of U.S. dollars)

	1997	2000	2005
Metals from soils	1–2	15–25	70–100
Metals from groundwater	0–0.4	0.4–1.0	1–3
Metals from wastewater	0–0.1	0.1–0.5	1–2
Radionuclides	0–0.5	4–10	40–80
Organics from groundwater	2–3	15–30	35–70
Other	4–8	20.5–36.5	67–115
Total	7–14	55–103	214–370

Source: D. Glass Associates, Inc.

These estimates assume very rapid growth as a result of dramatically increased technology adoption, particularly in the late 1990s. Such increased acceptance is dependent on a number of factors, including regulatory and public acceptance and the inherent capabilities of the technology (and the ability to prove such performance with believable data). We remain optimistic that phytoremediation will overcome these obstacles, and become an established, accepted treatment technology in the years to come.

International Markets

This chapter has limited its analysis to U.S. markets. The United States is the world's single largest environmental market, accounting for 35–45% of total world environmental spending (OECD, 1996). Generally speaking, the size of the remediation market in the entire European Union (EU) is just below half that of the United States, although most EU countries have only recently begun, or have yet to begin, conducting inventories of their hazardous waste sites, and enacting soil or groundwater protection legislation (Glass et al., 1995). Therefore, although the size of the market is equivalent, the forces driving remediation markets in Europe tend to lag behind those of the United States. On the other hand, the European market (and others around the world) may be more receptive to innovative technologies than has the United States. There is a small, generally anecdotal record of phytoremediation projects in Europe (e.g., Phytotech's activities in the Ukraine), leading to reason for optimism about phytoremediation's prospects, but at this point we feel that the market is too small to analyze or to make predictions for the future.

REFERENCES

Basta, N., and D. Veasey, 1996. Environmental firms react to a cooler market. *Chem. Eng.* (Jan.):37–43.

Boyd, V., 1996. Pint-sized plants pack a punch in fight against heavy metals. *Environ. Protect.* (May):38–39.

Cunningham, S. D., et al., 1996. Phtyoremediation of soils contaminated with organic pollutants. *Adv. Agron.* **56**:55–114.

Cunningham, S. D., and D. W. Ow, 1996. Promises and prospects of phytoremediation. *Plant Physiol.* **110**:715–719.

Glass, D. J., et al., 1995. International activities in bioremediation: Growing markets and opportunities. In *Applied Bioremediation of Petroleum Hydrocarbons*, R. E. Hinchee et al., eds., Battelle Press, Columbus, OH, pp. 11–33.

Glass, D. J., 1997a. Prospects for use and regulation of transgenic plants in phytoremediation. In *In situ and On-Site Bioremediation*, Vol. 4, B. C. Alleman and A. Leeson, eds., Battelle Press, Columbus, OH, pp. 51–56.

Glass, D. J., 1997b. Evaluating phytoremediation's potential share of the hazardous site remediation market. *Proc. 2nd Internatl. Conf. Phytoremediation.* IBC Library Service, Westborough, MA, pp. 7–39.

Glass, D. J., 1998a. The 1998 United States Market for Phytoremediation. D. Glass Associates, Inc., Needham, MA.

Glass, D. J., 1998b. Phytoremediation applications. *Genetic Eng. News*, September 15, 1998, p. 17.

Huang, J. W., et al., 1997. Phytoremediation of lead contaminated soils: Role of synthetic chelates in lead phytoextraction. *Environ. Sci. Technol.* **31**:800–805.

Medina, V. F., and S. C. McCutcheon, 1996. Phytoremediation: modeling removal of TNT and its breakdown products. *Remediation*, (Winter):31–45.

Mine Environment Neutral Drainage (MEND) 1997. Background document, available at http://www.nrcan.gc.ca/mets/mend/brief-e.htm (Oct. 1997).

Neidorf, R., 1996. Phytoremediation: Phytotech's hungry plants take up toxic metals. *EI Digest* (April):7–11.

Organization for Economic Cooperation and Development, 1996. *The Global Environmental Goods and Services Industry.* OECD, Paris.

Quinton, G. E., et al., 1997. A method to compare groundwater cleanup technologies. Paper presented at 4th Internatl. Bioremediation Symp., New Orleans, LA, April 30, 1997.

Raskin, I., 1996. Plant genetic engineering may help with environmental cleanup. *Proc. Natl. Acad. Sci.* (USA) **93**:3164–3166.

Rock, S., and S. Beckman, 1997. Phytoremediation field demonstrations in the U.S. EPA SITE Program. In *In situ and On-Site Bioremediation*, Vol. 3, B. C. Alleman and A. Leeson, eds., Battelle Press, Columbus, OH, p. 323.

Rugh, C. L., et al., 1996. Mercuric ion reduction and resistance in transgenic *Arabidopsis thaliana* plants expressing a modified bacterial merA gene. *Proc. Natl. Acad. Sci.* (USA) **93**:3182–3187.

Salt, D. E., et al., 1995. Phytoremediation: A novel strategy for the removal of toxic metals from the environment using plants. *Bio/Technol.* **13**:468–474.

Schnoor, J. L., et al., 1995. Phytoremediation of organic and nutrient contaminants. *Environ. Sci. Technol.* **29**:318A–323A.

The Hazardous Waste Consultant, 1996. Little prosperity in U.S. hazardous waste industry despite overall growth in environmental market. *Haz. Waste Consult.* (Nov./Dec.):1.16–1.17.

Thomson, B. M., and W. R. Turney, 1995. Minerals and mine drainage. *Water Environ. Res.* **67**(4):527–529.

U.S. Department of Energy, Office of Environmental Management, March 1995. *Estimating the Cold War Mortgage: The 1995 Baseline Environmental Management Report*, DOE/EM-0232.

U.S. Department of Energy, Office of Environmental Management, Jan. 1997. *Linking Legacies: Connecting the Cold War Nuclear Weapons Production Processes to Their Environmental Consequences*, DOE/EM-0319.

U.S. Environmental Protection Agency, Effluent Guidelines Division, Aug. 1984. Development document for effluent limitations guidelines and standards for battery manufacture, EPA 4401/1-84/067.

U.S. Environmental Protection Agency, Office of Solid Waste and Emergency Response, April 1997. Cleaning up the nation's waste sites: Markets and technology trends, EPA 542-R-96-005.

Watanabe, M. E., 1997. Phytoremediation on the brink of commercialization. *Environ. Sci. Technol.* **31**(4):182A–186A.

3

PHYTOREMEDIATION AND PUBLIC ACCEPTANCE

Robert K. Tucker and Judith Auer Shaw

Most plants are aesthetically pleasing. Gardening is one of the most popular activities for many people in the United States. The success of gardening magazines, attendance at flower shows, and prevalence of arboretums and botanical gardens all attest to the delight we take in the plant kingdom. Moreover, the relationships that humans have to plants long precedes our development as the modern species, *Homo sapiens*; the associations are rooted deep in our evolutionary past, and because of our dependence on plants—for food, medicine, and shelter—we have a deeply held innate appreciation of their value. Even in cities, many people long for parts of the "natural" world—witness the popularity of urban gardening and urban forestry. And of course landscaping, providing for house plants, and caring for the plants in many atria and offices of businesses across the land have become very substantial business enterprises.

In today's inner-city neighborhoods, we are likely to see the yards of abandoned buildings or factories with as much asphalt and rubbish as green. These sites—industrial "brownfields"—are abandoned properties, many of them contaminated, and few of them highly marketable. So they sit unattended, an eyesore to their community. When the cleanup of such sites begins, large earth-moving equipment such as bulldozers will be used for excavation or other treatment options, and this will result in long-term exposure to dust and other air pollutants.

Now take a look at a neighborhood in the center of Trenton, New Jersey. Next to an abandoned industrial facility, an orderly plot of mustard plants is under cultivation. It is thriving from the attention of the people from the neighborhood who watch over

Phytoremediation of Toxic Metals: Using Plants to Clean Up the Environment, edited by Ilya Raskin and Burt D. Ensley.
ISBN: 0-471-19254-6 ©2000 John Wiley & Sons, Inc.

it and water it regularly. Is this a risky behavior? No. This is phytoremediation, a gentler technology that recognizes that remediation is not just about cleaning up. It is also about impact on neighborhoods and about quality of life. This is a remediation that starts as a garden.

In contrast to the "dig and dump" or "pump and treat" technologies that require lots of equipment, create substantial dust and noise, and usually require the separation of the public from what is going on, phytoremediation can be done more gradually, on a human scale, and is an ideal technology for public involvement. Plants require attention and need nurturing. Most people relate to plants. Although the allegedly more gradual timeframes for phytoremediation have sometimes been cited as a disadvantage, we believe that the potential for public acceptance and community involvement in cleanup plans and remediation outweigh the drawbacks. Community acceptance is often the key to completing a cleanup in a timely fashion; opposition from the public has delayed some cleanups for years. When there is substantial community support, projects have overcome bureaucratic delay and have even jumped ahead in the priority list for permits and other requirements needed to begin. Thus, cleaning up a site with a method that the community supports will, more often than not, save time in the long run. The comparative economics of the cleanup processes, heavily in favor of phytoremediation, will also be a driving force for acceptance (Bishop, 1995).

Even apart from the ability of some plants to take up contaminants, the presence of ground cover provided by grasses, shrubs, and bushes on a site gives an element of safety, helping to shield people from direct contact with the soil and preventing the blowing of dust around the neighborhood. The project in Trenton was sponsored by a neighborhood organization led by Isles, Inc., a community development corporation. Further south in Camden, the New Jersey Environmental Federation has been experimenting with the use of compost in much the same way. Camden, one of the more contaminated urban areas in the state, has hundreds of abandoned sites with high lead levels. The simple addition of a layer of clean compost to these sites encourages plant growth to cover soil in areas contaminated with lead. The Federation, like Isles in Trenton, has reached out to citizens in the neighborhood and intimately involved them in Federation activities, recognizing a most important and fundamental policy to involve the community in every aspect of the process of identifying and cleaning up hazards in their neighborhoods (Goldsmith, 1997).

In urban America, we are in the midst of a severe environmental, economic, social, and cultural crisis. One of the nation's most pressing environmental problems is the state of its cities. This crisis is fundamentally an ecological one—in both the natural and the human sense. The recent good news for some cities that concerted civic efforts can bring urban areas back from the brink provides some encouragement. But this effort requires dealing with the full host of urban problems such as crime, poor schools, traffic and transportation, drugs, welfare, lack of employment, and a weak business community. Certainly one need is to render these areas environmentally safe and to clean up the legacy of toxic contamination. A valuable policy context for considering and discussing public acceptance of phytoremediation as a method of cleanup of contaminated sites is environmental justice and urban revitalization.

Environmental management leaders coined the term "brownfields" to describe the phenomenon of abandoned industrial sites, generally with some level of contamination from hazardous and toxic waste. The brownfields concept emerged as a natural outgrowth of groups and individuals seeking to reverse the tide of urban decay. Although not all brownfields are located in urban areas, the overwhelming thrust of the brownfields problem is decidedly urban in nature. The concentration of these old abandoned sites in what are now urban neighborhoods attracts other undesirable land uses such as junkyards, sewage plants, and other nonresidential businesses. The environmental impact of this convergence is significant and concentrates an unequal burden of municipal responsibility on a single geographic area (Lee, 1997).

The National Environmental Justice Advisory Council has characterized the brownfields issue as "yet another aspect of an intensifying set of systemic problems related to residential segregation, disinvestment of inner city areas, urban sprawl, degradation of the urban environment, and the polarization between urban and non-urban communities along lines of age, life style, race, socioeconomic status, and other spatially-related social divisions" (Lerner, 1996). The profusion of abandoned and contaminated industrial and commercial sites in many U.S. urban areas is a legacy of industrialization and patterns of growth that foster social decay by treating land, natural resources, communities, and populations as expendable and disposable commodities. The movement for environmental justice has suggested the need for redevelopment strategies that serve as instruments for ecologically sustainable and socially just development. With attention to involvement of the community in every step of the planning and implementation of the cleanup, phytoremediation can play an invaluable role in such strategies.

The process of addressing this inequity is directly tied to urban revitalization. The process of distributing the responsibilities of environmental management more equitably provides opportunities to introduce more sustainable and community-oriented activities into urban neighborhoods. It creates an opportunity for residents to become involved in enhancing the amenities that create nurturing environments where children can grow up healthy and in safety.

If society cannot find ways of revitalizing urban areas, development will necessarily encroach more and more on the nonrenewable resource of pristine natural space. In addition to realizing the crucial social role of healthy cities, the emergence of the brownfields issue signals recognition of the ecologically untenable nature of "greenfields" development or urban sprawl.

At the heart of all of these issues are the people who live in urban neighborhoods. They have to be involved in the decisionmaking process for cleanup and revitalization efforts to be effective and lasting. Involving communities in the cleanup process means that it is necessary for agencies as well as those in the private sector involved in accessing contaminant exposure or in cleanup procedures not only to institute policies and performance measures that encourage program personnel and policymakers to spend substantive time in neighborhoods as a regular part of their work to foster understanding of real problems, concerns, and aspirations of community residents but also to continually consult with and involve those residents in the decisionmaking process. The movement for environmental justice has increasingly established the concept that "people must speak for themselves" about their environment defined as the place where "they live, work, and play." It is essential that people in a community be involved in decisions about what to do with contamination. The many sites needing cleanup in urban areas will have special needs and concerns that come with working in city locations; however, there are also particular advantages in that regulatory agencies are aggressively pushing brownfields cleanup and in some cases providing financing or facilitating its procurement. The United States Environmental Protection Agency (EPA) has become an advocate of a policy framework based on environmental justice, and many states and cities have also realized the efficacy of this approach in many of their environmental programs (Fox, 1997). Environmental justice is predicated on the fact that the health of the members of a community, both individually and collectively, is a product of physical, social, cultural, economic, and spiritual factors. Environmental justice encompasses very clearly the inextricable link between these issues.

The process of engaging communities is challenging. It means understanding cultural differences and acknowledging voices that are not wholly focused on scientific aspects of cleanup. It means setting environmental cleanup in a social context—and recognizing that risk is not simply scientific or verifiable through risk assessments. In an inner-city community risk has very different parameters. Risk exists when a child cannot walk to school safely, or a family cannot afford health care, or when people become enraged when ill-prepared public-sector scientists who attend a meeting to discuss the need for remediation or when citizens won't believe or trust anyone who enters their neighborhoods to try to "fix" something. Risk in these circumstances is comprehensive—it includes risk from exposure to contaminants, but it is clearly much more than that.

This difference in definition complicates the approach to environmental cleanup and urban revitalization. It has to begin with a commitment to regarding cleanup in the context of community and as part of an overall plan to cleanup *and* create a better living environment. To be effective, revitalization must be accompanied by a high level of dialog where people who don't know about science and people who don't know about life in urban neighborhoods can become familiar with one another's stories and language. One reason phytoremediation is such an attractive technology is that it bridges that gap. Use of innovative technologies puts both the scientists and the residents on a common ground—neither is "certain" of the outcome—they are taking a risk together.

How does this work in practice? We believe the role and participation of adults and youth in the community is an important factor in education regarding exposure to toxic contaminants, development of strategies for community cleanup, and developing the political will and cohesion within the community to creatively deal with brownfields problems; it helps to speak with residents at public meetings and to speak with social studies and science students in the schools. The questions raised in these settings inform the discussion and identify concerns and questions the residents face in dealing with cleanup issues.

One example of concern about phytoremediation that might possibly be raised by some members of the public could involve the use of biotechnology for modification of some of the plants used. There has been some opposition to "genetic engineering" in the United States and especially in Europe, fueled by such advocates as Jeremy Rifkin, and by concern, for example, that the engineering of herbicide resistance in crop plants may be transferred to wild plants. Some of this concern has diminished as demonstrations have shown that some of the more lurid speculation was unfounded, but these issues cannot go unaddressed if there are concerns in the community about biotechnology or genetic engineering. Some of the strategies for risk communication that we discuss later in this chapter, and references to work in this area by other experts, will be helpful in dealing with public concern.

Particularly in urban areas, when residents are so close to sites, it is imperative to get the community involved in planning "from the very beginning." Communities must be involved in decisions about the investigation and cleanup of sites in their neighborhoods. Unless they have a place in the decisionmaking, the anger and frustration built up within the public may derail the cleanup process. Within the past 15 years (at the time of writing) agencies who deal with environmental health issues have become more adept at risk communication; building collaborative consultation with the public into overall risk management. The field of risk communication has grown rapidly and has become not only an area for research but an important tool in accomplishing cleanup of environmental hazards. The most important principle in risk communication is that it must be two-way; listening to the concerns, fears, and suggestions of the public is the first step in building a relationship of trust. This trust is crucial to the collaborative process—without it, the process follows an artificial path and ends up in conflict at a later date—after many hours are spent pursuing a solution that was never accepted by the public in the first place.

In 1988, the Division of Science and Research of the New Jersey Department of Environmental Protection (NJDEP), in collaboration with the Rutgers University Environmental Communication Research Program, interviewed academic experts, industry representatives, citizen leaders, and agency staff throughout the country in policy, technical, and community relations positions, to gain the experience and wisdom of those practitioners who had been "experimenting" for years with a variety of approaches to communicating risk.

The researchers concluded that unless agency personnel were trained in or sensitized to principles of risk communication, they are likely to view some public reactions to risk as irrational. Certainly public perceptions of risk are often at odds with those of scientists. Attempts to "lecture" the public about the "right" answers engender more anger than understanding. However, communities are quite capable of understanding the scientific aspects of risk assessment. Members of the public are quite capable of understanding technical detail. However, those who work with communities must take certain factors into account in order to understand how the public perceives and reacts to risk. NJDEP summarized these factors: *Voluntary risks* are accepted more readily than those that are imposed. Risks *under individual control* are accepted more readily than those under government control. Risks that *seem fair* are more acceptable than those that seem unfair. Risk information that *comes from trustworthy sources* is more readily believed than information from untrustworthy sources. Risks that seem *ethically acceptable* will be seen as less risky than those that are ethically objectionable. *Natural* risks seem more acceptable than artificial risks. Exotic risks seem more risky than familiar risks. Risks that are *associated with other, memorable events* are considered more risky. Risks that are *dreaded* seem less acceptable than those that carry less dread. Risks that are *undetectable* create more fear than detectable risks. Risks that are *well understood* by science are more acceptable than those that are not (Hance et al., 1988). Unless such factors are understood and taken into account, and unless communication is truly mutual, attempts to convey information may result in anger and outrage on the part of the community and frustration on the part of those who think they are trying to help in remediation activities.

Given the diversity in many U.S. urban areas, ethnic and cultural considerations are of utmost importance in working with local groups. Skills in communication, cultural sensitivity, the ability to work closely with those in the community such as ministers, local politicians, and recognized community leaders are crucial. It may be helpful to consult with anthropologists or other social scientists who have intimate knowledge of community organizations and structures before attempting to resolve complex risk issues.

It is important not only to establish genuine two-way dialog in the community but also to get citizens really involved in planning and decisionmaking for cleanup in their neighborhoods. Those who have become successful in working with citizens have developed various techniques for getting people involved. One of the most effective entrees is to work with community organizations that already have the trust and participation of the neighborhood. Churches, civic associations, and organized ethnic or other community groups should be contacted and cultivated. These organizations have recognized community leaders who can act as facilitators for local dialog. Caesar Chevez, for example, who went on to found and lead the United Farm Workers Union, was identified by the American Friends Service Committee when they set up a program to build housing in poor rural areas near Delano (in Kern County), California in the early 1950s. AFSC was able to take advantage of Chevez' natural leadership skills and supplement them with additional training and coaching in implementing their successful housing program. Chevez benefited from this early experience in his later efforts with the union.

In working with communities it is of utmost importance to develop a relationship of trust and then maintain that trust. This requires being open and honest with information. It means following up promptly to requests and suggestions. There are too many examples of recent actions by agencies that have caused people to distrust government; in fact, we are suffering a legacy of withheld information about environmental and public health risks that include effects of nuclear testing in Nevada after World War II, disinformation and lack of information about the manufacture and use of chemicals such as dioxin-containing Agent Orange during the Vietnam War, and the more recent series of disclosures about exposures that may be responsible for

Gulf War Syndrome, as well as a host of examples of lack of disclosure or delayed information about hazardous waste sites in many areas. Such actions can destroy trust and make the implementation of cleanup projects very much more difficult. Misinformation about risks, the types of toxic chemicals involved, or extent of exposure, for example, can engender outrage in a community. Such outrage has the potential to delay or derail remediation projects. Delay is the enemy of trust and community building. Thus it is important to keep momentum going once the project is off the ground. If delays are encountered, the ongoing dialog with the citizens ensures that correct information is circulated. Lack of dialog leads to assumptions that can be easily believed.

Another effective technique is to involve people from the community in work on the site. Lack of employment opportunities is a common problem in urban areas, so when there are jobs related to the project, they should be offered locally if possible. In other cases, like the phytoremediation project in Trenton, the oversight of the project was put in the hands of a corps of volunteers under the guidance of a technical staff person. In both cases the community has more ownership of the project. Ironically, there are many cases of remediation where contractors from outside the community bring in their own workers and proceed with little community participation.

Of course, safety considerations such as preventing worker exposure or off-site movement of contaminants is important in any remediation, but with community involvement that may extend to participation on site, it is especially important to have well-formulated safety protocols, not only to ensure that soil doesn't get tracked home but also so that such safety considerations can be discussed with the community so that they feel involved with this aspect of the project as well. Following good safety practices can be morale-boosting and educational for members of the community. When neighborhood residents can see that their health concerns are being addressed, they have confidence that the agency managers are respecting them and will in turn show more respect for the project.

THE MAGIC MARKER SITE IN TRENTON, NEW JERSEY: A CASE STUDY

As an example of a successful application of technology combined with community outreach and collaboration, phytoextraction is being used to clean up an old abandoned urban industrial site in Trenton, New Jersey. The Magic Marker site is typical of many of the so-called brownfield sites in New Jersey and across the United States, has been designated as a "Superfund" site. It has a history of industrial uses, including that of Gould, Inc., a battery manufacturer. That business left toxic lead that continues to contaminate the soil in the yard around the old buildings blighting the 400 block of Calhoun Street. Gould, which preceded the Magic Marker Corp, is a distant memory, and Magic Marker, the maker of writing implements, has come and gone. The neighborhood is left to cope as best they can with the industrial contamination left behind when the companies moved away. Researchers from Rutgers University devised a new technology and created a company, Phytotech, to implement it. They have been testing their new waste site cleanup technology. Starting with a 200×30-ft plot of freshly tilled earth, a rectangular patch of dirt hard by the northwest corner of the plant that looked like any other garden ready for seeding. They sowed the tract with the seeds of the Indian mustard plant, an innocuous-looking sprout that grows quickly and has the ability to accumulate lead from the soil in which it is planted. Citizens from the neighborhood were intimately involved in planning, implementation, and management of the site. At a public ceremony on April 22,1996, Phytotech's president, Dr. Bert Ensley, and the residents unveiled the garden with the help of officials from the city, the state, the federal Environmental Protection Agency, and Isles. Inc., a local community development corporation which played a pivotal role in getting the neighborhood involved. At the ceremony, James Rollins, a 72-year-old Dunham Street resident, said he moved in across the street from the plant in 1961, when Gould, the batterymaker, occupied the sprawling plant. At that time, fumes from weekend acid deliveries chased residents from their front porches. Rollins pointed out chalk-white acid stains that have permanently scarred the walls of the brick factory building. "Still, the residents have hope," Rollins said. "Isles kind of took us by the arm and led us to where we are, and now it's going good. All of the residents are cooperating and we're very energized. I hope you will clean our community," he told the gathered officials, "We will be with you."

Through Isles, Inc., residents who live near the factory have organized under the title "Northwest Community Improvement Association" and educated themselves about the cleanup process and forged new links to City Hall. From this collaboration, the City of Trenton expanded its involvement in neighborhood remediation and successfully garnered an EPA grant for $200,000 to provide funds for the next steps. It was that community involvement that ultimately led to Phytotech's garden, according to officials from the New Jersey Department of Environmental Protection, who have whole heartedly endorsed the research as well as the participation of the local neighborhood in the cleanup process. "It really was a community up kind of process," said NJDEP Assistant Commissioner Rick Gimello, at the ceremony. "The typical brownfields site tends to be economically driven instead of community driven. This was the reverse" (McGinty, 1996).

Dr. Michael Blaylock, a soil chemist with Phytotech, who would oversee the Magic Marker garden, explained that the mustard plants store the lead they pick up in their leaves, rather than their roots, which makes them easy to harvest and discard. In addition, he explained that the roots typically reach about 20 in. into the ground, well below the 6–8 in. depth where lead contamination usually resides (Blaylock, 1996).

With the plants reaching maturity in about 6 weeks, the researchers were able to grow and harvest several cycles in the garden during this spring and summer. In between each planting, they tested the lead levels in the soil and in the plants. Later in the year, they experimented with planting pumpkins on the site, and on October 22, 1996, just in time for Halloween, the researchers harvested a robust crop of the gourds (*The Trenton Times*, 1996). This was again done very publicly, involving Trenton and environmental agency officials and community members. At the meeting accompanying the pumpkin harvest, Phytotech announced that they had documented a reduction in lead to concentrations within established safety levels in 75% of the area they planted with mustard earlier in the year (Phytotech, Inc., 1997). Karen Waldron, a planner with the City of Trenton and coordinator of the city's brownfields projects, made the point that phytoremediation offers the promise of thorough, low-cost pollution cleanups that could help revitalize urban areas. The harvested plants can be incinerated, leaving an ash that is 40% metal or more. Researchers believe that when their cleanup technology becomes widely implemented, the incinerated ash containing this much metal would be valuable, like ore from a mine. But even if recovery of the metal weren't economically feasible, the economics of disposing of the metals in their much more concentrated state in the ash are substantially more favorable than "digging and dumping" the originally contaminated soil. The economics of phytoremediation are still being worked out. Dr. Ilya Raskin and his Phytotech colleagues estimate that using plants would cost $60,000–$100,000 to clean an acre of soil to a depth of about 20 in. In contrast, they estimate it would cost $400,000 to dig up the same amount of soil and haul it to a hazardous waste site.

Alan Mallach, Trenton's director of housing and development, and who has also played an instrumental role in moving the cleanup process forward, cautioned that the garden "Will not be, in itself, the solution to the Magic Marker site." But he readily agrees with Frank McLaughlin, a hydrogeologist with the NJDEP, that biotechnology clearly is "the future for brownfields sites which have low levels of contamination, compared to other, heavily polluted industrial sites." Jeanne Fox, a Regional Administrator for the EPA, feels that the experiment could have a far-reaching impact, saying,"if this works here, it will be used around the country."

That would fit nicely with the plans of Ensley, who has projected that his fledgling, 15-employee company will employ more than 170 workers and take in $50 million a year by 2001. "It's great." Ensley said. "You can build a company, create value and wealth, and do something good for the environment all at the same time."

The Magic Marker site in Trenton is a positive example where community involvement has definitely moved the process forward. Isles, Inc. built up a constituency and brought the community in from the very beginning. The residents selected the site they wanted to clean up: the location represented a risk to children in the neighborhood who managed to get onto the site. Not only was the site was near a school, at over seven acres; it was a major blight on the neighborhood (Peterson, 1996).

Citizens helped research the history of the site, going to the library, searching municipal records for tax information, and tracking down responsible parties and owners. Residents conducted a detailed four-page quality-of-life survey of the neighborhood. People spoke out about their needs and shared information about the neighborhood. They understood that the cleanup of a complex site such as this would be complicated and time-consuming. It was important to be up-front about the time involved and even the possibility of failure in the cleanup process.

The cleanup process could take more than 5 years, but the important thing is not the time. To the residents, the key is progress; they want to see that the process moves forward, and that the community itself is fully involved and in control of the process.

The agencies can't do it alone, City Hall is helpless by itself, and the neighborhoods need technical and financial help. No one entity can do it alone, but when all involved parties work together, the job gets done—coalitions are powerful!

The experience at Magic Marker shows that empowerment comes from collaboration. When sites are being cleaned up, the ancillary effects on the social and economic health of the neighborhood can be impressive. Improvement through site cleanup can indirectly affect solutions to other problems, especially by showing the community that by working together, their organized efforts produce results.

REFERENCES

Bishop, J. E., 1995. Pollution fighters hope a humble weed will help reclaim contaminated soil. *The Wall Street Journal* (Aug. 7) p. B1.

Blaylock, M., 1996. Cleanup at the Trenton, NJ brownfield site. Presentation to the American Chemical Society, Birmingham, AL, Sept. 10, 1996.

Fox, J., (Regional Administrator, Region II, U.S. EPA), 1997. Personal communication.

Goldsmith, A., (Executive Director of the New Jersey Environmental Federation), 1997. Personal communication.

Hance, B. J., et al., 1988. *Improving Dialogue with Communities.* New Jersey Department of Environmental Protection, Division of Science and Research, Trenton, NJ.

Lee, C., (Research Director of the United Church of Christ's Commission for Racial Justice), 1997. Personal communication.

Lerner, S., 1996. Brownfields of dreams. *The Amicus Journal* (Winter) p. 15.

McGinty, T., 1996. Plants weed out lead contamination. *The Trenton Times,* (April 24) p. A1.

Peterson, M., 1996. Lead-eating mustard plants. *New York Times,* (Oct. 23) p. B4.

Phytotech, Inc., 1997. *Selected Field Summaries of Phytoremediation, 1996 Site Demonstrations.* Phytotech, Inc., Monmouth Junction, NJ.

The Trenton Times, 1996. Reaping more than pumpkins. (Oct. 23) p. A8.

4

REGULATORY CONSIDERATIONS FOR PHYTOREMEDIATION

STEVE ROCK AND PHILIP G. SAYRE

INTRODUCTION

The regulator's view of phytoremediation is the same as for any proposed remediation technology and asks the basic questions: "Why do you think this technology will decrease risk to human health and the environment, and how will you show that it works?" "A treatment remedy must be "protective of human health and the environment, maintain protection over time, and minimize untreated waste" (40 CFR 300.430). Each application will be site-specific, and must be evaluated on a case-by-case basis by a regulator.

Interest in phytoremediation is on the increase because of favorable results of some apparently positive greenhouse results and field trials. The U.S. EPA has the dual role of seeking to protect human health and the environment associated with hazardous waste sites and encouraging development of innovative technologies that might be able to more efficiently clean up these sites.

Because of the potential utility of phytoremediation and its lower estimated costs, there is a great deal of interest on the part of site owners, managers, and their consultants and contractors in applying this technology to private, RCRA, Superfund, and brownfield sites. Balanced against this promise are questions regarding the true rate and extent of cleanup using this technology, as well as other associated risk concerns. Specific regulatory standards on the efficacy and potential risks for

Phytoremediation of Toxic Metals: Using Plants to Clean Up the Environment, edited by Ilya Raskin and Burt D. Ensley.
ISBN: 0-471-19254-6 ©2000 John Wiley & Sons, Inc.

phytoremediation have not been developed, and installations are currently approved on a site-by-site basis.

One persuasive indication of potential acceptance of a technique is previous successful applications on a similar sites. Because it is an innovative technology, phytoremediation does not have a long history of either completed cleanups or site closures. Results of studies done in greenhouses and on field test plots can be used to show proof of concept, and some of that data may be directly applicable. If time and funding permit, soil or water from the site should be used in lab or greenhouse studies. Such treatability studies can confirm the efficacy of the site-specific treatment.

LIMITS AND POTENTIAL OF PHYTOREMEDIATION AT HAZARDOUS WASTE SITES

Root contact is a primary limitation on phytoremediation applicability. Remediation with plants requires that the contaminants be in contact with the root zone of the plants. Either the plants must be able to extend roots to the contaminants, or the contaminated media must be moved to within range of the plants. This movement can be accomplished with standard agricultural equipment and practices, such as deep plowing to bring soil from 2 or 3 ft deep to within 8–10 in. of the surface for shallow rooted crops and grasses, or by irrigating trees and grasses with contaminated groundwater or wastewater.

Phytoremediation is also limited by the growth rate of the plants. More time may be required to phytoremediate a site than if more conventional cleanup technologies were used; excavation and landfill, or incineration takes weeks to months to accomplish, whereas phytoextraction or degradation may need several years. Therefore, for sites that pose acute risks for human and other ecological receptors, phytoremediation may not be the remediation technique of choice.

High concentrations of contaminants may inhibit plant growth and so may limit application on some sites or some parts of sites. This phytotoxicity could lead to a tiered remedial approach in which high-concentration waste is handled with expensive ex situ techniques that quickly reduce acute risk, whereas in situ phytoremediation is used over a longer period of time to clean the high volumes of lower concentrations of contamination.

Thus, sites with widespread, medium-level contamination within the root zone are candidates for phytoremediative processes.

REGULATORY ACCEPTABILITY

Often the most difficult sites to remediate technically and administratively are those abandoned facilities on the National Priority List (NPL), or Superfund sites. The Superfund evaluation and remedy selection procedures can serve as a model for the remediation processes adopted at other types of hazardous waste site. Therefore, an examination of how phytoremediation would be viewed when proposed for a Superfund site may prove helpful.

John Fagiolo, a remedial project manager (RPM) with EPA Region 5, explains that a Superfund site typically has three sets of interested parties or stakeholders: the site owner, users, or potentially responsible parties (PRPs); the federal and/or state and local regulators; and the neighbors and/or nonregulatory political entities. The needs and desires of the various groups are not always aligned. Often there are legal requirements or court orders involved.

On a Superfund site it is generally the task of the U.S. EPA RPM to choose a remedy and write the record of decision (ROD) for the site. The RPM's main task is to ensure the health and safety of the humans and the environment of the site and those affected by the contamination at the site. PRPs submit a specific remedy proposal (remedy) to U.S. EPA. Cost of cleanup is to be borne by the owners or those responsible for the site; thus the remedy recommended is typically the lowest-cost option.

The original and amended ROD, remedial investigation, feasibility study, remedial design (RD), technical memorandum, and other documents are available in the administrative record and information repository and should be consulted for in-depth details on the development and evaluation of the alternatives considered. This information is used in the evaluation of cleanup options to address contamination at the site.

Public input on remedy proposals and the information that supports these proposals is an important contribution to the cleanup remedy selection process. The public is encouraged to review and comment on the proposed plan.

The U.S. EPA uses nine criteria, which are required by law and described below, to evaluate these alternatives and determine the remedy preference. The evaluation criteria consisted of

1. *Overall protection of human health and the environment* determines whether the alternative eliminates, reduces, or controls threats to public health and the environment through institutional controls, engineering controls, or treatment.

2. *Compliance with applicable or relevant and appropriate requirements* (ARARs) evaluates whether the alternative meets federal and state environmental statutes, regulations, and other requirements that pertain to the site.

3. *Long-term effectiveness and permanence* considers the ability of the alternative to protect human health and the environment over time and the reliability of such protection, including the degree of certainty that the alternative will prove successful.

4. *Reduction of contaminant toxicity, mobility, or volume through treatment* evaluates the effectiveness of the alternative in reduction of the harmful effects of principal contaminants, reduction of the ability of the contaminants to move in the environment, and the reduction in amount of contamination present.

5. *Short-term effectiveness* considers the length of time needed to implement the alternative and the risks the alternative poses to workers, residents, and the environment during implementation.

6. *Implementability* considers the technical and administrative feasibility of implementation of the alternative, such as the practicability and difficulty of construction, and the availability of goods and services.

7. *Cost* considers the estimated capital and operation and maintenance costs, as well as present net worth costs. Present net worth is the total cost of the alternative over time in terms of current U.S. dollars.

8. *State acceptance* considers whether the state agrees with U.S. EPA analyses and recommendations of the studies and evaluations performed.

9. *Community acceptance* will be addressed in the ROD amendment. The ROD will include a responsiveness summary, which presents public comments and U.S. EPA responses to those comments. Acceptance of the recommended alternative will be evaluated after the public comment period.

RPMs may be cautious in approving the use of phytoremediation at a site as a result of the present lack of hard data. The field data that are available now are limited because of recent interest in phytoremediation and because the data that are available seldom are published in the peer-reviewed press. This situation should be corrected within the next 1–2 years (at the time of writing) through the efforts of programs such as SITE, the RTDF, and others.

RPMs often oversee numerous sites, and rely on the PRPs and others responsible for the sites to provide persuasive information that phytoremediation is adequate to address the risk concerns at a sites. Phytoremediation would not be selected for some sites because of the time required for the completion of cleanup.

For those sites for which RPMs may be willing to consider phytoremediation, at least three different assurances must be in place for final remedy selection:

1. Adequate containment of contaminated soils, groundwater, and sediments must be assured until the plants associated with phytoremediation have established themselves at the site to a point where they are either containing or degrading the contaminants of interests.

2. If phytoremediation is attempted and does not succeed, an adequate backup technology with a high chance of success must be ready and available for use at the site.

3. Evidence of the effectiveness of phytoremediation that is specific to the site matrix and contaminants must be presented. Since this technology is new, some RPMs may be willing to accept laboratory studies on the plants and contaminants of interest as the primary evidence to support the use of phytoremediation at the site. Such laboratory studies should at least show that the plants that would be used at the site are capable of remediating the site contaminants. With these three assurances in place and an adequate plan for

implementing phytoremediation and backup technologies at the site, some RPMs may be willing to allow phytoremediation to proceed at full scale to remediate the site.

Monitoring the efficacy of any innovative treatment may be more extensive than would be required for a more accepted technology. Monitoring needs to both address the decrease in the concentration of the contaminants in the media of concern and examine the fate of the contaminants. The monitoring plan must be tailored to the site and plants.

RESEARCH TO ADDRESS REGULATORY CONCERNS

To assess the appropriateness of any of the phytoremediation applications, regulators must have access to media and contaminant-specific field data that show the rate and extent of degradation or extraction or whatever the proposed mechanism is. The existing knowledge base is limited, and specific data on more plants, contaminants, and climate conditions are needed. Regulators also need standardized monitoring systems. Currently there is no industry or research consensus on which parameters are crucial to measure, and very few projects can afford to sample, analyze, or monitor a large number of parameters over the years needed for most phytoremediation projects.

The Superfund Innovative Technology Evaluation (SITE) program demonstrates field-ready technologies that are initiated and installed by the developer of the technology. SITE began evaluating phytoremediation projects in 1994. Currently three full demonstrations (including one at a Superfund site, the former Carswell AB with the USAF and ESTCP) and one Emerging Program project are under way using phytoremediation. Reports detailing the performance of the demonstrations will be published at the conclusion of the field work. Information on the SITE program or individual projects can be found on the internet at epa.gov/ORD/SITE.

In addition to U.S. EPA efforts, other federal agencies, universities, consultants, and remediation contractors have ongoing research on phytoremediation. All these projects expand the knowledge base of what plants can be expected to do consistently, and make the application of innovative technology more acceptable to regulators and consumers.

Continuing research and policy discussions in the related areas of determining possible risk-based alternative endpoints for cleanups, and measuring the intrinsic remediative capacity of a site (natural attenuation) will impact the applicability of many biological technologies, including plant-based systems.

Enhancements to the various phytoremediation process are continuing. Some applied research is directed at selecting and breeding plants that have more of an attractive quality such as hyperacummulation of metal, production of certain enzymes, and affinity or tolerance for contaminants. Research continues in genetic engineering of plants to combine positive traits, altering enzyme systems, or increasing a plant's natural range.

Plants used for phytoremediation are potentially regulable under several U.S. statutes. The United States Department of Agriculture (USDA) administers several statutes that could be used to regulate such plants, including the Federal Plant Pest Act (7 USC 150aa et seq.), the Plant Quarantine Act (7 USC 151 et seq.), and the Federal Noxious Weed Act (7 USC 2801 et seq.). Pertinent regulations are found at 7 CFR parts 319, 321, 330, 340, and 360, respectively. Under USDA authorities one type of plant (transgenic or naturally occurring) that would potentially be subject to review would be a plant considered as a plant pest. For additional guidance on USDA regulations pertaining to plants, please see the USDA Web site "http://www.aphis.usda.gov/bbep/bp/".

The U.S. EPA does not currently regulate plants intended for commercial bioremediation, although EPA believes the Toxic Substances Control Act (TSCA) gives EPA authority to do so, if such action is necessary to prevent unreasonable risk to human health or the environment. TSCA gives EPA authority to regulate "chemical substances." TSCA defines chemical substances broadly to mean all chemicals and mixtures of chemical substances.

Living organisms, such as plants, are mixtures of chemical substances and thus are subject to TSCA. Although TSCA could potentially be applied to plants used in bioremediation, EPA has not yet determined whether such action is necessary to protect the environment and human health. EPA to date has issued regulations for microorganisms only under Section 5 of TSCA (U.S. EPA, 1997). Further information on TSCA and biotechnology products can be found at the Web site http://www.epa.gov/opptintr/biotech/.

Certain plants engineered to contain sequences that afford the plant resistance to pests to enhance the remediation efficacy of the plant could be subject to review by EPA under its authority to regulate pesticides. EPA regulates pesticides under two statutes: the Federal Insecticide, Fungicide, and Rodenticide Act (FIFRA) and the Federal Food, Drug, and Cosmetic Act (FFDCA). Substances that plants produce to protect themselves against pests and disease are pesticides under the definition of FIFRA Section 2 (i.e., if they are "intended for preventing, destroying, repelling, mitigating any pest") regardless of whether the pesticidal capabilities evolved in the plants or were introduced by breeding or through the techniques of modern biotechnology. These substances, along with the genetic material necessary to produce them, are designated "plant pesticides" (U.S. EPA, 1994). Additional details about EPA plant pesticide regulations can be found at http://www.epa.gov/docs/fedrstr/EPA-PEST/1994/November/Day-23/. The U.S. Food and Drug Administration is responsible for food safety under the FFDCA. It is unlikely that FDA would be involved in the review of phytoremediation plants, however, since use of phytoremediation plants as food or food components is an unlikely scenario.

CONCLUSIONS

Phytoremediation is a relatively new category of treatment technologies. Some of the technologies have been proved on the greenhouse or laboratory scale and are being demonstrated and evaluated in the field. Because plants need a certain time and season to grow, results from field studies takes years to gather and assess. A number of studies under way should yield results over the next few years. As those data are harvested and compiled, regulators, site owners, and remediation specialists will be more equipped to decide whether phytoremediation is likely to be effective in reducing risk on any specific site with a given concentration of contaminant.

SECTION III

TECHNOLOGIES FOR METAL PHYTOREMEDIATION

5

PHYTOEXTRACTION OF METALS

MICHAEL J. BLAYLOCK AND JIANWEI W. HUANG

INTRODUCTION

The use of plants to extract toxic compounds from soils (phytoextraction) is being developed as a method for remediation of metal contaminated soils (Baker et al., 1994; Chaney, 1983; Raskin et al., 1994). Plants that can accumulate and tolerate unusually high concentrations of heavy metals in their tissue were responsible for drawing attention to the possibility of using plants in this manner. Accumulators of nickel (Ni) and zinc (Zn), for example, have been reported to contain as much as 5% of these metals on a dry-weight basis (Baker et al., 1994; Brown et al., 1994). Plants accumulating metals at a 5% (50,000 mg/kg) dry-weight concentration from a soil with a total metal concentration of 5000 mg/kg results in a 10-fold bioaccumulation factor. If the plant produces a significant amount of biomass while accumulating high concentrations, an important quantity of metal can be removed from the soil via plant accumulation. The metal-rich plant material can be collected and removed from the site using established agricultural practices, without the loss of topsoil associated with traditional remediation practices. The biomass can then be recycled to reclaim the metals that may have an economic importance. Alternatively, postharvest biomass treatments (i.e., composting, compaction, thermal treatments) can be employed to reduce the volume and/or weight of biomass for disposal as a hazardous waste if necessary. The metal bioaccumulation and concentration in the plant shoots above that of the soil concentration coupled with subsequent biomass reduction processes can greatly reduce the amount of contaminated material requiring disposal compared to

Phytoremediation of Toxic Metals: Using Plants to Clean Up the Environment, edited by Ilya Raskin and Burt D. Ensley.
ISBN: 0-471-19254-6 ©2000 John Wiley & Sons, Inc.

soil excavation, thereby decreasing the associated disposal costs. This process of extracting metals from the soil and accumulating and concentrating metals in the aboveground plant tissues enables plants to be used as part of a soil cleanup technology and is the fundamental basis for phytoextraction.

Three main factors influence or determine the ability of phytoextraction to effectively remediate a metals-contaminated site: (1) selection of a site conducive to phytoextraction, (2) metal solubility and availability for uptake, and (3) the ability of the plant to accumulate metals in the harvestable plant tissues. Each of these factors is addressed in relation to their role in the phytoextraction process.

METALS IN SOILS

Excessive toxic metal levels in soils pose significant hazards to human and animal health as well as to the ecosystem in general. Anthropogenic sources of heavy-metal deposition have increased as a result of the Industrial Revolution. Agriculture, mining, smelting, electroplating, and other industrial activities have resulted in the deposition of undesirable concentrations of metals such as As, Cd, Cr, Cu, Ni, Pb, and Zn in soil. Although trace metals are an important part of the soil ecosystem, the accumulation of these metals may be harmful to people, animals, plants, and other organisms contacting the soil or groundwater.

Distribution, Concentrations, and Regulatory Limits

Metal concentrations in soil range from < 1 mg/kg to as high as 100,000 mg/kg depending on the parent material and deposition events. Heavy-metal contamination by definition indicates that concentrations above background or native soil concentrations exist. From a regulatory or cleanup perspective, however, contamination is often defined as a concentration exceeding a particular value to which a level of risk has been assigned. Because risk varies with each metal and the associated exposure pathways, definition of the degree of contamination is specific for each contaminant. Moreover, regulatory limits for metal concentrations in soil vary considerably by state and even by site. Limits may be negotiated depending on site-specific factors and specific land use restrictions. The New Jersey Department of Environmental Protection, for example, has established soil cleanup criteria that differ for residential and nonresidential uses and are also subject to site-specific conditions that consider other environmental risks (NJDEP, 1996). Some limits are based on human health impacts from direct soil contact; others are based on ecological risk or secondary exposure pathways. Therefore, the criteria are used mainly as guidelines for further site evaluations based on background conditions and other risk factors to develop site-specific cleanup criteria. Soil cleanup criteria for some metal contaminants are presented in Table 5-1.

TABLE 5-1. Soil Cleanup Criteria[a] Established by the New Jersey Department of Environmental Protection (NJDEP, 1996)

Contaminant	Residential Direct-Contact Soil Cleanup Criteria	Nonresidential Direct-Contact Soil Cleanup Criteria
Antimony	14	340
Arsenic	20	20
Barium	700	47000
Cadmium	1	100
Copper	600	600
Lead	400	600
Nickel	250	2400
Selenium	63	3100
Silver	110	4100
Thallium	2	2
Vanadium	370	7100
Zinc	1500	1500

[a]The criteria provide total soil metal concentrations that may indicate the need for remedial action depending on additional site-specific factors. For additional details, please refer to Tables 3-1 and 7-1 from NJDEP's February 3, 1992 proposed rule entitled Cleanup Standards for Contaminated Sites, NJAC 7:26D.

Site Selection

Soil cleanup criteria are important in considering phytoremediation as a remedial option. The regulatory goals and timeline must be conducive to the application of phytoremediation. Because phytoremediation uses plants to extract a discrete mass of metal from the soil to reduce the measured concentration, the difference between the measured total metal concentration and the cleanup goal is a primary factor determining the feasibility of phytoremediation. For example, consider a soil contaminated with Ni at a concentration of 2800 mg/kg, uniformly distributed in the top 15 cm of the soil profile. Assume that a soil bulk density of 1.3 g/cm^3 yields 195 kg soil/m^2 and 546 g Ni/m^2 in the top 15 cm. In order to decrease the soil concentration to a 250 mg/kg residential cleanup standard, 497 g Ni/m^2 will have to be removed. A biomass yield of 1 kg/m^2 (10,000 kg/ha) containing 1% Ni (10,000 mg/kg) on a dry-weight basis will remove 10 g Ni/m^2 per crop, meaning that 50 crops would be required to achieve the residential standard. On the other hand, achieving a nonresidential goal of 2400 mg/kg is more realistic under the same conditions, requiring only 8 crops to meet the cleanup goal. This assumes of course that the crop will tolerate the Ni concentration and the Ni is in a form that is available or can be made available for plant uptake.

Additionally, it is important to remember that metal concentration is generally determined by analyzing the particle size fraction defined as soil (< 2 mm diameter). Particularly on disturbed sites, the amount of material defined as soil varies considerably, both laterally and vertically, because of rocks, gravel, and other debris.

LIVERPOOL JOHN MOORES UNIVERSITY
LEARNING & INFORMATION SERVICES

A simple mass balance calculation using an average soil bulk density value multiplied by the volume of soil and the measured concentration will only approximate the total mass of metal in the soil. The total metal removal required to bring the measured concentration into regulatory compliance may be much less than estimated and allow a faster cleanup than anticipated using phytoremediation. In the previous example, adjusting for a soil with 25% of material greater than 2 mm further reduces the mass amount of metal to be removed and the number of crops required to achieve the desired measured concentration by 25%. Determining the suitability of a site for phytoremediation must include the assessment of areas of soil with concentrations in a suitable range. For many sites phytoremediation may be used for large areas of the site with smaller areas or hot spots being treated by conventional methods. Where risk-based corrective action (RBCA) limits are imposed on the basis of metal bioavailability, phytoremediation may have improved applicability, as the plants will remove the most bioavailable metal first.

Metal Solubility and Availability

Metal solubility is dependent on soil characteristics and is strongly influenced by soil pH (Harter, 1983) and the degree of complexation with soluble ligands (Norvell, 1984). Metals in soil can exist as discrete particles or be associated with different soil components, including (1) free metal ions and soluble metal compounds in the soil solution, (2) exchangeable ions sorbed onto inorganic solid phase surfaces, (3) nonexchangeable ions and precipitated or insoluble inorganic metal compounds (e.g., oxides, hydroxides, phosphates, or carbonates), (4) metals complexed by soluble or insoluble organic material, and (5) metals bound in silicate minerals. Contamination events are usually indicated by discrete particles or elevated concentrations found in components 1–4; component 5 is indicative of background or indigenous soil metal concentrations (Ramos et al., 1994). The metals considered readily available for plant uptake are those that exist as soluble components in the soil solution or are easily desorbed or solubilized by root exudates or other components of the soil solution, often only a small portion of the total metal content of the soil. Because effective phytoextraction is dependent on a relatively abundant source of soluble metal to achieve significant uptake in the plant shoots, the soil conditions may need to be altered to increase metal solubility and availability.

Lead in soil, for example, is generally considered insoluble and unavailable for plant uptake with the solubility increasing as pH decreases (McBride, 1994). Predominant insoluble Pb compounds in the soil include Pb phosphates, Pb carbonates, and Pb (hydr)oxides. Table 5-2 presents some of the Pb minerals expected in soils and their respective formation constants.

In the soil, metal availability to plant roots increases substantially as the soil pH decreases below 5.5; however, plant growth may be inhibited or suppressed because of the increase in Al solubility and subsequent Al toxicity. Decreasing the soil pH through acidification treatments can enhance metal solubility, but the degree of pH adjustment is limited by the tolerance of the plant to acid conditions and in a more

TABLE 5-2. Formation Constants of Lead Compounds

$Pb_3(PO_4)_2 + 4H^+ \rightleftharpoons 3Pb^{2+} + 2H_2PO_4^-$	-5.26	
$Pb_4O(PO_4)_2 + 6H^+ \rightleftharpoons 4Pb^{2+} + 2H_2PO_4 + H_2O$	2.24	
$Pb_5(PO_4)_3OH + 7H^+ \rightleftharpoons 5Pb^{2+} + 3H_2PO_4 + H_2O$	-4.14	hydroxypyromorphite
$Pb_5(PO_4)_3Br + 6H^+ \rightleftharpoons 5Pb^{2+} + 3H_2PO_4 + Br^-$	-19.49	bromopyromorphite
$Pb_5(PO_4)_3Cl + 6H^+ \rightleftharpoons 5Pb^{2+} + 3H_2PO_4 + Cl^-$	-25.05	chloropyromorphite
$Pb_5(PO_4)_3F + 6H^+ \rightleftharpoons 5Pb^{2+} + 3H_2PO_4 + F^-$	-12.98	fluoropyromorphite
$PbCO_3 + 2H^+ \rightleftharpoons Pb^{2+} + CO_2 + H_2O$	4.65	cerussite
$PbSO_4 \rightleftharpoons Pb^{2+} + SO_4^{2-}$	-7.79	anglesite
$PbS \rightleftharpoons Pb^{2+} + S^{2-}$	-27.51	galena

Source: Lindsay (1979).

practical sense by the cost. Optimally, the soil pH would be adjusted to the point where plant dry matter yields due to pH effects begin to decrease unacceptably. Obviously, this point will vary with soil type and plant species.

Lead can also be solubilized through the addition of chelating or complexing agents. The chelate complexes the free metal ion in solution, allowing further dissolution of the sorbed or precipitated phases until an equilibrium between the complexed metal, free metal, and insoluble phases occurs (Norvell, 1991). The quantity of Pb solubilized and maintained in the soil solution will be a function of the chelate concentration, affinity of the chelate for Pb and competing ions, and the solubility of Pb compounds in the soil. Lead carbonates and Pb phosphates, for example, have limited solubility and are not readily available as a source of Pb for plant uptake. The addition of EDTA, however, should solubilize the Pb from the carbonate phase and provide a soluble source of Pb to the plant root.

Predictions regarding the ability of a chelating agent to solubilize a metal in the soil can be made using equilibrium constants found in the literature and assuming that the metal solubility is controlled by known solid phases. For example, Lindsay (1979) approximates the Pb^{2+} activity in soils at $10^{-8.5}$ M for soils between pH 5.5 and 7.5 where Pb and phosphate solubilities are controlled by chloropyromorphite and hydroxyapatite, respectively. Reported formation constants of Pb–EDTA [log K = 19 (Martell and Smith, 1974, 1982)] and stability diagrams (Norvell, 1991; Sommers and Lindsay, 1979) indicate that EDTA should solubilize Pb under these conditions. If the Pb solubility is controlled strictly by galena (log K = –27.5), EDTA would not be very effective in solubilizing Pb. The extreme variability associated with the contaminant form as well as the soil properties on disturbed sites makes it difficult to accurately predict all the competing reactions and conditions. Much of the work, therefore, relies on empirical data to determine the effectiveness of an applied chelator to increase metal solubility.

Treatability studies are often conducted to evaluate a particular soil for its suitability for a remediation treatment, in this case phytoextraction. An important step in the phytoextraction treatability study is the evaluation of metal solubility. Sequential extraction techniques have been employed to estimate chemical relationships and metal speciation. The sequential extraction uses successive extractions to selectively remove operationally defined soil fractions. The metal concentration in the resulting solution is determined and reported as being associated with the particular fraction extracted. Many variations of procedures exist, but they generally follow an order similar to that of Tessier et al. (1979). The initial extraction is designed to extract the soluble–exchangeable fraction using a neutral salt. That extraction is followed in order by a carbonate extraction (buffered solution at pH 5), Fe and Mn oxides (extracted with a reducing agent such as 0.3 M hydroxylamine hydrochloride), organic matter (oxidizable with acidic hydrogen peroxide), and finally a total digestion of the remaining material to constitute the residual fraction. The amount of metal available for phytoremediation is then estimated on the basis of the distribution of metal between the fractions. The results should be interpreted with the understanding that the extracted fractions are operationally defined and not necessarily specific soil components. For example, the carbonate fraction consists of soluble compounds at pH 5 and is not limited solely to carbonate compounds.

Bioavailability or plant availability is expected to be greatest for the soluble–exchangeable fraction and decrease with each succeeding fraction. Metal associated with the residual fraction is not considered available for removal by phytoextraction. Chelating agents have been used to estimate metal bioavailability and is the basis for the DTPA (diethylenetrinitrilopentaacetic acid) soil test for micronutrient and heavy-metal availability (Lindsay and Norvell, 1978; Amacher, 1996). By extracting Pb-contaminated soils with EDTA before and after conducting sequential extractions, we can see the relationship between chelate extractable metal and the associated soil chemical fractions. In several Pb-contaminated soils tested, EDTA consistently removed a large portion of the Pb from the fractions considered bioavailable (soluble–exchangeable, carbonate, oxide, and organic) with very little Pb extracted from the residual fraction (Fig. 5-1). The ability to make metal or radionuclide contaminants in these fractions available for plant uptake through the use of chelators or other soil amendments is vital to the success of any phytoextraction operation. The next limitation will then be the ability of the plant to accumulate the metal from the soluble forms. Determining the most effective soil amendment and effective concentration for the specific metal to enhance the plant availability is much of the art of phytoremediation. If the metal concentration in the residual fraction exceeds the regulatory criteria, however, it may be impossible to remove enough metal through phytoextraction to meet the cleanup criteria.

METAL ACCUMULATION IN PLANTS

The success of using plants to extract metals from contaminated soils requires a better understanding of the mechanisms of metal uptake, translocation, and accumulation by plants. The understanding of metal transport characteristics could help us to design a better cultivation protocol to enhance metal accumulation in plants. The efficiency of metal phytoextraction is a function of a number of factors. The major factors are plant species, metal availability to plant roots, metal uptake by roots, metal translocation

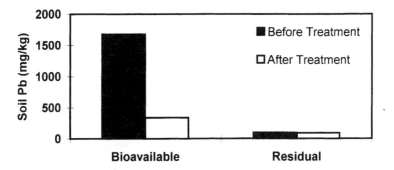

Figure 5-1   Total Pb extracted in the bioavailable and residual fractions from a Pb-contaminated soil before and after extraction with 2.5 mmol/kg EDTA (25 mL of 250 µM EDTA equilibrated with 2.5 g soil). The bioavailable fraction is the sum of Pb extracted sequentially in the soluble–exchangeable, carbonate, oxide, and organic fractions (Tessier et al., 1979). The residual fraction is defined as the Pb remaining (extractable with concentrated nitric acid) after the previous extractions (unpublished results).

from roots to shoots, and plant tolerance to toxic metals. Our research has been focused on optimizing these factors to maximize metal accumulation in plant shoots.

Plant Species Variation in Metal Accumulation

Plant species vary significantly in the ability of accumulating metals from contaminated soils. There is a small number of plant species endemic to metalliferous soils that can tolerate and accumulate high levels of toxic metals. These plants, termed *metal hyperaccumulators*, can accumulate more than 0.1% of Pb, Co, Cr, or more than 1% of Mn, Ni, or Zn in plant shoots when growing in their natural habitat (Brooks et al., 1977; Baker and Brooks, 1989). To date, there are approximately 400 known metal hyperaccumulators in the world (Baker and Walker, 1989). Some of these hyperaccumulators have been tested for phytoextraction of metals from contaminated soils (Baker et al., 1994, Brown et al., 1994). Field trials with selected metal hyperaccumulators (*Thlaspi caerulescens, Alyssum murale, A. lesbiacum, A. tenium*, and other species) indicated that these plant species accumulated high levels of Zn and Cd in the shoots (Baker at al., 1994); however, the remediation potential may be limited by these plants because of the slow growth rate and low biomass produced by the plants. Additional studies using *T. caerulescens* and *Silene vulgaris* to remediate Zn- and Cd-contaminated soils have produced similar results (Brown et al., 1994). Recently, our laboratories and several other laboratories have carried out extensive research to screen for metal hyperaccumulating plants from high biomass species (Kumar et al., 1995; Blaylock et al., 1997; Huang and Cunningham, 1996; Huang et al., 1997b). Results from these studies demonstrated that metal hyperaccumulation can be achieved with selected high biomass agronomic crops in conjunction with soil amendments application to the contaminated soils (see discussion later in this chapter).

Metal Absorption by Roots

A number of toxic metals are also essential plant elements (Zn, Ni, Mn, and Cu), and others are not known essential elements for plants (Pb, Cd, U, and ^{137}Cs). These elements enter the plants primarily via root absorption of the cation from the soil solution. Since root cells of many plant species have plasma membrane potentials ranging from -100 to -200 mV (Huang et al., 1992; Wang et al., 1994), there exists a very large electrical chemical potential gradient across the plasma membrane. The electrical chemical potential gradient across the plasma membrane of root cells can drive these metals into root cells. Furthermore, the metal activity of the listed elements in the cytoplasm of root cells must be maintained at very low levels because of the chemical characteristics of cytoplasm. The negative membrane potentials and the low intracellular metal activity could result in a large driving force for the influx of these metals into root cells. There is circumstantial evidence in the literature suggesting that Zn, Cu, and Ni are transported into the root cell via a common transporter (Kochian, 1993). Using radiotracer techniques, plant physiologists have studied voltage-gated cation channels in the plasma membrane of root cells, and identified strongly voltage-gated Ca channels operating in the root-cell plasma membrane (Huang et al., 1994; Marshall et al., 1994). The voltage-gated channels are characterized by increased Ca transport activities as the transmembrane electrical potential is depolarized. These authors also found that the activity of the voltage-gated divalent cation channels was rapidly inhibited by Cu, Cd, and Pb. The inhibition of the channel activity by Pb, Cu, and Cd could arise from the blockage of the channel by these metals, or competitive transport of Pb and Cd through the cation channels. Using Fura-2 to monitor Pb entry into isolated bovine chromattin cells, Tomsig and Suszkiw (1991) have detected permeation of Pb through a Ca channel. They also found that voltage-gated Pb transport was blocked by nifedipine (a Ca channel blocker) and stimulated by BAY K8644 (a Ca channel agonist). These results suggest that Pb, Cd, and other divalent cations can be transported into root-cell via voltage-gated cation channels. Environmental factors could have a significant effect on the gating of these cation channels, thus altering the metal absorption by roots.

Metal Translocation from Roots to Shoots

For practical reasons, the metal concentration in plant shoots is the most important parameter for phytoextraction. Currently, the harvested portion of plants at most contaminated sites are limited to the aboveground parts; therefore, the metal absorbed by roots must be translocated to the shoots for successful phytoextraction. A number of physiological processes are involved in the long-distance transport of metals, including metal unloading into root xylem cells, long distance transport within the xylem to the shoots, and metal reabsorption from the xylem stream by leaf mesophyll cells.

The general model for metal unloading into the xylem vessels involves the absorption of metals from soil solution into the root symplasm, then the absorbed metals are unloaded from the xylem parenchyma into mature xylem vessels (Kochian, 1991). After metals are unloaded into the xylem vessels, they are transported to the shoots by the transpiration stream. For most toxic metals, the rate of metal translocation from roots to shoots is much lower compared than the rate of uptake of these metals. For example, using shoot:root concentration ratio to estimate Pb translocation from roots to shoots, Huang and Cunningham (1996) found that for any fixed time period, the amount of Pb translocated to the shoots is less than 30% of the Pb absorbed by the roots. Because of the slow translocation of Pb from roots to shoots, roots have been considered to be the main barrier for the translocation of Pb and other heavy metals. To test whether root removal would alter Pb translocation, Huang et al. (1997a) investigated Pb accumulation in excised shoots of four species (corn, goldenrod, ragweed, and sunflower) that produce high biomass. In that study, plants were grown in nutrient solution in the absence of Pb, and the shoots were excised under the solution surface at the root–shoot junction. The excised shoots were then transferred into the nutrient solution with 5 cm of the shoots immersed in the nutrient solution containing 20 μM Pb. For the portion of excised shoots immersed in the solution, the Pb concentration reached a value similar to the concentration found in intact roots for the same plant species. However, for all species tested, Pb concentration in the portion of shoots not immersed in solution was only 2–8% of the portion immersed in the solution. The results indicate that plant shoots have the capacity to accumulate Pb and probably other heavy metals; however, the limiting step is the long-distance translocation from roots to shoots.

Why is metal translocation in plants generally lower? For the metals pertinent to phytoremediation, most of them are divalent cations. As previously discussed, once these cations are transported into root cells, they are either precipitated in the cell or chelated with an organic compound. For long-distance transport, the toxic metals have to be chelated with organic compounds available inside the cell. Experimental data show that by adding synthetic chelates to nutrient solution, Pb translocation is rapidly increased (Fig. 5-2). Furthermore, the chelate-enhanced Pb translocation was pH-dependent. At pH 5.5 without added EDTA, there was very little Pb translocated to the shoots; however, following the addition of 0.2 mM EDTA to the nutrient solution containing 0.2 mM Pb, the shoot Pb concentration increased to 0.8 mg/kg. When the nutrient solution pH was decreased to 3.5, shoot Pb concentration increased to 16 mg/kg for the same Pb and EDTA concentration. The data clearly demonstrated that chelate additions could significantly increase Pb translocation from roots to shoots.

THE PHYTOEXTRACTION PROCESS

Phytoextraction utilizes the roots of plants to absorb, translocate, and concentrate toxic metals from the soil to the aboveground harvestable plant tissues. The concentration process results in a reduction of contaminated mass and also a transfer of the metal from an aluminosilicate-based matrix (soil) to a carbon-based matrix

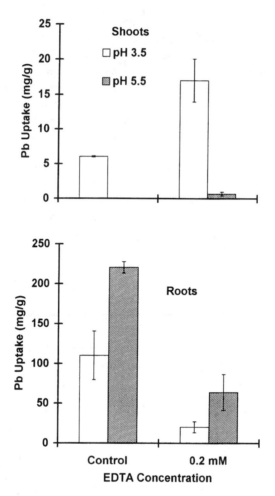

Figure 5-2   Effects of applying EDTA (2 mM) chelated Pb (2 mM) to nutrient solution on Pb accumulation in shoots (*a*) and roots (*b*) of *B. juncea* grown in nutrient solution at pH 3.5 and pH 5.5. [From Blaylock et al. (1997).]

(plants). The carbon in the plant material can be oxidized to carbon dioxide, further decreasing (and concentrating) the mass of material to be treated, disposed of, or recycled. The phytoextraction process is dependent on the metal being first accumulated by the plant roots and then translocated to harvestable plant tissues. The harvestable portions are generally regarded as the aboveground plant material, although roots of some crops may also be harvestable.

Role of Soil Amendments in Metal Phytoextraction

Plants grown on heavy-metal-contaminated soils generally do not accumulate high levels of the targeted metals in the plant tissue with the exception of certain metal hyperaccumulators, such as Zn or Ni hyperaccumulators. As we have discussed previously, these hyperaccumulators are generally not suitable for commercial phytoextraction because of the slow growth and lower biomass production of these plants. The goal of commercial metal phytoextraction is to reduce the levels of toxic metals in the contaminated soils within a reasonable timeframe (1–3 years). To achieve this goal, we need to use plants that are able to accumulate greater than 1% of targeted metals in shoots and produce more than 20 metric tons of shoot biomass/ha per year. Previous results from both greenhouse and field experiments clearly indicate that shoot metal concentrations for plants grown on the contaminated soils were far below the level targeted for commercial phytoextraction. However, with the application of soil amendments to the contaminated soils, we are now able to achieve the targeted metal concentration for commercial phytoextraction (Blaylock et al., 1997; Huang and Cunningham, 1996; Huang et al., 1997b). The application of soil amendments (such as synthetic chelates and organic acids) increases metal desorption from soil to soil solution and metal translocation from roots to shoots.

It has been demonstrated that most heavy metals are rapidly accumulated in the roots if the metals are bioavailable in the plant growth media; however, only a small portion of the absorbed metals is translocated to the shoots (Jones et al., 1973; Kumar et al., 1995; Huang and Cunningham, 1996; Ebbs and Kochian, 1997). Two major limitations to the phytoextraction of heavy metals are the low metal bioavailability in the soil and the poor metal translocation from roots to shoots. For example, for most Pb-contaminated soils that we have studied, Pb in soil solution is usually less than 0.1% of total soil Pb. Furthermore, for plants grown on the Pb-contaminated soils, Pb translocation from roots to shoots was less than 30% for the plant species showing the highest rate of Pb translocation from roots to shoots (Huang and Cunningham, 1996). A key to the success of metal phytoextraction is to increase and maintain metal concentrations in the soil solution. Chelates and other chemical compounds have been used to increase the solubility of metals in plant growth media, and could significantly increase metal accumulation in plants (Wallace et al., 1977; Norvell, 1991; Albasel and Cottenie, 1985). Data from our laboratories and others demonstrated that chelates, organic acids, and certain chemical compounds can be used to trigger metal hyperaccumulation in a number of agronomic crops with high biomass production. In the following section, we discuss the use of soil amendments to enhance phytoextraction of Pb, U, and ^{137}Cs from contaminated soils.

Lead   Results from hydroponic studies indicate that shoot Pb concentrations in plants increased dramatically as Pb levels in the nutrient solutions increased (Kumar et al., 1995; Huang and Cunningham, 1996). This leads us to use soil amendments to increase Pb desorption from soil to soil solution. For the soil amendments tested so far, certain synthetic chelates are very effective in increasing Pb levels in the soil solution. For example, within 24 h after the application of a chelate (HEDTA) to a Pb-contaminated soil (total soil Pb 2500 mg/kg), the Pb concentration in soil solution increased from 17 µM for the control (without added HEDTA) to 19000 µM for the HEDTA (6 mmol/kg) treatment (Huang et al., 1997b). The increase of Pb concentration in soil solution was positively correlated with the increase of Pb concentrations in plants (Fig. 5-3). To search for an ideal chelate to enhance Pb phytoextraction, researchers have compared the relative efficiency of five synthetic chelates (CDTA, DTPA, HEDTA, EDTA, and EGTA) in enhancing soil Pb desorption and plant Pb accumulation (Blaylock et al., 1997; Huang et al., 1997b). For the chelates tested, EDTA was the most efficient chelate in increasing Pb desorption from soil to soil solution, and Pb accumulation in plant shoots. We also examined the correlation between Pb concentrations in plants and Pb levels in soil solution. Lead concentrations in plant shoots increased linearly with increasing Pb levels in the soil solution, accomplished by applying chelates to Pb-contaminated soils (Huang et al., 1997b).

Uranium   Surface soil contamination with U has resulted from the development of the nuclear industry, which involved the mining, milling, and fabrication of various U products. Because there are large areas of U contaminated soils in the world, engineering-based remediation such as excavation requires millions of tons of soils to be disposed of as low-level radioactive waste. We have been developing phytoextraction techniques for the cleanup of U-contaminated soils (Huang et al., 1998). To search for ideal soil amendments to enhance soil U desorption from soil to

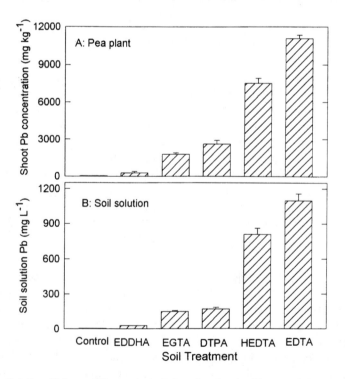

Figure 5-3   Relative efficiency of five synthetic chelates in enhancing Pb accumulation in plants grown on the chelate-treated Pb-contaminated soil (*a*) and Pb desorption from soil to soil solution (*b*). Control denotes the Pb-contaminated soil (total soil Pb 2500 mg/kg) without chelate treatment. The level of each chelate used was 0.5 g/kg soil. [Adapted from Huang et al. (1997b).]

soil solution, the U-contaminated soils were treated with a number of selected soil amendments (synthetic chelates, inorganic and organic acids, and sodium and potassium bicarbonates). The efficiency of soil amendments in enhancing soil U desorption was determined by examining U concentration in the soil solution in response to the addition of the soil amendments to the contaminated soils. Among the amendments tested, citric acid was most effective in enhancing U desorption from soil to soil solution (Huang et al., 1998). For example, application of citric acid (20 mmol/kg) to a U-contaminated soil (total soil U 280 mg/kg) increased U concentration in soil solution from 1.2 to 240 mg/L, which represents a 200-fold increase of U in soil solution. Addition of citric acid to the contaminated soil transiently reduced soil pH by 0.5–1.0 unit. The application of nitric acid at the same concentration as citric acid also reduced the soil pH in a similar magnitude; however, the increase in soil U desorption by applying nitric acid was much less than that by the citric acid (Huang et al., 1998). The results indicate that the reduction of soil pH contributed only part of the enhanced soil U desorption, and the driving force for the citric-acid-enhanced soil U desorption was probably the chelation between U and citric acid.

On the basis of the soil U desorption results, we investigated the potential of applying a number of different soil amendments to the contaminated soils in U phytoextraction. Similar to soil U desorption, citric acid was the most effective in triggering U hyperaccumulation in plants. For most plant species tested, there was a dramatic increase in shoot U concentration (Fig. 5-4). For the species tested so far, the stimulation in shoot U concentration was highest in *B. juncea* (cultivar 426308). Compared to the control (without soil amendment), the shoot U concentration in *B. juncea* increased more than 1000-fold within a week. Using the *B. juncea* cultivar, we also examined time-dependent kinetics of U accumulation. Citric-acid-triggered U hyperaccumulation could be achieved within 24 h after the amendment application, and the shoot U concentration reached a steady state 3 days after the amendment application (Huang et al., 1998).

Cesium-137 It has been reported that application of NH_4^+ and K^+ to soil can significantly alter Cs desorption from soil to soil solution and Cs accumulation in plants (Lasat et al., 1998). Besides U, radioactive Cs and Sr are radionuclides that pose major health risk to humans and animals. We have studied the potential of phytoremediation for the cleanup of ^{137}Cs-contaminated soils. Using similar strategy for the phytoextraction of Pb and U from contaminated soils, we tested the effects of applying synthetic chelates, organic acid, ammonium, and potassium fertilizers on ^{137}Cs desorption from soil to soil solution. For the soil amendments tested, ammonium fertilizers (both ammonium sulfate and ammonium nitrate) are very effective in enhancing ^{137}Cs desorption from soil to soil solution. For example, the application of 20 mmol/kg $(NH_4)_2SO_4$ to contaminated soil (total soil ^{137}Cs of 370 pCi/g) increased the soil solution ^{137}Cs from 0.4 to 4.5 pCi/g (unpublished results). Synthetic chelates and organic acids did not have significant effects on ^{137}Cs desorption from soil to soil solution. On the basis of the soil extraction results, we tested the effect of applying $(NH_4)_2SO_4$ on ^{137}Cs accumulation in a number of plant species. As shown in Figure 5-5, shoot ^{137}Cs concentration could be significantly increased by applying

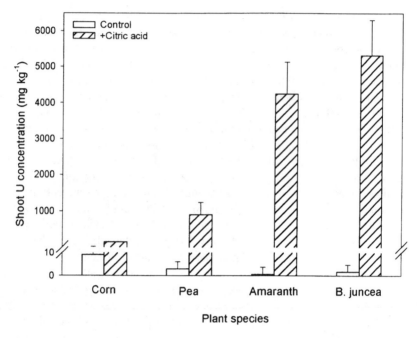

Figure 5-4   Uranium accumulation in shoots of four plant species grown on a U-contaminated soil (total soil U 750 mg/kg) in response to the addition of citric acid (20 mmol/kg soil). Plants were grown in the contaminated soil for 4 weeks before applying citric acid, and were harvested one week after the citric acid application (unpublished results).

ammonium fertilizers. Furthermore, the simulation of shoot ^{137}Cs accumulation was significantly varied among plant species tested. Corn and wheat plants showed better response to the ammonium stimulated ^{137}Cs accumulation in shoots (Fig. 5-2). Application of $(NH_4)_2SO_4$ (10 mmol/kg) increased ^{137}Cs concentration in corn shoots from 100 to 1200 pCi/g. The results suggest that we could use ammonium fertilizer to enhance phytoextraction of ^{137}Cs from contaminated soils.

Mechanisms of Soil-Amendment-Triggered Metal Hyperaccumulation

Major limiting factors for phytoextraction of metals from contaminated soils are the lower metal bioavailability in the soil and poor metal translocation from roots to shoots. Application of soil amendments could partially eliminate these limiting steps in the metal phytoextraction. There are several mechanisms involved in the soil amendment triggered metal hyperaccumulation in plants. First, the increase in metal level in soil solution is required. This is clearly demonstrated in the case for Pb, U, and [137]Cs. The addition of soil amendments increased these metals in the solution by more than 10-fold for ^{137}Cs and 100-fold for Pb and U (Huang et al., 1997b; 1998). The main driving force for soil-amendment-induced surge of metal concentration in soil solution could be the chelation between metals and the chelating compounds. In

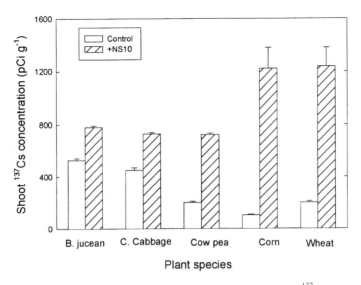

Figure 5-5   Cesium-137 accumulation in shoots of five plant species grown on a ^{137}Cs-contaminated soil (total soil [137]Cs 370 pCi/g) in response to the addition of ammonium sulfate (10 mmol/kg soil). Plants were grown in the contaminated soil for 3 weeks before applying the ammonium fertilizer, and were harvested 2 weeks after the application of ammonium sulfate (unpublished results).

the case of ^{137}Cs, ion exchange between NH_4^+ and or K^+ with [137]Cs$^+$ plays a major role in increasing [137]Cs$^+$ level in soil solution. Second, chelating compounds could buffer metal activity near the root surface and thus maintain a constant supplying of free metal to the uptake sites of roots. Third, the complex of metal chelate could be directly absorbed by roots and translocated to shoots. For example, in the case of EDDHA application, a purple color (EDDHA solution color) was observed in leaves of corn and pea plants within 12 h after applying the EDDHA solution to the contaminated soils (Huang et al., 1997b). These observations indicate that the chelate was rapidly absorbed by roots and translocated to the shoots. Recently, we have directly detected EDTA in the xylem sap collected from the plants grown on the contaminated soil treated with EDTA (unpublished results). Finally, it may also be possible that chelating compounds at higher levels alter plant ion transport systems and/or root-cell membrane structure such that metal uptake and translocation are facilitated. Further research is needed to elucidate the exact mechanism involved.

FIELD PRACTICES

Contaminated soils at industrial sites may have been exposed to numerous activities, including compaction from heavy traffic, disposal of waste products, buildings, and excavations, which disrupt the soil structure and alter the arable properties presenting unique challenges for the application of phytoremediation in the field. Techniques to restore the land to a condition conducive to plant growth is the first step before phytoremediation can occur. The application of phytoremediation in the field requires some creativity as well as solid scientific, environmental, and agronomic skills. Because of the extreme variability of soil properties at contaminated sites, debris, and regulatory concerns, one must be able to distinguish those tasks and sites that are difficult to manage from those that are impractical to improve the soil quality, fertility, and metal uptake. Phytoextraction in the field becomes somewhat of an art that draws heavily on experience from other disciplines.

Phytoextraction is still a new technology, and very little published information exists on successful field techniques. Several examples of field applications have been reported, however. Excessive concentrations of selenium in irrigated agricultural soils are being successfully managed through plant removal of selenium (Banuelos et al., 1993). Recently, some promising results in remediating Pb contaminated soils have also been reported. Phytoextraction was used at a Pb contaminated site in a residential area of Boston, Massachusetts to reduce the total soil Pb to a target concentrations of 400 mg/kg (Blaylock, 1999). The initial soil Pb ranged from 640 to 1900 mg/kg with an average of 984 mg/kg. The subsurface soil exhibited lower total Pb levels than the surface, averaging 538 mg/kg at the 15–30 cm depth and 371 mg/kg at the 30–45-cm depth. The soil the total soil Pb concentrations in the surface soil were reduced from an average of 984 to 644 mg/kg in one growing season using phytoextraction (Blaylock, 1999). A similar reduction was also reported for a site in New Jersey.

Field operations consist of an initial site visit and walkover to assess physical characteristics followed by an initial sampling and survey for a treatability study. The use of portable X-ray fluorescence spectrometers can greatly aid in the identification of the magnitude and distribution of surface contamination. On completion of the treatability study and establishment of regulatory goals and expectations, the site is prepared for cropping using standard agronomic practices. Crop selection is based on site conditions, the contaminant of concern, and the geographic location. The crop is produced and metal accumulated through addition of appropriate soil amendments with attention to minimizing any potential downward migration of the contaminant through control of irrigation. When the crop has reached its optimum metal content (product of biomass × metal concentration), it can be harvested and disposed and the process repeated. Biomass disposal options are site-dependent but focused on reducing the amount of material to be disposed of through various biomass treatment options. The process can then be repeated with successive crops until the desired regulatory goal is achieved.

Phytoextraction has the potential to remediate many metals- and radionuclide-contaminated sites. It should be realized that not every site is conducive to phytoremediation as a result of excessively high contaminant concentrations or unsuitable conditions for plant growth. However, in many sites phytoextraction may fill the role of a less expensive and invasive form of treatment.

REFERENCES

Albasel N., and A. Cottenie, 1985. Heavy metal uptake from contaminated soils as affected by peat, lime, and chelates. *Soil Sci. Soc. Am. J.* **49**:86–390.

Amacher, M. C., 1996. Nickel, cadmium, and lead. In *Methods of Soil Analysis, Part 3—Chemical Methods*, D. L. Sparks, ed., Madison, WI.

Baker, A. J. M., and R. R. Brooks, 1989. Terrestrial higher plants which hyperaccumulate metallic elements—A review of their distribution, ecology and phytochemistry. Biorecovery **1**:81–126.

Baker, A. J. M., and P. L. Walker, 1989. Ecophysiology of metal uptake by tolerant plants. In *Heavy Metal Tolerance in Plants: Evolutionary Aspects*. A. J. Shaw, ed., CRC Press, Boca Raton, FL, p. 155–176.

Baker, A. J. M. et al., 1994. The possibility of in situ heavy metal decontamination of polluted soils using crops of metal-accumulating plants. *Res. Conserv. Recycl.* **11**:41–49.

Banuelos, G. S. et al., 1993. Boron and selenium removal in boron-laden soils by four sprinkler irrigated plant species. *J. Environ. Qual.* **22**:786–792.

Blaylock, M. J. et al., 1997. I. Enhanced accumulation of Pb in Indian mustard by soil-applied chelating agents. *Environ.Sci. Technol.* **31**:860–865.

Blaylock, M. J. 1999. Field demonstrations of phytoremediation of lead contaminated soils. In: *Phytoremediation of Trace Elements*, G. S. Banuelos and N. E. Terry, eds., Ann Arbor Press, Ann Arbor, MI.

Brown, S. L. et al., 1994. Phytoremediation potential of Thlaspi caerulescens and bladder campion for zinc- and cadmium-contaminated soils. *J. Environ. Qual.* **23**:1151–1157.

Chaney, R. L., 1983. *Land Treatment of Hazardous Wastes.* Noyes Data Corp., Park Ridge, NJ, pp. 50–76.

Ebbs, S. D., and L. V. Kochian, 1997. Toxicity of zinc and copper to *Brassica* species: Implication for phytoremediation. *J. Environ. Qual.* **26**:776–781.

Harter, R. D., 1983. Effect of soil pH on adsorption of lead, copper, zinc and nickel. *Soil Sci. Soc. Am. J.* **47**:47–51.

Huang, J. W. et al., 1992. Aluminum effects on calcium fluxes at the root apex of aluminum-tolerant and aluminum-sensitive wheat cultivars. *Plant Physiol.* **98**:230–237.

Huang, J. W. et al., 1994. Voltage-dependent Ca^{2+} influx into right-side-out plasma membrane vesicles isolated from wheat roots: Characterization of a putative Ca^{2+} channel. *Proc. Natl. Acad. Sci. USA* **91**:3473–3477.

Huang, J. W., and S. D. Cunningham, 1996. Lead phytoextraction: Species variation in lead uptake and translocation. *New Phytol.* **134**:75–84.

Huang, J. W. et al., 1997a. Phytoextraction of lead from contaminated soils. In *Phytoremediation of Soil and Water Contaminants*, E. L. Kruger et al., eds., The American Chemical Society, pp. 283–298.

Huang, J. W. et al., 1997b. Phytoremediation of lead-contaminated soils: Role of synthetic chelates in lead phytoextraction. *Environ. Sci. Technol.* **31**:800–805.

Huang, J. W. et al., 1998. Phytoremediation of uranium-contaminated soils: Role of organic acids in triggering uranium hyperaccumulation in plants. *Environ. Sci. Technol.* **32**:2004–2008.

Jones, L. H. P. et al., 1973. Lead uptake from solution by perennial ryegrass and its transport from roots to shoots. *Plant Soil* **38**:403–414.

Kochian, L. V., 1991. Mechanisms of micronutrient uptake and translocation in plants. In *Micronutrients in Agriculture,* 2nd ed. J. J. Mortvedt et al., eds., Soil Science Society of America, Madison, WI, pp. 229–296.

Kochian, L. V., 1993. Zinc absorption from hydroponic solutions by plants roots. In *Zinc in Soils and Plants*, A. D. Robson, ed., Kluwer Academic Publishers, Boston, pp. 45–57.

Kumar, N. P. B. A. et al., 1995. Phytoextration: The use of plants to remove heavy metals from soils. *Environ Sci. Technol.* **29**:1232–1238.

Lasat, M. M. et al., 1998. Phytoremediation of a radiocesium-contaminated soil: evaluation of cesium-137 bioaccumulation in the shoots of three plant species. *J. Environ. Qual.* **27**:165–169.

Lindsay, W. L., 1979. *Chemical Equilibria in Soils*, Wiley, New York, pp. 329–342.

Lindsay, W. L., and W. A. Norvell, 1978. Development of a DTPA soil test for zinc, iron, manganese, and copper. *Soil Sci. Soc. Am. J.* **42**:421–428.

Marshall, J. et al., 1994. Membrane potential-dependent calcium transport in right-side-out plasma membrane vesicles from *Zea mays* L. roots. *Plant J.* **5**:683–694.

McBride, M. B., 1994. *Environmental Chemistry of Soils.* Oxford University Press, New York, pp. 336–337.

NJDEP, 1996. *Soil Cleanup Criteria.* New Jersey Department of Environmental Protection Proposed Cleanup Standards for Contaminated Sites, NJAC 7:26D.

Norvell, W. A., 1984. Comparison of chelating agents as extractants for metals in diverse soil materials. *Soil Sci. Soc. Am. J.* **48**:1285–1292.

Norvell, W. A., 1991. Reactions of metal chelates in soils and nutrient solution. In *Micronutrients in Agriculture,* 2nd ed., J. J. Mortvedt et al., eds., Soil Science Society of America, Madison, WI, pp. 187–227.

Ramos, L. et al., 1994. Sequential fractionation of copper, lead, cadmium and zinc in soils from or near Doñana National Park. *J. Environ. Qual.* **23**:50–57.

Raskin, I. et al., 1994. Bioconcentration of heavy metals by plants. *Curr. Opin. Biotechnol.* **5**:285–290.

Sommers, L. E., and W. L. Lindsay, 1979. Effect of pH and redox on predicted heavy metal-chelate equilibria in soils. *Soil Sci. Soc. Am. J.* **43**:39–47.

Tessier, A. et al., 1979. Sequential extraction procedure for the speciation of particulate trace metals. *Anal. Chem.* **51**:844–850.

Tomsig, J. L., and J. B. Suszkiw, 1991. Permeation of Pb^{2+} through calcium channels: fura-2 measurements of voltage- and dihydropyridine-sensitive Pb^{2+} entry in isolated bovine chromaffin cells. *Biochim. Biophys. Acta* **1069**:197–200.

Wallace, A. et al., 1977. Some interactions in plants among cadmium, other heavy metals, and chelating agents. *Agron. J.* **69**:18–20.

6

PHYTOSTABILIZATION OF METALS

WILLIAM R. BERTI AND SCOTT D. CUNNINGHAM

INTRODUCTION

The remediation of metal-contaminated sites often involves excavation, stabilization of the soil with cement or similar material, and landfilling of the contaminated soil; soil washing; or capping. These practices, which are based on civil engineering, are environmentally invasive and expensive. This may prohibit their use at many sites, especially very large contaminated areas and small sites within residential areas. At many of these sites, no action is taken to adequately assess and, when necessary, reduce the risks to human health and the environment. In some cases, the party responsible for a contaminated site no longer exists and the costs of remediation are transferred to the public. In other cases, the contaminated areas are not high priorities on remediation agendas. Without some remediation effort, contaminated areas that cannot support plant growth remain barren, and the metal-bearing soil remains exposed to human contact and to erosion that may carry contaminants off site. Because of the limitations of current remediation technologies, there is an increasing need for remediation techniques that are low cost, low input, and environmentally benign, yet equally protective of human health and the environment.

One developing alternative remediation technique for metal-contaminated sites is phytostabilization, also called *inplace inactivation* or *phytorestoration*. Phytostabilization is a site stabilization technique that reduces the risk of soil contaminants through the use of soil amendments that induce the formation of insoluble contaminant species. Less soluble forms of metal contaminants are less

Phytoremediation of Toxic Metals: Using Plants to Clean Up the Environment, edited by Ilya Raskin and Burt D. Ensley.
ISBN: 0-471-19254-6 ©2000 John Wiley & Sons, Inc.

likely to leach through the soil profile, and are less likely to interact biologically with humans, animals, or plants. Plants are used to cover the soil surface to prevent erosion, reduce water percolation, and serve as a barrier to prevent direct contact with the soil. Phytostabilization does not remove the contaminants from the soil, but reduces the hazards to human health and the environment by a different yet equally protective strategy. This strategy may be adaptable to a wide range of sites, including large abandoned sites and those in urban areas.

The premise of phytostabilization is not entirely new. Phytostabilization builds on a broad base of knowledge and experience in fundamental soil chemistry and agricultural and conservation practices. Traditional agronomic practices, coupled with disturbed land revegetation and land application of sludges, provide valuable insight on how to optimize plant growth in harsh or suboptimum conditions. Simultaneously, they have served as a tool for gaining a better understanding of the chemistry of common soil amendments and how heavy metals behave in soil systems and are influenced by common soil amendments. For example, the use of soil amendments such as lime, gypsum, and fertilizers has been an integral part of traditional agriculture to improve the physical and chemical properties of soil for plant growth. The practice of mine spoil revegetation as well as the management of sewage-sludge-amended soils also provides a basis for phytostabilization. Many mine spoils are characterized by disturbed soil structure, poor supply of plant nutrients such as phosphorus, and high levels of heavy metals, and may exhibit extremely low soil pH. These conditions necessitate the use of fertilizers and lime to encourage plant growth by either improving soil fertility or reducing metal toxicity to plants (Pitchell et al., 1994). Management guidelines have been developed for the land application of sewage sludges, which not only contain valuable plant nutrients for recycling but also may contain heavy metals of potential environmental concern. Sludge-amended soils are often maintained at near neutral to slightly alkaline pH using liming agents. Under these conditions, many heavy metals precipitate, becoming less soluble and less mobile (Alloway, 1995). Additionally, the organic matter in the sludge has a high affinity for some metals (e.g., Cu, Pb) (Alloway, 1995), which also may help to reduce the mobility of these metals.

Phytostabilization is similar to establishing a pasture using agricultural equipment, planting schemes, and soil amendments. Before planting, the soil is plowed to prepare a seed bed and to incorporate lime, fertilizer, and amendments for inactivating metal contaminants. Plants are either seeded or transplanted. Irrigation is used as necessary to establish plant growth. The amendments used in phytostabilization may be similar to those used in agriculture (e.g., lime, phosphorus fertilizers, organic materials). The application rates required to inactivate metal contaminants, however, are usually much greater than the rates used in agriculture to adjust soil fertility. Agriculture soil tests are useful for determining the need for lime, fertilizer, and organic matter. Soil tests may also detect other limitations to plant growth, such as high sodium levels. Because phytostabilization is similar to some agricultural practices, it is also inexpensive when compared to current remediation practices (Cunningham and Berti, in press).

Phytostabilization is an adaptable technique that can help minimize various hazards and eliminate exposure pathways to organisms of soil metal contaminants (Chaney and Ryan, 1994), although the most effective amendments and applications rates may vary for individual metal contaminants. For example, in the United States the primary concern of Pb-contaminated soils is ingestion by young children or pregnant women. Children and the developing fetus are particularly sensitive to health impairment from exposure to Pb, which is determined by measuring blood Pb concentrations. Phytostabilization of Pb contaminated soil may address this risk pathway by enhancing the formation of soil Pb species that are insoluble in the digestive tract of young children and women of childbearing age. Similar approaches may be used for other metals, such as As, Cd, Cr, Hg, or Ni (Waldron, 1980; Merian, 1991), where human health is the primary concern. Where plant toxicity or food-chain contamination are additional concerns, reducing plant availability and uptake of metal contaminants may be another goal of phytostabilization. The specific concerns for contamination may vary from site to site. With all metal contaminants in soils, degradation of environmental quality is also an important consideration, although this may be difficult to assess. Whatever the risk pathway or target organism, the amount of risk posed by contamination is related to the exposure rate and duration to the organism of concern, and the form of the contaminant in the soil. The form of the contaminants in the soil can by altered with soil amendments that promote the formation of biologically inactive metal species.

THE USE OF SOIL AMENDMENTS

Soil amendments for phytostabilization should inactivate metal contaminants rapidly following incorporation, preventing leaching, plant uptake, and reducing biological availability (Fig. 6-1). The chemical alterations of the contaminant in the soil should be long lasting if not permanent. Additionally, soil amendments should (1) be inexpensive, (2) be easy to handle and apply, (3) be safe to the workers handling the amendment, (4) be compatible with and nontoxic to the plants selected for revegetation, (5) be readily available or easy to produce, and (6) not cause additional environmental impact to the site. Some amendments may have secondary benefits such as supplying plant nutrients or increasing soil moisture holding capacity (e.g., phosphate fertilizers, organic materials). Soil amendments that have little to "negative" economic value are preferred to more expensive materials. Such materials are usually waste products or byproducts that must be disposed or recycled. Material generators will sometimes pay others to take them. These materials include biosolids from wastewater treatment plants, animal and plant manures, and byproducts from industrial processes. The most promising soil amendments to reduce the availability of metal contaminants to sensitive organisms are phosphate fertilizers, organic matter or biosolids, iron (Fe) or manganese (Mn) oxyhydroxides, natural or artificial (synthetic and byproduct) clay minerals, or mixtures of these amendments. The mechanisms by which different amendments alter the form of metal contaminants in soil may vary with different soil amendments and contaminants, and, in some cases,

AVRIL ROBARTS LRC

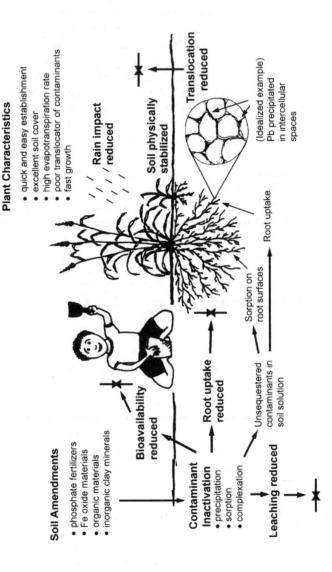

Figure 6-1 The role of soil amendments and plants in the phytostabilization of heavy-metal-contaminated soil.

are largely speculative. Suggested mechanisms include precipitation, humification, sorption, and redox transformations, (Table 6-1) (Cunningham and Berti, in press; Green, 1997).

Phosphate amendments may enhance the formation of highly insoluble forms of metal contaminants, especially Pb. Phosphate amendments may include phosphoric acid, phosphate fertilizers, or byproducts high in phosphates. In the case of Pb-contaminated soils, phosphate addition may produce pyromorphites (Ma et al., 1993; Zhang et al., 1997), which are Pb phosphate minerals that are extremely insoluble even under very acidic conditions (e.g., the conditions in the stomach of a fasting human). Phosphate changes the chemistry of some heavy metals very rapidly. Changes in Pb leachability, solubility, or bioavailability appear to occur rapidly (Laperche et al., 1997). Additionally, the changes seem to be stable and long-lasting. The application rates necessary to achieve desirable changes in contaminant solubility are typically much higher than the fertilizer rates used in agriculture (e.g., 5000 mg P/kg for Pb inactivation compared to 15–30 mg P/kg for some agricultural crops). For Pb-contaminated soils, at least, phosphate amendment appears to be one of the more

TABLE 6-1.   Soil Amendments in Phytostabilization, Suggested Applicability to Metal Contaminants, and Possible Modes of Contaminant Inactivation

Amendment Type		Possible Target Contaminants	Suggested Mode of Inactivation
Phosphate materials	H_3PO_4, apatite, calcium orthophosphates, Na_2HPO_4, KH_2PO_4, other phosphate fertilizers, high-phosphate byproducts	Pb	Formation of insoluble metal phosphate minerals, such as Pb pyromorphites
Hydrous Fe oxides	Iron Rich or other byproducts containing hydrous Fe oxides, isolated hydrous Fe oxides	As, Cd, Cu, Ni, Pb, Zn	Sorption of contaminants on oxide surface exchange sites, coprecipitation, or formation of contaminant–Fe compounds
Organic materials	Manures, composts, sludges, and other biosolids	As, Cd, Cu, Pb	Sorption of contaminants on exchange sites, or incorporation into the organic material
Inorganic clay minerals	Synthetic zeolites, natural aluminosilicates, or aluminosilicate byproducts from burning of coal refuse	As, Cd, Cu, Mn, Ni, Pb, Zn	Sorption of contaminants on mineral surface exchange sites, or incorporation into the mineral structure

Source: Adapted from Cunningham and Berti (in press) and Green (1997).

promising forms of phytostabilization. Without careful planning and management, however, the treatment of some soils with phosphates may produce undesirable effects. Added phosphates may cause soil As to become more mobile (Berti, unpublished data). Additionally, excess soil phosphate may pose an environmental concern in areas with surface waters that may be sensitive to nutrient inputs.

Materials containing hydrous Fe oxides have a great capacity to sorb heavy metals such as As, Cd, Pb, Ni, and Zn, and reduce their availability or mobility in soils. Several research groups are investigating the use of one material, Iron Rich, which is a byproduct from TiO_2 production and contains 50–60% hydrous Fe oxide by weight (Berti and Cunningham, 1997; Brown and Chaney, personal communication, 1996). Other research has investigated the use of pure hydrous Fe oxides for immobilizing Pb and Cd in contaminated soils. Sellstone et al. (personal communication, 1997) found that hydrous Fe oxide added at 2.5–5.75% (w/w) to a Pb-contaminated soil reduced Pb solubility in an extraction mimicking the fasting conditions of child's stomach by 34–50%. These researchers found that a combination of hydrous Fe oxides and phosphoric acid or triple superphosphate (a type of calcium phosphate fertilizer) produced a greater reduction in Pb solubility in this extraction than any of these materials alone.

Organic materials such as composts, manures, sludges, or byproducts from burning fossil fuels, may inactivate metals such as As, Cd, Cu, Pb, or Zn by sorption at exchange sites or by complexation. Most of these amendments are byproducts or waste products that may be readily available and relatively inexpensive to obtain, but their properties may vary from one batch to another. Some organic amendments have been shown to reduce metal solubility (Brown and Chaney, 1996; Berti and Cunningham, 1994); however, research on Pb-contaminated soils suggested that a reduction in Pb solubility with the addition of some biosolids did not necessarily translate to reduced biological availability (Brown and Chaney, 1996).

Other amendments include synthetic, natural, or byproduct aluminosilicates (Mench et al., 1994; Gworek, 1992; Vangronsveld et al., 1993). One study demonstrated that some synthetic zeolites may have a high capacity to sorb metals tightly to the zeolite surface (Gworek, 1992). In some cases, contaminants may also be incorporated into the zeolite structure with time, possibly resulting in a more permanent reduction in metal solubility. It may be possible to select specific zeolites for their affinity for specific metals. Synthetic zeolites, however, may be expensive to produce for phytostabilization purposes. Other studies have investigated the use of a modified aluminosilicate (Beringite) from the burning of coal refuse (Mench et al., 1994; Vangronsveld et al., 1993). Beringite appeared to have a high affinity for Cd, Cu, Pb, and Zn in one contaminated soil, and it reduced the solubility and plant availability of these metals. In addition to the metals listed above, the treatment of contaminated soils with zeolites or modified aluminosilicates also appears effective for Mn, Ni, and some forms of As (Green, 1997).

Liming agents have been used to maintain a neutral soil pH in sludge-amended soils and to overcome plant growth limiting factors in the reclamation of acid mine spoils (Albesel and Cottenie, 1985). The ability of liming agents to inactivate metal contaminants in soils, however, may be limited. Liming agents must be periodically reapplied to maintain the desired soil pH at which metal contaminants are insoluble. In some cases, the pH at which metal inactivation is optimum may be too high for healthy plant growth, reducing plant micronutrient availability and necessitating further site input. Additionally, the forms of metal (e.g., Pb) that may be present under limed conditions may not have reduced biological availability. Liming may reduce plant availability of heavy metals (Albesel and Cottenie, 1985), but may not reduce the solubility of contaminants (e.g., Pb) under very acidic environments such as the human stomach (Brown and Chaney, 1996). Because of its limitations, liming alone does not appear to be a suitable long-lasting treatment for phytostabilization. Lime, as a metal inactivation agent, may be beneficial as a component of a treatment matrix to enhance the effects of other materials (Green, 1997).

THE ROLE OF PLANTS

Plants play an important secondary role in phytostabilization by physically stabilizing the soil with dense root systems to prevent erosion and by protecting the soil surface from human contact and rain impact with a dense canopy (Fig. 6-1). Plant roots also help to minimize water percolation through the soil, further reducing contaminant leaching. Plant roots may also provide surfaces for sorption or precipitation of metal contaminants (Laperche et al., 1997). Some plants sequester in their roots any soil metal contaminants that remain in the soil solution. Currently, the role of plants in phytostabilization does not extend to chemically altering the form of the contaminants. It is conceivable that some plants could contribute to contaminant inactivation by directly inducing the formation of insoluble metal compounds inside plant tissues or on root surfaces. This and similar possibilities have not received much attention and are so far relatively unexplored.

There are some important considerations when selecting plants for phytostabilization. Unlike the plants chosen for phytoextraction, plants chosen for phytostabilization should be poor translocators of metal contaminants to aboveground plant tissues that could be consumed by humans or animals. Plants should also be tolerant of the soil metal levels as well as the other initial site conditions (e.g., soil pH, salinity, soil structure, water content). The plants must grow quickly to establish ground cover, have dense rooting systems and canopies, and have relatively high transpiration rates to effectively dewater the soil. Additionally, the plants must be easy to establish and care for, and have a relatively long life or be able to self-propagate. Some plant species or plant ecotypes have been found growing on contaminated soils, seemingly without much difficulty. Although these plants may exhibit some desirable site tolerance characteristics, many do not grow quickly or are small and cannot provide sufficient ground cover. At this time, the plants that appear most suited for phytostabilization are plants with well-known agronomic characteristics, such as those used for soil conservation and management.

DETERMINING HAZARD REDUCTION

Some measurement of hazard reduction must be made to validate the effectiveness of phytostabilization because the contaminants are not removed from the soil. Currently, there is no regulatory or scientific consensus on what types of measurement best indicate a hazard reduction. A variety of techniques may be used to demonstrate the effect of soil treatments on soil metal contaminants. These techniques include chemical, spectroscopic, microscopic, or biological tests that may provide information regarding the solubility, mineralogy, or biological availability of the contaminants.

Chemical tests are often used to determine changes in the relative solubility of a metal after the incorporation of amendments. These tests include U.S. regulatory tests such as the toxicity characteristic leaching procedure (TCLP) (U.S. EPA, 1990) and the simulated precipitation leaching procedure (SPLP) (U.S. EPA, 1995), which were designed to measure the leachability of contaminants under landfill conditions (TCLP) or acid rain conditions (SPLP). Another type of test is sequential chemical extractions. The methods for sequential chemical extraction vary for each metal and differ among researchers (Berti et al., 1997). Sequential chemical extraction often begins with a mild extractant, such as pure water or a dilute salt, and continues with progressively harsher extractants until the last step, which is often a total digestion of the soil by boiling in strong acids. This type of test does not provide conclusive evidence regarding the metal species present in the soil, but can help categorize the forms of metal that may be present in the soil based on differences in solubility. Solubility "profiles" of the soil can be compared before and after treatment with soil amendments to show shifts of the contaminants toward more tightly bound or unavailable fractions.

Other chemical tests have been used to characterize a contaminant's potential bioavailability to a target organism (e.g., human). One such test is known as the *physiologically based extraction test* (PBET) (Ruby et al., 1993, 1996). It was designed to simulate the conditions in the stomach and intestines of a young fasting child (e.g., pH, digestive enzymes, organic acids, temperature, and residence time). The PBET was originally designed to estimate risks for Pb and As. For these metals, soil ingestion by children is most often the primary concern. Results from the PBET appear to correlate well with swine and rat dosing studies for Pb ($r^2 = 0.93$ for rats (Ruby et al., 1996), and $r^2 = 0.85$ for swines (Medlin, 1997), Fig. 6-2). For As, the correlation between PBET results and animal models [swine (Medlin, 1997), rabbit (Cotter-Howells et al., 1994), and monkey (Manceau et al., 1996), Fig. 6-3) is not as good as with Pb. For both As and Pb, however, bioavailability as measured using animal models varies significantly from the default values of 30% for soil Pb and 100% for soil As (Figs. 6-2 and 6-3). Currently, animal dosing studies (e.g., swines) are more widely accepted for estimating Pb and As availability to children, although chemical extraction tests have been accepted as a measure of bioavailability in a few specific cases (Ruby, personal communication).

Spectroscopic and microscopic techniques are useful to determine the chemical forms of metal contaminants in the soil. These tests include X-ray diffraction (XRD),

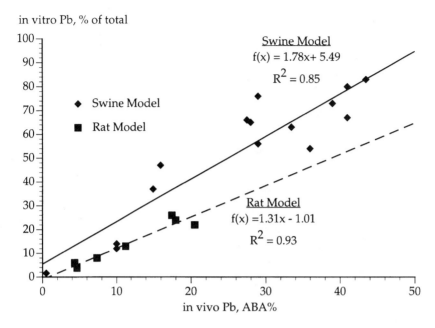

in vitro Pb, % of total

Figure 6-2   In vitro relative Pb bioaccessibility measured using the PBET procedure at pH 1.5 versus EPA swine model blood Pb relative bioavailability (RBA) [diamonds (Medlin 1997)] and in vitro relative Pb bioavailability measured using the PBET procedure at pH 2.5 versus Sprague–Dawley rat model blood Pb RBA [squares (Ruby et al., 1996)].

X-ray absorption spectroscopy (XAS), and scanning electron microscopy with energy dispersive X rays (SEM/EDX). Currently, the use of XRD and XAS on whole soils or soil components has limitations, and may still require further development. XRD usually requires metal concentrations greater than 1% and is applicable only where metals are in a crystalline form. XAS has been a valuable tool in describing the metal chemistry of relatively simple systems, such as isolated minerals, by providing information regarding the oxidation state and types of chemical bond present for an element. This technique quickly becomes difficult as the matrix becomes more complex (i.e., becomes more like whole soil). Some recent research using whole Pb-contaminated soils, however, has produced promising results (Cotter-Howells et al., 1994; Manceau et al., 1996). SEM provides qualitative information concerning the physical or chemical nature of soil metal contaminants. This technique is well developed and widely used in soil chemistry, and has proved useful in detecting soil Pb minerals with low solubility (Davis et al., 1992, 1993).

The third category of techniques to measure hazard reduction include bioassays that target specific organisms, such as plant assays, earthworm assays, animal dosing studies (e.g., swine or rat studies), or human dosing studies. Because most bioassays are organism-specific, their results may have limited applicability to other organisms. One common exception, however, is the use of animal dosing studies to estimate contaminant bioavailability to humans. The value of a bioassay in determining a

in vitro As, % of total

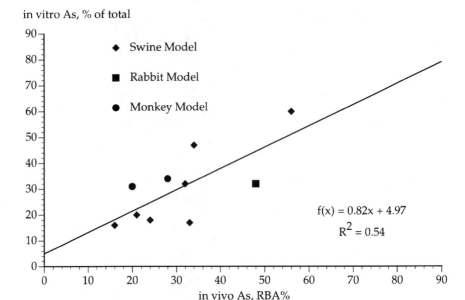

Figure 6-3   In vitro relative As bioaccessibility measured using the PBET procedure at pH 1.5 vs. EPA swine model blood As relative bioavailability (RBA) [diamonds; (Medlin, 1997)], and in vitro relative Pb bioaccessibility measured using the PBET procedure at pH 2.5 versus rabbit (squares) and *Cynomolgus* monkey model (circles) blood As RBA (Ruby et al., 1996).

hazard reduction depends on the selection of an appropriate target organism. The target organism may be the organism most sensitive to the contaminants at a site, or the organism chosen as the most important to protect. For instance, at sites contaminated with Pb or As, swine or rat dosing studies have been used to estimate bioavailability to humans. These tests are expensive and time-consuming, which prohibits their frequent use. Physiologically relevant tests, such as the PBET mentioned above, may provide an easy and inexpensive alternative to animal dosing studies.

BIOAVAILABILITY AND RISK ASSESSMENT

The techniques described above measure different physical and chemical aspects of metal contaminants in soils that can sometimes be related to metal bioavailability. The term *bioavailability*, under its broadest definition, refers to the portion of metal in soil that is capable of interacting with macro- and microorganisms (Peijnenburg et al., 1997). When human health is the primary concern at a site, bioavailability is more narrowly defined as the portion of the soil contaminant that can enter the human circulatory system following ingestion (Hrudley et al., 1996). Other pathways into the body (e.g., inhalation or dermal contact) may also be considered, but are minor for most inorganics in soils. The issues surrounding bioavailability are still evolving. There continue to be many unanswered questions regarding the best ways to measure bioavailability for the many organisms at potential risk from heavy-metal contamination. For Pb, much research has been conducted to elucidate the factors that affect bioavailability in humans (e.g., soil chemistry, human nutrition, and stomach pH). For many other metal contaminants, however, relatively little information has been gathered regarding bioavailability and how it can be measured. Although there are many uncertainties surrounding bioavailability, it is recognized as an important component of risk, and may be valuable in determining remediation requirements. It is also a key factor in the theory behind phytostabilization.

Our current understanding of how bioavailability contributes to human risk is captured in the integrated exposure and uptake biokinetic (IEUBK) model (U.S. EPA, 1994a, 1994b). The IEUBK model was developed by the U.S. EPA to estimate the percentage of children who are at risk in areas contaminated with Pb. It incorporates potential Pb exposure from multiple sources, including air, food, water, soil, and dust. The model also considers many other parameters such as age, nutritional status, and Pb bioavailability. IEUBK predicts the blood Pb concentrations of children in an exposed population. Using default assumptions, including that the total soil Pb is 30% bioavailable to a child, soils cannot contain more than about 400 mg Pb/kg before more than 5% of the children in an exposed population exhibit blood Pb above a critical level of 10 µg Pb/dL blood. By changing only the default soil Pb bioavailability value in the model from 30% to 10% for a soil containing 2000 mg

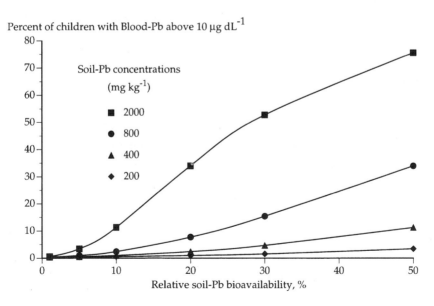

Figure 6-4   Influence of soil Pb bioavailability on the percentage of children with blood Pb levels above 10 µg/dL, as determined using the IEUBK model (U.S. EPA, 1994a, 1994b).

AVRIL ROBARTS LRC

Pb/kg, we see that the percent of children predicted to be at risk falls from over 50% to 11% (Fig. 6-4). This scenario may still present an unacceptable risk to many even when soil Pb bioavailability is 10%. The demonstration, however, shows that in some risk scenarios, bioavailability may have an influential role in determining the human and environmental hazards associated with metal contaminants. The demonstration also suggests that remediation technologies that target contaminant bioavailability, such as phytostabilization, should have a dramatic effect on risk reduction.

There may be different ways to determine or model bioavailability depending on the target organism (i.e., the organism that is to be protected), or metal of concern. The IEUBK model is specific for ingestion of Pb by children of up to 7 years old. At many Pb-contaminated industrial sites where children are not likely to be present, the exposure of pregnant site workers and the level of Pb in fetal blood is a primary concern. In these cases, another risk assessment approach for adult exposures has been developed and proposed by the U.S. EPA Technical Review Workgroup for Lead (U.S. EPA, 1996). At this time, very few models are available for metal contaminants other than Pb that include bioavailability as an important component in assessing hazard and risk. The development of models that include bioavailability may prove valuable for site and risk assessment regardless of whether phytostabilization is considered as a remediation option. The development of such models will most likely depend on simultaneous development of tests to measure bioavailability.

GENERAL CONSIDERATIONS FOR PHYTOSTABILIZATION

Phytostabilization offers some attractive advantages when compared to other site stabilization techniques. Adding amendments such as fertilizers or organic materials and establishing a vegetative cover may be less disruptive to the site than some current remediation practices that involve excavation or stabilization with cement. Phytostabilization can be adapted to different sites with a variety of conditions (e.g., soil pH, salinity, soil texture, metal levels, and contaminant types) through the careful selection of amendments, rates, and plants. Because it is an agronomically based technique, phytostabilization is relatively inexpensive and is the least expensive alternative to traditional site stabilization techniques. Phytostabilization may be an appropriate remediation choice at many types of sites (e.g., industrial sites, shooting ranges, smelting sites, mining areas, and urban areas). This technique may be advantageous where decontamination strategies are impractical because of the size of the area or the funds available for remediation. Phytostabilization may also serve as an interim strategy to reduce risk at sites where complications delay the selection of the most appropriate technique for the site. If phytostabilization is not accepted as the final strategy for these sites, it can at least be used to protect barren contaminated areas from erosion or leaching until an acceptable approach is determined. A simplified version of phytostabilization could be implemented by homeowners who have slightly elevated levels of heavy metals in the soil to help reduce the potential risks to their families. For instance, a homeowner whose soil may contain 300–400 mg Pb/kg may reduce health hazards by applying phosphate fertilizers and maintaining a healthy lawn over the area of concern. Sites such as small areas around homes may not be considered top priorities on some remediation agendas; phytostabilization may allow property owners to further reduce any actual or potential hazard of soil Pb.

Some technical and logistical aspects of phytostabilization must be addressed through research. To date, most research in phytostabilization has been conducted on a small scale in the laboratory, in the greenhouse, or on small plots. The transfer of this research to field plots has recently begun to demonstrate this technique to be both practical and effective at a full-scale level. More field trials would help strengthen regulatory and public support for this technique. Additionally, much of the research has been conducted primarily with Pb-contaminated soils, most of which have contained less than 4000 mg Pb/kg. Phytostabilization must be examined for a variety of soils with a variety of contaminants to determine its applications and limitations regarding contaminant types and concentrations. So far, soil amendments and rates have been chosen empirically from fundamental knowledge of soil chemistry. For instance, phosphate addition to Pb-contaminated soils is often based on the molar ratios of the desired Pb species to be formed. As the development of phytostabilization continues, this approach should eventually blend with experience in using different amendments and rates in research.

The environmental impacts of the application of metal-inactivating amendments to soils must also be investigated along with the effects of these amendments on the metals of concern. For instance, large applications of phosphate fertilizers may effectively inactivate heavy metals such as Pb, but could cause nutrient imbalances in nearby sensitive surface waters, possibly resulting in eutrophication. The potential impacts of soil amendments on environmental quality must be well understood so that they can be properly managed to prevent further degradation of the environment.

Phytostabilization appears to be adaptable to many types of site. Despite its adaptability, this technique may have some limitations. At some sites, conditions may be too harsh to support plant growth without extensive efforts, resources, and expense to ameliorate the existing site conditions and improve soil fertility. It may be possible to incorporate soil amendments to depths below that of typical agricultural plows, although this may be labor-intensive, require specialized equipment, and may increase the costs associated with remediation. Phytostabilization at sites contaminated with several heavy metals, or with a combination of metal and organic contaminants, may require a multistep approach. In these cases, different types of soil amendments could be applied together or in stages to the site to inactivate different types of contaminants. Alternatively, phytostabilization could be included with other techniques as part of a comprehensive remediation strategy.

REGULATORY CONSIDERATIONS

Regulatory acceptance of a developing remediation strategy is crucial to the establishment and use of any new technique. Regulatory and public acceptance often requires thorough demonstration of the technique to prove that it is practical and equally protective of human health and the environment compared to current techniques. The phytostabilization field trials that have been recently initiated, if successful, should help raise support for this technique. The risk reduction afforded by phytostabilization, however, may be difficult to convey to nonscientific communities. For this technique, risk reduction is based on technical concepts of contaminant solubility and bioavailability. Not only is the chemistry of the technique complex; the issues concerning bioavailability remain matters of debate. It is possible that until phytostabilization is accepted widely as a viable remediation technique for metals-contaminated sites, the technique could be implemented in certain situations as a site maintenance technique to help establish plant growth on barren sites, or to help reduce risk until an acceptable remediation strategy is determined. Using phytostabilization in this way may also help to provide additional demonstration of the efficacy of this technique.

ECONOMIC CONSIDERATIONS

For any emerging remediation technique to become widely practiced, it should be economical to implement. A brief economic comparison of phytostabilization with traditional site stabilization and decontamination strategies is provided in Table 6-2 (Cunningham and Berti, in press). Phytoextraction is included in the comparison because it is another promising remediation alternative. This comparison considers the costs to implement these techniques on sites contaminated with greater than 0.2% Pb to a depth of 30 cm. The soil is classified as a characteristic hazard because the Pb exceeds the U.S. regulatory TCLP limit of 5 mg Pb/L. Landfilled soil must be placed either in an industrial landfill with cement stabilization to reduce the TCLP to below 5 mg Pb/L or in a hazardous waste landfill without stabilization. For phytoextraction, the soil was assumed to be contaminated with 0.14% Pb. Annual harvests were assumed to be at least 40 tons of plant biomass per hectare at 1% Pb per dry weight of plant tissue, which would reduce the soil Pb level to 0.04% (400 mg Pb/kg) within 10 years (or 20 tons of biomass per hectare per year in 20 years). At those sites where remediation requires more than one year (i.e., phytoextraction, phytostabilization, and soil capping), the costs are adjusted to reflect an annual inflation rate (3%) and an annual discount rate (12%).

Site decontamination techniques include (1) site excavation and landfilling of the soil in a regular landfill, (2) soil washing by particle separation and placement of the fines in a hazardous waste landfill, and (3) phytoextraction using Na_4EDTA (1 mol EDTA to 1 mol Pb) with harvests as described above. The economic ranking of these techniques is excavation and landfilling ($1.6 million/ha) > soil washing ($790,000/ha) > phytoextraction ($279,000/ha). Site stabilization techniques include (1) asphalt capping for site end use as a parking lot, (2) soil capping with 60 cm of uncontaminated off-site soil and a vegetative cover, and (3) phytostabilization with triple superphosphate (a calcium phosphate fertilizer) at 90 tons/ha and Iron Rich at 400 tons/ha, with fertilizer, lime, and four mowings per year for 30 years. Of these techniques, asphalt capping is most expensive ($160,000/ha), soil capping is moderately expensive ($130,000/ha), and phytostabilization is least expensive

TABLE 6-2   Economic Comparison of Selected Site Decontamination and Stabilization Techniques[a]

Remediation Alternative	Description	Net Present Cost (U.S. $/ha)
	Site Decontamination	
Excavation and landfilling	Soil is excavated to 30 cm, stabilized with cement, and placed in an industrial landfill	1,600,000
Soil washing by particle separation	The soil is excavated to 30 cm, processed in a soil washing facility to remove fine material (20% of whole soil); fine material is stabilized with cement and placed in a hazardous waste landfill	790,000
Phytoextraction	EDTA is applied at a rate of 7400 kg/ha (1 mol EDTA to 1 mol Pb), along with lime and fertilizer; annual plant harvest is 40 tons/ha, with plant concentrations of 1% Pb; soil Pb is reduced from 0.14 to 0.04% in 10 years	279,000
	Site Stabilization	
Asphalt capping	The asphalt cap will be used as parking lot; the cap consists of a 20-cm base layer, 25-cm subgrade layer, and a 4-cm top layer; drains and curbs are included	160,000
Soil capping	Uncontaminated off-site soil is placed over the contaminated area to a thickness of 60 cm; vegetation is established on the cap surface, but no other barriers or geotextiles are included	130,000
Phytostabilization	Lime and nitrogen and potassium are added to improve soil fertility; triple superphosphate (a calcium orthophosphate) is applied at 90 tons/ha and Iron Rich is applied at 400 tons/ha; grass is sown, and the site is mowed 4 times per year for 30 years	60,000

[a]Economic comparison summarized from Cunningham and Berti (in press).

($60,000/ha). In their respective categories, both phytoextraction and phytostabilization are the least expensive techniques to implement. Most would prefer to use a decontamination strategy to remove the cause for concern to avoid limitation of future landuse options. This is not always possible, however, for practical or economic reasons. Where a site stabilization technique is acceptable, phytostabilization may be an attractive option.

AVRIL ROBARTS LRC

CONCLUSION

Phytostabilization is an emerging alternative stabilization technique that appears to be adaptable to a range of site conditions and remediation needs. It is an attractive technique because it is less expensive than current site stabilization practices, probably less environmentally invasive, and easy to implement. This technique may be particularly well suited to sites where stabilization is an appropriate approach to remediation, or where a short-term risk-reducing practice is acceptable until the final remediation of a site can be accomplished. Currently, some technical and regulatory issues must be resolved before phytostabilization can become a widely practiced technique. The types and rates of different inactivating amendments must be optimized for the most effective reduction in risk. Longevity of treatments and limits associated with contaminant type and concentration require further investigation. Questions regarding bioavailability and how it should be measured still remain, and are perhaps the most crucial questions for the acceptance of phytostabilization. The technical development of phytostabilization may be inherently important apart from the establishment of an alternative remediation technique. The scientific knowledge gained from the intensive study of the chemistry of heavy metals in soils and their interactions with other soil components will be a valuable asset to the hazard and risk assessment of contaminated soils. Additionally, the development of a technique such as phytostabilization brings together researchers from many disciplines. This interdisciplinary approach to solving problems of soil contamination will promote greater interaction among these areas, which may lead to additional insights and solutions.

ACKNOWLEDGMENTS

The authors wish to thank Gary Quinton for his invaluable assistance in the economic analysis of remediation alternatives, and Ellen Cooper for her excellent help in preparing and editing the chapter.

REFERENCES

Albesel, N., and A. Cottenie, 1985. Heavy metals uptake from contaminated soils as affected by peat, lime, and chelate. *Soil Sci. Soc. Am. J.* **49**:386–390.

Alloway, B. J., 1995. *Heavy Metals in Soils*, 2nd ed. Blackie Academic and Professional, New York.

Berti, W. R., and S. D. Cunningham, 1994. Remediating soil with green plants. In *Trace Substances, Environment and Health*, D. R. Cothern, ed., Science Reviews, Northwood, UK, pp. 43–51.

Berti, W. R., and S. D. Cunningham, 1997. Inplace inactivation of Pb in Pb-contaminated soils. *Environ Sci Technol.* **31**:1359–1364.

Berti, W. R. et al., 1997. Sequential chemical extraction of trace elements: development and use in remediating contaminated soils. In *Proc. 3rd International Conf. Biogeochemistry of Trace Elements.*

Brown, S. L., and R. L. Chaney, 1996. Personal Communication. Further information on the Joplin site being considered for field plots for the IINERT study of in situ inactivation of soil Pb.

Chaney, R. L., and J. A. Ryan, 1994. *Risk Based Standards for Arsenic, Cadmium, and Lead in Urban Soils.* Dechema, Frankfurt, Germany.

Cotter-Howells, J. D. et al., 1994. Identification of pyromorphite in mine-waste contaminated soils by ATEM and EXAFS. *Eur. J. Soil Sci.* **45**:393–402.

Cunningham, S. D., and W. R. Berti, in press. Phytoextraction or phytostabilization: technical, economic, and regulatory considerations of the soil-lead issue. In *Proc. 4th International Conf. Biogeochemistry of Trace Elements*, Terry, N., ed.

Davis, A. et al. 1992. Bioavailability of arsenic and lead in soils from the Butte, Montana, Mining District. *Environ. Sci. Technol.* **26**:461–468.

Davis, A. et al. 1993. Micromineralogy of mine wastes in relation to lead bioavailability, Butte, Montana. *Environ. Sci. Technol.* **27**:1415–1425.

Green, A. S., 1997. *Report from the In Situ Soil Remediation Technology Workshop of the International Lead Zinc Research Organization, Inc.* (ILZRO).

Gworek, B., 1992. Lead inactivation by zeolites. *Plants and Soil* **143**:71–74.

Hrudley, S. E. et al., 1996. *Bioavailability in Environmental Risk Assessment*. CRC Lewis Publishers, New York, p. 7.

Laperche, V. et al., 1997. Effect of apatite amendments on plant uptake of lead from contaminated soil. *Environ. Sci. Technol.* **31**:2745–2753.

Ma, Q. et al., 1993. Immobilization of soil Pb by apatite. *Environ. Sci. Technol.* **27**:1803–1810.

Manceau, A. et al., 1996. Direct determination of lead speciation in contaminated soils by EXAFS spectroscopy. *Environ. Sci. Technol.* **30**:1540–1552.

Medlin, E. A., 1997. *An In Vitro Method for Estimating the Relative Bioavailability of Lead in Humans*. M.S. thesis, University of Colorado, Boulder.

Mench, M. J. et al., 1994. A mimicked in-situ remediation study of metal-contaminated soils, with emphasis on cadmium and lead. *J. Environ. Qual.* **23**:58–63.

Merian, E., ed., 1991. *Metals and Their Compounds in the Environment: Occurrence, Analysis, and Biological Relevance.* VCH, New York.

Peijnenburg, W. J. G. M. et al., 1997. A conceptual framework for implementation of bioavailability of metals for environmental management purposes. *Ecotox. Environ. Safety* **37**:163–172.

Pitchel, J. R. et al., 1994. Comparison of amendments and management practices for long-term reclamation of abandoned mine lands. *J. Environ. Qual.* **23**:766–772.

Ruby, M. V. et al., 1996. Estimation of lead and arsenic bioavailability using a physiologically based extraction test. *Environ. Sci. Technol.* **30**:422–430.

Ruby, M. V. et al., 1993. Development of an in vitro screening test to evaluate the in vivo bioaccessibility of ingested mine-waste lead. *Environ. Sci. Technol.* **27**:2870–2877.

U.S. Environmental Protection Agency, 1990. Hazardous waste management system; identification and list of hazardous waste toxicity characteristic revision; final rule, Part II, *Fed. Reg.* **55**(61):11,798–11,877; Part V, *Fed. Reg.* **55**(126):26,986–26,998.

AVRIL ROBARTS LRC

U.S. Environmental Protection Agency, 1994a. *Guidance Manual for the Integrated Exposure Biokinetic Uptake Model for Lead in Children.* EPA/540/R-93/081,PB93-963510. U.S. Environmental Protection Agency, Office of Emergency and Remedial Response, Research Triangle Park, NC.

U.S. Environmental Protection Agency, 1994b. *Technical Support Document: Parameters and Equations Used in the Integrated Exposure Biokinetic Uptake Model for Lead in Children.* (v. 099d). U.S. Environmental Protection Agency, Office of Emergency and Remedial Response: Research Triangle Park, NC.

U.S. Environmental Protection Agency, 1995. *Test Methods for Evaluating Solid Waste. Physical/Chemical Methods,* 3rd ed., 4 vols. Proposed update EPA 530/SW-846, Chapter 6.

U.S. Environmental Protection Agency, 1996. *Recommendations of the Technical Review Workgroup for Lead for an Interim Approach to Assessing Risks Associated with Adult Exposures to Lead in Soil.* Adult Lead Risk Assessment Committee, Technical Review Workgroup for Lead, New York.

Vangronsveld et al., 1993. Reclamation of a bare industrial area contaminated by non-ferrous metals: *in situ* metal immobilization and revegetation. *Environ. Pollut.* **87**:51–59.

Waldron, H. A., ed., 1980. *Metals in the Environment.* Academic Press, New York.

Zhang, P. et al., 1997. Pyromorphite formation from goethite adsorbed lead. *Environ. Sci. Technol.* **31**:2673–2678.

7

PHYTOFILTRATION OF METALS

SLAVIK DUSHENKOV AND YORAM KAPULNIK

INTRODUCTION

Pollution of surface water and groundwater with heavy metals and radionuclides is a problem humanity can not afford to ignore. Enormous amounts of heavy metals have been mobilized in the past century as a result of global, industrial, and, in particular, metalliferous mining (Kelly, 1988) as well as smelting, agricultural, and waste disposal activities (Ross, 1994). Today, radioactive contamination remains one of the most menacing legacies of the Cold War. Severe soil and water contamination have occurred in areas of radioactive material mining and reprocessing. During the period of airborne and surface nuclear arms testing, a significant amount of anthropogenic isotopes such as ^{137}Cs, ^{239}Pu, and ^{240}Pu were dispersed into the global environment. In the second half of the twentieth century, nuclear power plant accidents contributed to the described contamination. The total activity of all the radioactive material released in the Chernobyl nuclear power plant accident (in the Ukraine in 1986) is today estimated to have been 3.7 10^{18} (Askbrant et al., 1996) to 12×10^{18} Bq (IAEA, 1996).

The presence of heavy metals or radionuclides (generally referred in this chapter as *toxic metals*) in water often jeopardizes the ecosystem stability and poses serious danger to human health (Eisenbud, 1983; Gogolev and Wilke, 1997; Prohl and Muller, 1996; Ross and Kaye, 1994). A variety of methods to remove toxic metals from water, based on ion exchange or chemical and microbiological precipitation, has been developed and used with some success (Janson et al., 1982; Palmer et al., 1988). These technologies have distinct efficiencies for different metals and may be very costly if

Phytoremediation of Toxic Metals: Using Plants to Clean Up the Environment, edited by Ilya Raskin and Burt D. Ensley.
ISBN: 0-471-19254-6 ©2000 John Wiley & Sons, Inc.

large volumes, low metal concentrations, and high cleanup standards are involved. There is a great need for reliable and inexpensive technologies that can reduce toxic metal concentrations to environmentally acceptable levels. Such technologies might also be effectively used in pollution prevention and waste reduction programs.

Recently there has been an increasing interest in the use of living and nonliving bacteria, fungi, and algae for the bioremediation and recovery of heavy metals from aqueous streams (Summers, 1992; Veglio and Beolchini, 1977; Wilde and Benneman, 1993). Live or dead cultured cells of a higher plant, Angel's trumpet (*Datura innoxia* Mill.) have been used to remove Ba^{2+} from solution (Jackson et al., 1990). Commercial applications of this research are still restrained by the prohibitive cost of growing pure cultures of cells and microorganisms, and by the need for their immobilization and/or separation from the aqueous stream. Metal-accumulating fungi (Tobin et al., 1994) and two species of aquatic fern, *Azolla filiculoides* Lam. (Sela et al., 1989) and *Azolla pinnata* R. Br. (Jain et al., 1989), have also been tested as metal biosorbents capable of remediating industrial effluents. Aquatic flower plants have also been utilized for water purification. Water hyacinth [*Eichornia crassipes* (Mart.) Solms] (Falbo and Weaks, 1990; Kay et al., 1984; Turnquist et al., 1990), pennywort (*Hydrocotyle umbellata* L.) (Dierberg et al., 1987), and duckweed (*Lemna minor* L.) (Mo et al., 1989) can remove various heavy metals from water. However, the efficiency of metal removal by these plants seemed to be low because of their small size and small, slow-growing roots. The high water content of aquatic plants also complicates their drying, composting and incineration.

In contrast, terrestrial plants develop much longer, fibrous root systems covered with root hairs that create an extremely high surface area (Dittmer, 1937), and these roots are easily dried in the open air. Hydroponically cultivated roots of several terrestrial plants were recently discovered to be effective in absorbing, concentrating, or precipitating toxic metals from polluted effluents. This process was termed *rhizofiltration* (Dushenkov et al., 1995). It was recently found that hydroponically grown seedlings of some terrestrial plants could be also used for metal removal from solution (Salt et al., 1997). Preliminary estimates indicate that, in many cases, the efficiency of rhizofiltration compares favorably with that of currently employed water treatment technologies. In this chapter we will summarize these recent efforts and evaluate the potential of this technology as an effective phytoremediation tool.

MECHANISMS OF PHYTOFILTRATION

Many interactions between nutrients and plant roots have been noted (Marschner, 1995). Possible mechanisms of toxic metal removal by plant roots include extracellular precipitation, cell wall precipitation and adsorption, and intracellular uptake followed by cytoplasmic compartmentalization or vacuolar deposition (Fig. 7-1). The precise removal mechanisms are largely unknown and are not necessarily similar for different metals. Intracellular uptake of toxic metals may employ the same mechanisms that are responsible for the uptake of essential ions such as K^+, Ca^{2+}, Mg^{2+}, NO_3^-, and SO_4^{2-}. Solute transport across membranes may be both a passive

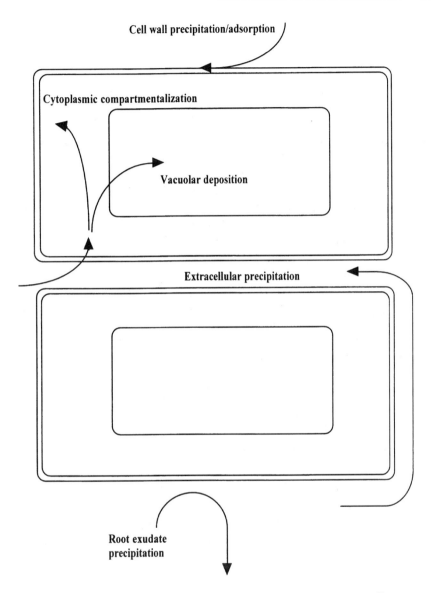

Figure 7-1 Mechanisms of metal removal from the solution by plant roots or seedlings.

process along the concentration gradient or a process linked directly or indirectly to energy-consuming mechanisms (Marschner, 1995). Membrane transport systems such as aqueous pores, ion efflux pumps, ion selective channels, and proton–anion contraports may fail to discriminate among different ions that have similar ionic radii and the same ionic charge. It is well established that Sr is an analog of Ca in living organisms (Kabata-Pendias and Pendias, 1989), and the effect of K on ^{137}Cs uptake in plants is well documented (Seel et al., 1995). The inhibitory effect of Ca^{2+}, Zn^{2+},

Mn^{2+}, and Cu^{2+} on Cd^{2+} uptake suggests that Cd is accumulated intracellularly via the same transport mechanism (Salt et al., 1997). Phytochelatins play an important role in the accumulation and detoxification of Cd in the vacuole (Salt et al., 1995, 1997). Phytochelatins appear to be the primary metal-binding polypeptides of plants that are constitutively expressed and play a central role in accumulating and/or detoxifying excess metals in plant cells. This phenomenon can further assist phytofiltration technology as it would increase specificity of metals to binding domains in the plant roots. Moreover, plants can contain several phytochelatins to increase the multibinding capacity of metals. Different metals induce phytochelatin synthesis; yet, binding of metals by phytochelatins has been demonstrated for Cu, Zn, Pb, and Cd only. Unfortunately, comparative binding affinities for various combinations of metals and peptide lengths are not yet available. There is strong evidence that the free amino acid, histidine may also play an important role in nickel transport in several nickel hyperaccumulators in the genus *Alyssum* (Krämer et al., 1996).

In the phytofiltration concept, uptake of ions by plants is defined as those ions retained after water rinses. The movement of ions or salts across the plant membrane by a carrier is dependent on metabolic energy and is determined to be by active transport. However, entry and association of ions with plant cells may also occur by a number of physical processes, including diffusion, ion exchange, mass flow, and adsorption. Precipitation and exchangeable sorption, which may be two major processes involved in removal of metals, tend to form insoluble compounds. Plant appoplast is influenced by the properties of the cell wall and middle lamella and is found to be negatively charged. Therefore, cations from the external solution can accumulate in a nonmetabolic step in the free space, where anions are "repelled." Plant species as well as root segment age differ considerably in cation-exchange capacity (number of cation-exchange sites) located in cell walls. This aspect should be considered when selecting a plant for phytofiltration. Moreover, there is little relationship between cation adsorption and cation absorption, and some of the cations remain attached to form an ion-exchange-like surface. The development of anion-exchange reactions (that characterized by short uptake time periods, a rate-limiting step, and an apparent selectivity of one ion in preference to another) is one important component of phytofiltration. Although the formulation of such a reaction would be similar to that employed for kinetic analysis in a carrier theory, it has a diffusion-limited step rather than carrier turnover.

Removal of U from solution is associated with a combination of such physical and chemical processes as chelation, ion exchange, and chemical precipitation. Positively charged U ions may interact with carboxyl groups of polygalacturonic acid and other negatively charged binding sites within plant cell walls (Dushenkov et al., 1997b). It was found that lead is precipitated from solution by sunflower roots mostly as lead phosphate (Dushenkov et al., 1995). Microscopic analysis of corn roots, exposed to soluble Pb showed the presence of similar electron-dense deposits inside and outside cells (Malone et al., 1974). Analysis of *B. juncea* roots exposed to Pb showed the formation of precipitates that contained substantial amounts of Pb carbonates in the cell walls (Raskin et al., 1997). Lead can also bind to exchangeable anionic binding sites in the cell wall (Broyer et al., 1972). Cell wall fractionation studies have shown that the largest amount of Pb reversibly binds to the pectic acid fraction (Lane et al., 1978). Extended X-ray absorbance fine-structure (EXAFS) analysis of *B. juncea* roots exposed to $Pb(NO_3)_2$ solution revealed that Pb was primarily bound to carboxyl groups, suggesting the involvement of polygalacturonic acid (Raskin et al., 1997). A histochemical study of Pb and Cd distribution in corn (*Zea mays* L.) seedlings implied that an endodermal barrier may prevent heavy-metal penetration into the stele (Seregin and Ivanov, 1997). In roots exposed to sublethal concentrations of Cd and Pb, those metals were found to be primarily associated with cell walls in the root cap, rhizodermis, and cortex.

RHIZOFILTRATION

The biofilter formed by biologically active, high-surface-area plant roots can be remarkably effective in sorbing pollutants from water (Dushenkov et al., 1995). The ideal plant for rhizofiltration should exhibit characteristics that provide the maximum toxic metal removal from a contaminated stream, in conjunction with easy handling, low maintenance cost, and a minimum of secondary waste requiring disposal. It is also desirable for a plant used for a rhizofiltration system to produce hydroponically significant amounts of root biomass or surface area. Sunflower plants grown in a specially designed rhizofiltration system can produce up to 50 g dry weight/m^2 per day. A rhizofiltration plant should be able to accumulate significant amounts of the contaminant of concern and to tolerate high levels of a toxic metal (at least temporarily); it should also preferably have a high root:shoot ratio and grow safely in controlled environments. Until the ideal rhizofiltration plant can be genetically engineered, it may be necessary to compromise among the properties described above in order to achieve the optimum performance.

Once a suitable plant for rhizofiltration is identified, a specifically constructed rhizofiltration unit should be designed to provide conditions for plant growth and for removal of toxic metals (Fig. 7-2). Because interactions between mineral nutrients supplied to support plant growth might chemically interact with the polluted water (yielding precipitating complexes), it is important to establish a dense root system in the contaminated water and to provide the plant with adequate nutrition without the addition of nutrients to the treated water. To meet these requirements, foliar nutrient application can be considered; however, establishment of a shell feeding zone in the portion of the root system is more efficient. Nutrients can be supplied only through the artificial soil mixture located on the top of the hydroponic system (feeder layer). A dense net of fine lateral roots will then develop in the feeder layer (Fig. 7-2), and the majority of the roots grow through the net in a search for water and to create a rhizofilter in the treated medium. This type of general setup provides the plant with enough nutrients and simultaneously avoids the addition of chemicals to the aqueous stream. Rhizofiltration units may be organized according to differing engineering designs to accommodate specific site conditions (Fig. 7-3).

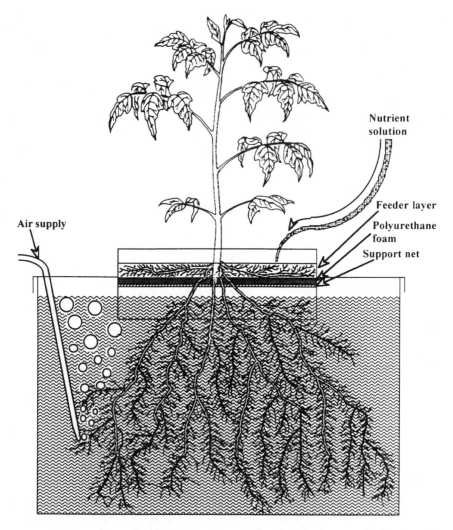

Figure 7-2 Lateral section of the rhizofiltration unit. The majority of roots are exposed to the contaminated aerated solution. This part of the root system provides water for the plant and ensures toxic metal uptake. Nutrients are supplied only through the feeder layer, an artificial soil mixture located on top of the hydroponic system. A dense net of fine lateral roots is developed in the feeder layer. This type of general setup provides the plant with enough nutrients and simultaneously avoids addition of chemicals to the aqueous stream.

Different plant species, both dicotyledonous and monocotyledonous, were evaluated for the ability to remove lead from water (Dushenkov et al., 1995). Lead concentrations in roots ranged from 5.6 to 16.9% on a dry-weight basis (Fig. 7-4). Among the fast-growing crop plants, root capacity to accumulate lead declined in the order sunflower > Indian mustard > tobacco > rye > spinach > corn. It has been shown that Indian mustard plants are capable of removing Pb^{2+} from aqueous solutions

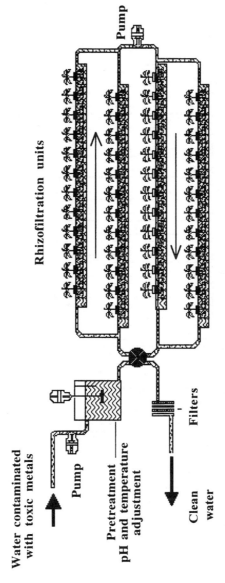

Figure 7-3 Schema of a flow-through rhizofiltration system.

95

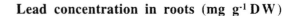

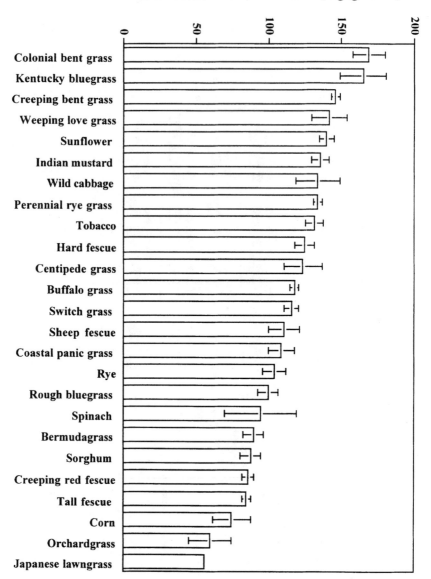

Figure 7-4 Lead accumulation in roots of different plants. Hydroponically grown 4-week-old sunflower plants were treated with continuously aerated solution of 300 mg/L Pb supplied as Pb(NO$_3$)$_2$. Vertical bars denote S.E. ($n = 4$).

having a wide range of concentrations, from 4 to 500 mg/L (Dushenkov et al., 1995). The roots of Indian mustard were also effective in the removal of Cd^{2+}, Cr^{6+}, Cu^{2+}, Ni^{2+}, and Zn^{2+}. The bioaccumulation coefficient, that is, the ratio of metal concentration in root tissue (μg/g DW) to the initial metal concentration in the solution (mg/L), determined after 24 h of metal treatment, varied significantly for different metals. At the concentrations used, Pb had the greatest bioaccumulation coefficient, 563, while Zn had the lowest, 131 (Dushenkov et al., 1995).

Miniature rhizofiltration units were used for growth chamber experiments (Dushenkov et al., 1997b). Sunflower plants tested in batch experiments in a growth chamber significantly reduced water concentrations of Cd^{2+}, Cr^{6+}, Cu^{2+}, Mn^{2+}, Ni^{2+}, and Pb^{2+} within the first hour of treatment (Fig. 7-5). In a 24-h experiment, the plants were much less efficient in removing anionic species such as AsO_2^- and SeO_4^{2-} from solution (Fig. 7-6). It was observed that most of the cationic species of toxic metals were removed from solution, at least initially, much more rapidly than anionic ones.

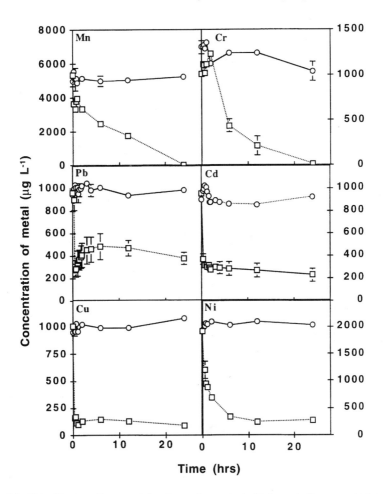

Figure 7-5 Rhizofiltration of heavy metals using sunflower plants. Hydroponically grown plants were put on top of a 13-cm-deep plastic jar that contained 750 mL of continuously aerated media. The total volume of the solution was kept constant by adding deionized water to compensate for water lost through plant transpiration, sampling, and evaporation. Control treatments did not contain plant material. Salts of $Cd(NO_3)_2 \cdot 4H_2O$, $K_2Cr_2O_7$, $Cu(NO_3)_2 \cdot 3H_2O$, $Ni(NO_3)_2 \cdot 6H_2O$, $Mn(NO_3)_2 \cdot 4H_2O$, and $Pb(NO_3)_2$ were used as the metal sources.

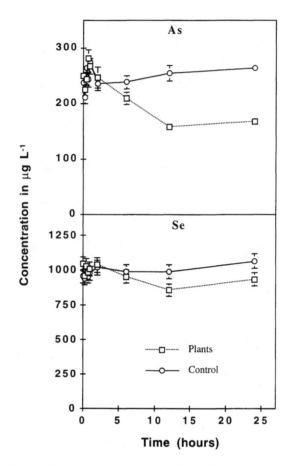

Figure 7-6   Rhizofiltration of heavy metals using sunflower plants. Hydroponically grown plants were put on top of a 13-cm-deep plastic jar that contained 750 mL of continuously aerated media. The total volume of the solution was kept constant by adding deionized water to compensate for water lost through plant transpiration, sampling, and evaporation. Control treatments did not contain plant material. Salts of $NaAsO_2$, and Na_2SeO_4 were used as metal sources.

Only a few terrestrial plant species were tested for the removal of radionuclides from water. Sunflower was found to be superior to Indian mustard and bean for uranium removal (Dushenkov et al., 1997b). Terrestrial plants were screened for their ability to concentrate ^{137}Cs and ^{90}Sr from water in a pond near the Chernobyl nuclear power plant (Dushenkov et al., 1997a). The highest bioaccumulation coefficient for ^{137}Cs was measured in sunflower roots, and that for ^{90}Sr in sunflower shoots (Dushenkov et al., 1997a). Sunflower roots accumulated up to 8 times more ^{137}Cs than the roots of timothy (*Phleum pratense* L.) or foxtail (*Alopecurus pratensis* L.). Sunflower shoots contained 2.5×10^6 Bq of ^{90}Sr per kilogram dry weight—a quantity of radioactivity sufficient for this material to be held in quarantine in the Chernobyl exclusion zone (Sorochinsky, 1997).

In batch experiments with hydroponically grown sunflower plants (Dushenkov et al., 1997b), it was shown that the treatment of Cs-, U-, or Sr-contaminated water with roots dramatically reduced the concentration of metals within a few hours (Fig. 7-7). The most rapid removal kinetic was demonstrated for U. Uranium concentration decreased 10-fold in 1 h. After 48 h, an equilibrium was reached at 10 μg/L. For Cs there was no reduction in concentration within the first hour of treatment, but after 6 h a noticeable reduction in Cs concentration was measured. Within 24 h, almost all of the Cs was removed, leaving a final concentration of less than 3 μg/L. Strontium concentration was reduced to 35 μg/L within 48 h. The concentration of Sr continued to decline for the next 2 days, reaching a level of 1 μg/L.

Uranium movement in plants is very limited under normal conditions, which explains the low U concentration in aboveground matter, even in plants grown on U-enriched soil (Kabata-Pendias and Pendias, 1989). Uranium levels in the sunflower shoots from the rhizofiltration systems generally remained low (Dushenkov et al., 1997b), and slightly elevated U concentrations in shoots (5 μg/g) were observed only in plants used to treat a solution containing 2430 μg/L of U at pH 5.0. The U concentrations in roots showed a positive linear correlation with the U solution concentration. Sunflower roots had a higher relative affinity for U at pH 5 than at pH 7.

Sunflower roots concentrated uranium from solution by up to 10,000-fold. Bioaccumulation coefficients were similar for various U concentrations used in the experiments. However, a lower pH resulted in a significant increase of U bioaccumulation. In the concentration range from 10 to 2430 μg/L, average bioaccumulation coefficients for U in sunflower roots were 6624 ± 870 and 3379 ± 430 at pH 5 and 7, respectively (Dushenkov et al., 1997b).

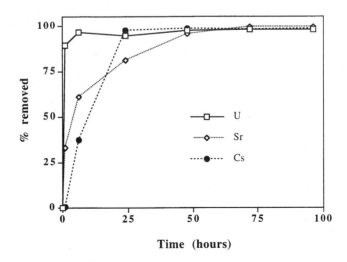

Figure 7-7   Rhizofiltration of U, Sr, and Cs. Four-week-old sunflower plants (cultivar 187) were used to treat 750 mL of solution. Root DW were, respectively, 1.21 ± 0.20 g for U, 1.13 ± 0.08 g for Sr, and 0.94 ± 0.21 g for Cs. The mean of three replicates is shown.

Scientists from the Institute of Cell Biology and Genetic Engineering in Kiev (Ukraine) (Grodzinsky et al., 1997; Sorochinsky et al., 2000) evaluated the use of plants to treat water contaminated with ^{137}Cs and ^{90}Sr in the Ukraine. A natural pond 1 km west of the Chernobyl nuclear power plant was selected as the water source. Radionuclide concentrations in this water body were close to 80 and 1200 Bq/L for ^{137}Cs and ^{90}Sr, respectively. Roots of 8-week-old sunflower plants were submerged in the water for 48 h, and the plants were then replaced with fresh plants. This batch protocol was repeated for a total of 12 days and resulted in the removal of 90% of the ^{137}Cs and 80% of the ^{90}Sr from the pond water (Grodzinsky et al., 1997).

BLASTOFILTRATION

It was recently discovered that when grown in aerated water (aquacultured), young plant seedlings may be as effective as roots in removing heavy metals from water (Salt et al., 1997). The technology of using plant seedlings to remove toxic metals from water was termed *blastofiltration* (*blasto* means seedling in Greek) (Raskin et al., 1997). Many plant species including *Brassica juncea* L. Czern., *Brassica napus* L., *Brassica rapa* L., *Medicago sativa* L., and *Oryza sativa* L. can germinate and grow for up to 10 days in aerated water in the absence of light and nutrients. Five-day-old seedlings of *B. juncea* were effective in accumulating various metals from artificially contaminated water containing Ca^{2+}, Mg^{2+}, K^+, SO_4^{2-}, and NO_3^- at levels resembling those found in ground water (Salt et al., 1997). They were able to concentrate the divalent cationic metals Pb, Sr, Cd, and Ni by a factor of 500–2000 over the concentration in solution. The monovalent cation, Cs^+ and the hexavalent Cr anion (CrO_4^{2-}), were accumulated less strongly, with bioaccumulation coefficients of 100–250 (Salt et al., 1997).

Blastofiltration may represent an alternative plant-based water treatment technology. It takes advantage of the fast increase in surface:volume ratio that occurs after germination and exploits the fact that some germinating seedlings also absorb or adsorb large quantities of toxic metal ions. This property makes seedlings perfectly suitable for water remediation. To produce seedling cultures to be used for blastofiltration, seeds, water, and air are the only components required.

FIELD PERFORMANCE OF RHIZOFILTRATION SYSTEMS

In naturally polluted water, a wide range of solutes is expected to be found and, potentially, interfere with the phytofiltration process. Validation of the rhizofiltration concept was carried on site with polluted water sources. A pump-and-treat rhizofiltration system was recently tested for cleaning groundwater contaminated with uranium in Ashtabula, Ohio, USA (Dushenkov et al., 1997b). A pilot rhizofiltration system with a total volume of 1500 L operated for 6 months. The incoming uranium concentration, ranging from 20 to 1800 µg/L, was pumped from a 5-m-deep drilled well and water was adjusted to pH 5.5. A total of more than 200 m^3 of contaminated water at the Ashtabula site was treated in this manner with 6-week-old sunflower plants. Figure 7-8 illustrates the performance of the pilot rhizofiltration system.

Flow rates of 0.15–1.60 L/min were used to evaluate the dynamic characteristics of the rhizofiltration system. The pilot-scale rhizofiltration system reduced the U concentration from several hundred micrograms per liter to below the regulatory discharge level of 20 µg/L. When the flow rate was increased, thus reducing the residence time in the tanks, the highest flow rate that provided acceptable system operation was found to be approximately 1 L/min. Reductions in soluble U concentrations correlated with the increased U concentration in the roots. The highest U concentration in roots was observed in the first tank of the system. Under continuous flow, roots were able to accumulate more than 1.0% of U on a dry-weight (DW) basis, producing a bioaccumulation coefficient of 30,000 (Dushenkov et al., 1997b). Longer operations could be permitted when a 2-week plant replacement interval could be adapted. It is noteworthy that the rhizofiltration system was able to treat spikes of high uranium concentration and to provide reliable removal of uranium from contaminated water to below the site specific EPA groundwater standard of 20 µg/L.

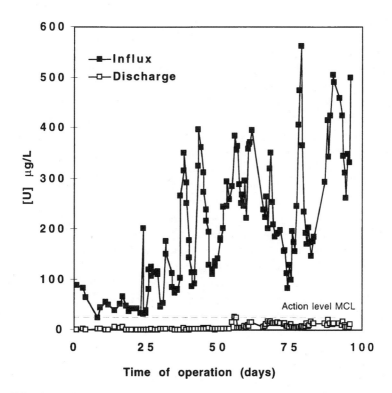

Figure 7-8 Pilot rhizofiltration system performance at the Ashtabula, OH site. Sunflower plants were grown hydroponically for 6 weeks and then were transferred to the rhyzofiltration system for 2 weeks. Plants were periodically harvested and replaced with fresh plants from the nursery. The pilot rhizofiltration system was assembled in the greenhouse. A surge tank was used for the pretreatment of the source water. The water temperature was maintained at 24 °C and the pH was automatically adjusted to 5.5 ± 0.3.

TABLE 7-1   **Bioaccumulation Coefficient (Concentration of Radionuclide in Dried Plant Tissues Divided by Radionuclide Concentration in Water) for Various Plants Grown in Pond Water at Chernobyl**

| | Bioaccumulation Coefficient | | | |
| | ^{137}Cs | | ^{90}Sr | |
Plant Species	Shoots	Roots	Shoots	Roots
Common timothy (*Phleum pratense* L.)	576	987	NA[a]	110
Meadow foxtail (*Alopecurus pratensis* L.)	2713	2796	920	NA
Indian mustard (*Brassica juncea* L. Czern.)	1973	NA	NA	NA
Peas (*Pisum sativum* L.)	382	863	NA	NA
Sunflower (*Helianthus annuus* L.)	452	4933	2016	799
Concentration in water (Bq/L)	82		1250	

[a]Data not available

   In situ rhizofiltration systems were tested in a small artificial pond with a surface of 50 m², 1 km west of the Chernobyl nuclear power station that contained ^{137}Cs and ^{90}Sr. Up to 80% of the water surface was covered with floating rhizofiltration units (Sorochinsky et al., 2000). Each unit contained sunflower plants grown on 1 × 1-m rafts constructed from Styrofoam. A 0.5-m² cavity in the center of the raft permitted plant roots to extend into the pond water.

   Sunflower roots concentrated ^{137}Cs 25,000-fold compared with its activity in the pond water. Transport of ^{137}Cs to shoots was limited (Table 7-1). In contrast to ^{137}Cs, ^{90}Sr was located mostly in the shoots; the bioaccumulation coefficient for ^{90}Sr in shoots reached 3400. Sunflower produced averages of 190 g of shoots and 51 g of roots per m². The rhizofiltration system removed more than 60,000 Bq/m of ^{137}Cs activity and more than 268,000 Bq/m of ^{90}Sr activity. Preliminary calculations showed that in just one season, plants removed total amounts of ^{137}Cs and ^{90}Sr equal to the amounts of these radionuclides presented in the pond water in the spring. Natural removal of activity was compensated by the release of ^{137}Cs and ^{90}Sr from sediments, and by radionuclides moving into the pond with runoff water. The final autumn concentration of radionuclides in the pond water was only a few percent less than that in the spring. It might be possible to use ponds on radioactive-contaminated territory as radionuclide traps. Rhizofiltration could be used in this case to remove activity from the ecosystem continuously, by concentrating radionuclides from four to five orders of magnitude in plants.

MODELING AND ECONOMICAL EVALUATION OF THE RHIZOFILTRATION PROCESS

Modeling various aspects of rhizofiltration processes is an important tool for decisionmaking, economic evaluation, and system performance under different conditions. A mathematical model of the rhizofiltration process was developed, based on the laboratory and field results (Fleisher et al., 1996). The key component of the model is an expression describing the concentration reduction by plant uptake in a rhizofiltration tank. To obtain this expression, potential variables involved within a single tank in the rhizofiltration system were identified (Fig. 7-9).

A strong relationship was found between the observed concentration reduction by plant activities (C_e') and the metal concentration available to the plants [$C_a = (C_w, C_{in}')$]. Subsequent experiments confirmed that the Michaelis-Menten equation was an appropriate choice to describe this relationship.

The model incorporates the Michaelis-Menten equation for C_e' into the mass balance of the rhizofiltration system. This approach allows prediction of the tank concentrations for any part of the rhizofiltration system and for any number of days of operation. The model is suitable for tank concentrations (C_w) below 600 µg/L; for tank concentrations above 600 µg/L, modification by a tuning factor is required for accurate predictions. The model was tested at different flow rates and concentrations and yielded good correlations with observed data in all cases. The model requires fine-tuning for the specific contaminant and site conditions.

Preliminary cost estimations were based on the results obtained during operation of a pilot-scale rhizofiltration system at a former uranium processing plant in Ashtabula, OH (Dushenkov et al., 1997b). The actual inlet concentrations at Ashtabula ranged from 21 to 874 µg/L, which showed no adverse effect on the performance of the system. For purposes of cost evaluation, a constant inlet concentration of 350 µg/L of uranium was assumed at a flow rate of up to 1.05 L/min per trough. The capacity for the full-scale system was assumed to be 3800 L/min, for 365 days per year, annually treating a total volume of 2 billion L of U-contaminated water. For these specified parameters, 25 kg per day of loaded biomass would require treatment and disposal. The roots of the plants contain up to 90% water at harvest, and

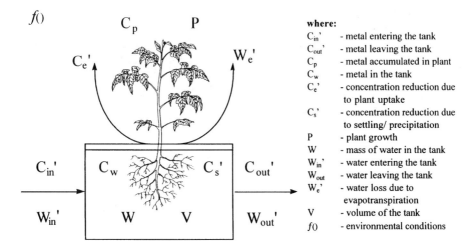

Figure 7-9 Variables of the rhizofiltration process in a single tank.

the mass of the dried roots can be reduced by up to an additional 90% by roasting or other thermal treatment, or by advanced composting processes. The roots can also be extracted by an acid solution to yield a nonradioactive biomass and a concentrated acid solution of uranium. The total cost per 1000 L treated was calculated at $1.50.

PERSPECTIVES

Treatment of polluted water is a series of interrelated, interdependent processes that can be resolved by the phytofiltration concept. The plant roots (or seedlings) provide an efficient, practical, and inexpensive conceptual solution for large water volumes. Rhizofiltration may find a market niche as a water treatment technology that is complementary to both ion exchange and constructed wetlands. Further development and commercialization of rhizofiltration and blastofiltration will be determined by economics as well as by such technical advantages as applicability to many toxic metals, ability to treat high volumes, reduced need for toxic chemicals, reduced volume of secondary waste, possibility of recycling, and the likelihood of regulatory and public acceptance.

REFERENCES

Askbrant, S. et al., 1996. Mobility of radionuclides in undisturbed and cultivated soils in Ukraine, Belarus and Russia six years after the Chernobyl fallout. *J. Environ. Radioactiv.* **31**:287–312.

Broyer, T. C. et al., 1972. Some aspects of lead in plant nutrition. *Plant Soil* **36**:301–313.

Dierberg, F. E. et al., eds., 1987. *Aquatic Plants for Water Treatment and Resource Recovery.* Magnolia Publishing, Orlando, FL, pp. 497–504.

Dittmer, H. J., 1937. A quantitative study of the root and root hairs of a winter rye plant (*Secale cereale*). *Am. J. Bot.* **24**:417–420.

Dushenkov, S. et al., 1997a. Phytoremediation: A novel approach to an old problem. In *Global Environmental Biotechnology*, D. L. Wise, ed., Elsevier Science B. V., Amsterdam, pp. 563–572.

Dushenkov, S. et al., 1997b. Removal of uranium from water using terrestrial plants. *Environ. Sci. Technol.* **31**:3468–3474.

Dushenkov, V. et al., 1995. Rhizofiltration: The use of plants to remove heavy metals from aqueous streams. *Environ. Sci. Technol.* **29**:1239–1245.

Eisenbud, M. 1983. *The Human Environment—Past, Present, and Future*. National Council of Radiation Protection and Measurements, Bethesda, MD.

Falbo, M. B. and T. E. Weaks, 1990. A comparison of *Eichhornia crassipes* (Pontederiacea) and *Sphagnum quinquefarium* (Sphagnaceae) in treatment of acid mine water. *Econ. Bot.* **44**:40–49.

Fleisher, D. H. et al., 1996. *Decision Support Software for Phytoremediation Systems*. ASAE paper no. 965042, ASAE. Phoenix, AZ.

Gogoley A., and B. M. Wilke, 1997. Combination effects of heavy metals and fluoranthene on soil bacteria. *Biol. Fertil. Soils* **25**:274–278.

Grodzinsky, D. et al., 1997. Rhizofiltration method for removing radionuclides from contaminated water. *Dokl. Nacion. Akad. Nauk Ukraini* **8**:189–192.

IAEA, 1996. One decade after Chernobyl: Summing up the consequences of the accident. In Agency IAE, ed. *International Conference on One Decade After Chernobyl: Summing up the Consequences of the Accident.* Jointly sponsored by the European Commission, International Atomic Energy Agency, World Health Organization, in cooperation with the United Nations (Dept. Humanitarian Affairs). International Atomic Energy Agency, Vienna, Austria, p. 555.

Jackson, P. J. et al., 1990. The removal of barium ion from solution using *Datura innoxia* Mill. suspension culture cells. *J. Environ. Qual.* **19**:644–648.

Jain, S. K. et al., 1989. Removal of some heavy metals from polluted water by aquatic plants: Studies on duckweed and water velvet. *Biol. Wastes* **28**:115–126.

Janson, C. E. et al., 1982. Treatment of heavy metals in wastewaters: What wastewater-treatment method is cost-effective for electroplating and finishing operations? Here are the alternatives. *Environ. Prog.* **1**:212–216.

Kabata-Pendias, A., and H. Pendias, 1989. *Trace Elements in Soils and Plants.* CRC Press, Boca Raton, FL.

Kay, S. H. et al., 1984. Effect of heavy metals on water hyacinths [*Eichornia crassipes* (Mart.) Solms]. *Aq. Toxicol.* **5**:117–128.

Kelly, M., 1988. *Mining and the Freshwater Environment.* Elsevier Science Publisher, London and New York.

Krämer, U. et al., 1996. Free histidine as a metal chelator in plants that accumulate nickel. *Nature* **379**:635–638.

Lane, S. D. et al., 1978. Lead toxicity effect on indole-3-ylacetic acid-induced cell elongation. *Planta* **144**:79–84.

Malone, C. et al., 1974. Localization of lead accumulated by corn plants. *Plant Physiol.* **53**:388–394.

Marschner, H., 1995. *Mineral Nutrition of Higher Plants*, 2nd ed. Academic Press, San Diego.

Mo, S. C. et al., 1989. Uptake of mercury from aqueous solution by duckweed: The effect of pH, copper, and humic acid. *J. Environ. Sci. Health* **A24**:135–146.

Palmer, S. A. K. et al., 1988. *Metal/Cyanide Containing Wastes. Treatment Technologies.* Noyes Data Corporation, Park Ridge, NJ.

Prohl, G., and H. Muller, 1996. Radiation exposure from radionuclides in ground water: An uncertainty analysis for selected exposure scenarios. *Radiat. Environ. Biophys.* **35**:205–218.

Raskin, I. et al., 1997. Phytoremediation of metals: Using plants to remove pollutants from the environment. *Curr. Opin. Biotechnol.* **8**:221–226.

Ross, S. M., 1994. Sources and forms of potentially toxic metals in soil-plant systems. In *Toxic Metals in Soil-Plant Systems*, S. M. Ross, ed., Wiley, Chichester, U.K., pp. 3–25.

Ross, S. M., and K. Kaye, 1994. The meaning of metal toxicity in soil-plant systems. In *Toxic Metals in Soil-Plant Systems*, S. M. Ross, ed., Wiley, Chinchester, U.K., pp. 27–61.

Salt, D. E. et al., 1995. Mechanisms of cadmium mobility and accumulation of Indian mustard. *Plant Physiol.* **109**:1427–1433.

Salt, D. E. et al., 1997. Metal accumulation by aquacultured seedlings of Indian mustard. *Environ. Sci. Technol.* **31**:1636–1644.

Seel, J. F. et al., 1995. Uptake of ^{137}Cs In vegetable crop grown on a contaminated lakebed. *Health Phys.* **68**:793–799.

Sela, M. et al., 1989. The accumulation and the effect of heavy metals on the water fern *Azola filiculoides. New Phytol.* **1989**:7–12.

Seregin, I. V., and V. B. Ivanov, 1997. Histochemical investigation of cadmium and lead distribution in plants. *Russ. J. Plant Physiol.* **44**:791–796.

Sorochinsky, B., 1997. Using plants to remove ^{137}Cs and ^{90}Sr from water. Personal communication.

Sorochinsky, B. et al., 2000. Decontamination of shell water bodies of 10 kilometers ChNPP zone using rhizofiltration. In *Problemi Chernobilskoi Zoni Otchuzdeniya.* Administration of Chernobyl Exclusion Zone. Nuclear Engineering, Kiev.

Summers, A. O., 1992. The hard stuff: Metals in bioremediation. *Curr. Opin. Biotechnol.* **3**:271–276.

Tobin, J. M. et al., 1994. Metal accumulation by fungi: Applications in environmental biotechnology. *J. Industrial Microbiol.* **13**:126–130.

Turnquist, T. D. et al., 1990. Nickel uptake by the water hyacinth. *Environ. Sci. Health* **A25**:897–912.

Veglio, F., and F. Beolchini, 1977. Removal of metals by biosorption: A review. *Hydrometallurgy* **44**:301–316.

Wilde, E. W., and J. R. Benneman, 1993. Bioremoval of heavy metals by the use of microalgae. *Biotech. Adv.* **11**:781–812.

8

THE USE OF PLANTS FOR THE TREATMENT OF RADIONUCLIDES

M. Cristina Negri and Ray R. Hinchman

INTRODUCTION

Radioactive contamination has become problematic since the development of nuclear technology in the second half of the twentieth century. Because of the types of activity involved, governmental institutions or organizations in close relationship with government agencies, own most of the radioactively contaminated sites. Causes for radioactive contamination include accidental spills and emissions from all of the operations typical of the nuclear fuel cycle, especially mining (^{222}Rn) and milling (^{238}U, ^{230}Th, ^{226}Ra, ^{210}Pb); fallout from nuclear testing (^{131}I, ^{90}Sr, ^{137}Cs, Pu); and accidents, such as the Chernobyl disaster, which is the most recent and probably the most serious. Other sources of radioactive contamination involving naturally occurring radionuclides are extractive operations, including oil drilling, which may bring to the surface naturally occurring radioactive materials (NORMs), notably uranium, radon, radium, and thorium. Globally dispersed radionuclides are estimated to contribute to more than one-quarter of the normalized collective effective dose equivalent commitment, the second largest contributor after mining and milling operations (United Nations, 1988).

The challenges associated with the remediation of soil, groundwater, and wastewater from radionuclides are similar to those associated with the remediation of other inorganic contaminants, with the added health risk deriving from these elements' radioactivity. Also, the radioactive decay component may influence,

Phytoremediation of Toxic Metals: Using Plants to Clean Up the Environment, edited by Ilya Raskin and Burt D. Ensley.
ISBN: 0-471-19254-6 ©2000 John Wiley & Sons, Inc.

especially for shorter-lived radionuclides, the selection of the most appropriate technology. While soil excavation and transfer to an approved landfill (as is or after stabilization in concrete or glass matrices) is the most common treatment for radioactively contaminated soils, size separation by soil washing, extraction with complexing agents, ion exchange, flocculation, and reverse osmosis–ultrafiltration are other technologies that are currently available at different stages of demonstration or deployment for the removal of radionuclides from soil and water. Site remediators and stakeholders are becoming increasingly interested in phytoremediation—broadly defined as the engineered use of higher plants to remove pollutants from a contaminated medium—because of its potential economic competitiveness and intrinsic ability to leave the soil undisturbed and fertile for subsequent revegetation with more ecologically appropriate plant species.

Phytoremediation is particularly well suited for the treatment of wide surfaces of low-level contamination. In that sense, phytoremediation of radionuclide contamination is theoretically the ideal case, because activity levels of radionuclides of environmental concern often translate into minute amounts in terms of mass. The United States Department of Energy (U.S. DOE) sponsored an international workshop in Slavutych, Ukraine, in the winter of 1998 at which scientists evaluated the feasibility and research needs to phytoremediate the exclusion zone around Chernobyl and use the obtained biomass as a feedstock for a biomass-to-energy facility. Although final decisions on the program are pending, it was evident that if an active cleanup program were to be implemented, phytoremediation would be the only practical alternative to consider for the decontamination of such a large impact area (Chernobyl Phytoremediation and Biomass Energy Conversion Workshop, Slavutych, Ukraine, Feb. 1998, unpublished).

Phytoremediation has not been extensively applied to the remediation of radionuclides. Although this technology has not been fully deployed for this particular application, several bench and field studies have been performed, and interest in this technology is growing. Phytoremediation has been tested for the remediation of uranium-contaminated wastewater and soil at the Ashtabula site and of uranium-contaminated soil from the Fernald site, both in Ohio. Other experiments include feasibility studies of the removal of cesium-137 (^{137}Cs) from soil at Brookhaven National Laboratory, NY. Both ^{137}Cs and ^{90}Sr removal have been also studied for application at a pond near Chernobyl in Ukraine. Tests are also under way to assess the feasibility of using phytoremediation to remove ^{137}Cs from soil at the Argonne National Laboratory (ANL) West site, within the Idaho National Engineering and Environmental Laboratory (INEEL). Table 8-1 lists these studies, and the following subsections provide more detail on them, as well as a discussion on the behavior and soil–plant transfer ratios of several of the most environmentally relevant radionuclides, to help elucidate the factors that may have a key role in determining phytoremediation potential.

As in the case of heavy metals, phytoremediation has been proved to be most effective and at a more advanced stage of development for treating readily available contaminants and therefore to treat surface water and groundwater contamination.

TABLE 8-1   Completed and Ongoing Phytoremediation Studies for Removal of Radionuclides

Radionuclide	Site	Type of Study	Status (December 1998)	Reference
Uranium	Ashtabula, OH wastewater	Pilot rhizofiltration	Completed	Dushenkov et al. (1997a)
Uranium	Ashtabula, OH soil	Bench, phytoextraction	Completed	Huang et al. (1998a)
Uranium	Fernald, OH soil	Greenhouse, bench	Ongoing	Cornish et al. (1995a)
^{137}Cs	Brookhaven National Lab, NY, soil	Bench, greenhouse; field	Completed on-going phase II	Cornish et al. (1997); Lasat et al. (1997)
^{137}Cs	Argonne National Laboratory-West, soil	Bench, greenhouse; field	Completed planned 1999	Idaho Dept. Health and Welfare (1998)
^{137}Cs, ^{90}Sr	Chernobyl, Ukraine, surface water	Greenhouse; field	Completed	Dushenkov et al. (1997b)
^{3}H	Argonne National Laboratory, IL groundwater	field	planned 1999	Argonne National Laboratory, 1998

Soil-adsorbed radionuclides have typically been more difficult to treat. Success in soil treatment at this stage depends on the development of specific amendments and treatments that can increase the rate of transfer of the radionuclide into plant-available forms, without further dispersing radionuclides into the environment. The most difficult challenge associated with remediating radionuclide-contaminated soil is that the contamination is typically a legacy of past activities, and as time has passed, the contaminant in the soil has been redistributed on the basis of its characteristics and soil properties. Both strong adsorption to clay minerals (such as in the case of ^{137}Cs) and coprecipitation with oxides (U, Pu) are believed to continue slowly over the years and to decrease the accessibility of the contaminant to remediation technologies (e.g., phytoremediation). Because only small fractions of the radionuclides are typically available for biological uptake, and the replenishment of this quota is strongly limited over time, phytoremediation appears to be more feasible when the total amount of contaminant to be removed (difference between the actual contamination levels and the remediation goal) is small. Successful phytoremediation depends largely on soil characteristics.

Although the economic advantages of phytoremediation over other technologies are still largely unquantified, other factors may make it worth pursuing for the remediation of radionuclides, in that it may prove more effective than other technologies in reducing the dose rate to workers (very little human presence is expected at the contaminated site after the installation) and in reducing potential resuspension of surface-deposited contamination by wind erosion and soil excavation because the soil surface is relatively undisturbed and eventually covered by a uniform vegatative cover. On the other side, concerns have been raised over the potential impact on the food chain of standing contaminated plant biomass available to animals as feed. To prevent this, several strategies have been proposed, such as favoring the use of nonpalatable plant species, taking advantage of root accumulation, and minimizing the standing time for contaminated biomass before harvesting. Biomass disposal is also more problematic because of the radioactive contamination.

CESIUM-137 (^{137}Cs)

Cesium-137 (half-life 32 years) is among the most relevant fission-products constituent of fallout. It is also commonly found as a consequence of spills and accidents of which Chernobyl is the most recent example. For this reason, ^{137}Cs has been widely studied in plant–soil systems. Direct studies focusing on phytoremediation applications have also been conducted or are currently under way.

Radiocesium has long been known to tightly bind to soils. Sequential extractions conducted by Salbu et al. (1994) on Chernobyl-impacted soils showed that 60–95% of the ^{137}Cs was found in the nitric acid and residual fractions, both considered unavailable for plant uptake.

Very small fractions of radiocesium are usually in readily reversible ion-exchange sites in mineral soil. Although typical exchange sites on organic matter and planar clay sites, referred to as regular exchange sites, RESs, contribute to the sorption of cesium, micaceous clays (such as illite and vermiculite) are responsible for its tightest retention. Radiocesium is preferentially sequestered by a small number of sites, referred to as frayed edge sites (FESs), which are located at the edge-interlayer positions of micaceous clays (Fig. 8-1). Poorly hydrated alkali metal ions (such as K^+, Rb^+, Cs^+, NH_4^+) are selectively sorbed by these sites preferentially over larger, more hydrated cations (Sweeck et al., 1990; Wauters et al. 1994). Wauters et al. (1994) found that ^{137}Cs fixation by clay soils was also positively correlated with soil concentrations of Ca and Mg, probably because of the steric effect of these bulkier ions, which allows Cs to migrate faster through the expanded edge interlayers. Although no direct effect of liming and pH was found on Cs bioavailability, this phenomenon may have consequences in the indirect effect of liming and pH on the bioavailability of Cs in the long term (Wauters et al., 1994).

Organic soils are believed to retain ^{137}Cs in a more reversible way, thus actually acting as temporary sinks of plant-available ^{137}Cs. Valcke and Cremers (1994) concluded, on the basis of their studies on ^{137}Cs sorption and desorption dynamics in soils, that soils with very high organic matter (>80%) bound Cs in a readily reversible form in the organic matter RES. This phenomenon was considered less substantial at lower concentrations of soil organic matter (<40%) when FES sites on micaceous clays were present in sufficient amounts to intercept the ^{137}Cs. In support of this conclusion, Kirton et al. (1990) studied cesium cycling after the Chernobyl accident

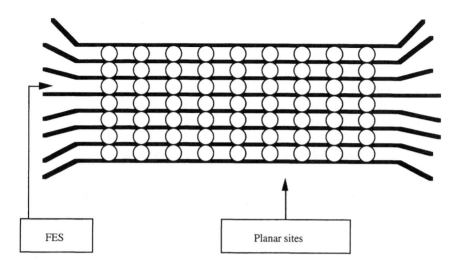

Figure 8-1 Schematic view of regular exchange sites (RES) and frayed-edge sites (FES) of a mica particle for Cs adsorption. [From Wauters et al. (1994), reproduced with permission.]

in an upland site in Cumbria and reported that Cs recycling from soil and root mat into fresh growth of vegetation was quite effective. In the case of a particular soil sample, the cumulative removal of ^{137}Cs by the vegetation approached 25% of that initially estimated to be present in the original soil. Soil microorganisms also appear to play a significant role in cesium cycling in organic soil. Microflora at different forest sites in Germany contained an average of 13% and a maximum of 56% of the total ^{137}Cs found in the organic layers of the soils (Bruckmann and Wolters, 1994). Cesium fluxes in undisturbed forest soil were also found to be correlated with dissolved organic carbon fluxes, and both were highly correlated with soil temperature (Tegen and Dorr, 1996).

Both K^+ and NH_4^+ ions have been shown to displace radiocesium from both the RES and FES exchange complexes. Cesium bound by organic soils was readily and almost quantitatively desorbed by K^+ and NH_4^+ ions (Valcke and Cremers, 1994). Their relative effectiveness is, however, controversial, with some authors finding that K^+ displaces more Cs (Wauters et al., 1994), whereas other authors (Prozorov et al., 1997; Lasat et al., 1997) find that NH_4^+ alone or in combination with strong acids is more effective at displacing Cs. Sodium ions can desorb radiocesium from nonselective sites, but they cannot compete with cesium for selective sites. Both ammonium and potassium may, once bioavailable, compete with ^{137}Cs for plant uptake, so the resulting uptake or discrimination of these three ions is probably the cumulative result of a number of concomitant factors, such as plant-specific growth patterns, nutrient requirements and uptake rates, respective base saturation, chemical equilibria in the soil solution, and soil properties. Radiocesium displacement appears to be rate-controlled by the buildup of the Cs in the liquid phase (Wauters et al., 1994). Kinetic studies conducted by Prozorov et al. (1997) revealed that ^{137}Cs leaching with acid–salt solutions is a slow process, with equilibration times of approximately 20 days (diffusion-controlled exchange reaction). It is also unclear whether the displacement of Cs by other cations can be sustained over time or decreases after peaking under the optimum conditions. Although a continuous, slow release of K from illitic clays is a known effect of weathering, very little information is available about the release of Cs from illitic clays by the same mechanism.

The addition of potassium before planting significantly decreased Cs uptake by lettuce, turnips, okra, eggplant, corn, and beans. Seel et al. (1995) found a significant inverse correlation between mean Cs concentration in plants and mean K concentration in the underlying soil. Robison and Stone (1992) also reported an inverse relationship between Cs in plant (coconuts) and K in coral soil. In both cases, the soil was poor in micaceous clays, so the relationship may be an example of more straightforward cases in which the FES adsorption mechanism has little influence. On a practical note, however, Seel et al. (1995) noted that a decrease in the amount of biomass caused by a deficiency of K offset the increased Cs concentration of the no-K treatment, so the final Cs inventory in the plant harvest was stable at 0.3% of the soil and not significantly different between K-fertilized and nonfertilized treatments. Grauby et al. (1990) found that although the effect of ammonium ions was always in favor of an increased radiocesium uptake from spiked soil, the effect of K additions depended on the soil cationic exchange capacity (CEC). When K additions were made to a low-CEC soil, a fast decrease of ^{137}Cs uptake was followed by an increased transfer factor. High-CEC soils always showed significant reductions in transfer factors following additions of K. Prister et al. (1992), in discussing practical countermeasures to limit the transfer of radiocesium to crops in the Chernobyl-affected zone, stated that applying ammonium sulfate or nitrate increased the accumulation of radiocesium in crops up to threefold.

Wallace et al. (1979) found that spiking soil with ^{137}Cs mixed with a stable cesium carrier dramatically increased plant uptake, most probably by saturating the adsorption sites at the disadvantage of the radiocesium, which was then more available for uptake. Schulz et al. (1960) conducted extensive studies on the fixation and displacement of radiocesium by a number of mono- and divalent cations and concluded that Na$^+$ and most divalent cations were poor exchangers for steric reasons, whereas K$^+$, NH$_4^+$, and Cs$^+$ are able to release part of the ^{137}Cs because they can substitute for it in the crystal. CsCl removed approximately 30% of the fixed radiocesium in illite, 5% in vermiculite, and 20–60% of the fixed radiocesium in other soils of different mineralogic composition. Although K$^+$ and NH$_4^+$ were better at removing radiocesium from Aiken clay, CsCl was the best extractant for Vina clay loam and Yolo loam. In a study conducted by Nishita et al. (1960), potassium added to an artificially contaminated soil already containing relatively high levels of potassium did not significantly affect radiocesium uptake by ladino clover, but after multiple croppings reduced the K levels in the soil, additions were effectively reducing radiocesium uptake. On the other side, additions of even small amounts of stable cesium to the same soil markedly increased the radiocesium uptake by plants, even when the Cs was added at levels seriously injurious to plants (approximately 4.5 meq Cs/g soil). The effectiveness of stable cesium in displacing radiocesium in soil with aged contamination has not been investigated. Adding chelating agents or soluble organic acids did not prove effective in increasing plant uptake of ^{137}Cs (Essington et al., 1962; Lasat et al., 1997).

As a consequence of its quick sorption by soil components, ^{137}Cs from fallout is typically found in the upper layers of mineral soils, leaving ploughing and other physical disturbances the main factors in its distribution downwards into the soil profile (Salbu et al., 1994; Horrill and Clint, 1994). In forest soil on the other hand, cesium migration appears to be more common. Migration from the litter layer to the mineral layer appeared to occur over the years 1987–1994, with a progressive impoverishment of the more surficial, organic undecomposed layers; an enrichment of the deeper mineral layers; and a no-net change of the in-between humified Oh horizon (Rühm et al., 1996).

Plant concentration factors commonly found in the literature for radiocesium vary between 0.01 and 1 (Menzel, 1965; Øhlenschlaeger and Gissen-Nielsen, 1989; Kirchman et al., 1993). Differences in Cs uptake among plant species are reportedly similar in magnitude to those caused by different soil types (Broadley and Willey, 1997; Andersen, 1967a). In agreement with the previously reported references regarding cesium mobility in different soil types, plants grown in organic soil have been shown to accumulate Cs in higher amounts than plants grown in mineral soil [see Paasikallo (1984), in Entry et al. (1996)]. Coral soil in the Marshall Islands (which has relatively elevated levels of organic matter and lack of clay) has shown concentration ratios of 5 (Robison et al., 1992 [see also Koranda (1978) in Kirchman et al. 1993)].

Irrigation may have an effect on Cs uptake in that, together with evapotranspiration, it allows for physicochemical solute transport and because irrigation has a general positive effect on plant growth, vigor, and uptake (Wadey et al., 1994). Evapotranspiration from physiologically active plants provides a major vertical driving force for purely physicochemical solute transport (Wadey et al., 1994). On the other side, however, excessive irrigation may dilute the soil solution to levels that are unfavorable to plant uptake.

Inoculation with vesicular arbuscular mychorrizae has been shown to increase the amount of ^{137}Cs accumulated by the mychorrizal plants sweet clover (*Melilotus officinalis*) and sorghum (*Sorghum sudanense*) (Rogers and Williams, 1986). Mychorrizal heather was found to translocate more Cs from roots to shoots than nonmychorrizal plants (Horrill and Clint, 1994). However, the highest ^{137}Cs removals reported in the literature were found in nonmychorrizal plants, such as those in the Chenopodiaceae, beet (*Beta vulgaris*), quinoa (*Chenopodium quinoa*), and Russian thistle (*Salsola kali*) (Broadley and Willey, 1997; Arthur, 1982; Blanchfield and Hoffman, 1984). Cesium-137 concentration factors in Russian thistle grown on a buried waste disposal pit averaged 35 (Arthur, 1982). Redroot pigweed (*Amaranthus retroflexus*), a close relative of the Chenopodiaceae, also proved to be very effective in taking up Cs (Lasat et al., 1997). Broadley and Willey (1997) found that taxa in the Chenopodiaceae had significantly higher mean-shoot concentrations of Cs than those in the Gramineae. Also, they recorded a strong linear relationship between mean shoot Cs and Rb (simulant for K) concentrations within the Chenopodiaceae, which accumulated Cs and Rb (K) in a ratio of about 1:9 (compared with 1:80 for the

Gramineae), thus discriminating quite less between Cs and Rb. Fast-growing plants (Chenopodiaceae) reportedly accumulated higher radiocesium concentrations than did slower growing plants (Gramineae), indicating a potential relationship between plant growth strategies and radiocesium uptake (Willey and Martin, 1997).

When grown in conditions of known, elevated bioavailability, many plants have been shown to take up and translocate Cs rather efficiently. Lasat et al. (1997) found high bioaccumulation rates in hydroponically grown plants. Concentration factor values were between 38 and 165, with dicotyledon species accumulating 2–4 times as much Cs as grasses. In other experiments, water hyacinth (*Eichornia crassipes*) was found to deplete Cs in 37 Bq/mL–3.7 kBq/mL (10^{-3}–10^{-1}-μCi/mL) solutions. Uptake attained equilibrium over a contact period not exceeding 5 h. In these experiments, a 60-fold increase in medium activity resulted in a 17-fold increase in accumulation levels, and 85–95% of the Cs was found in the root system (Jayaraman and Prabhakar, 1982). Ponderosa pine and Monterey pine seedlings grown on spiked perlite-peat medium were able to take up approximately 6–8% of the supplied ^{137}Cs in 4 weeks; accumulation was evenly divided between shoots and roots (Entry et al,. 1993). The same authors found that the duration of exposure was curvilinearly related to the total accumulation of ^{137}Cs, and, similarly, the relationship between accumulation and radiocesium concentration was also curvilinear. Dushenkov et al. (1997a) found dramatic reductions of Cs by hydroponically grown sunflowers within a few hours.

Cesium appears to be highly mobile in plant tissue. Zehnder et al. (1995) studied foliar absorption of soluble ^{137}Cs by grapevine plants and concluded that radiocesium was easily taken up through the leaf surface, transported to other plant tissues, and released from the roots to the soil in measureable amounts. In the soil, it would then undergo fixation by clays, thus becoming less available to plants. Indirect evidence suggests also that ^{137}Cs can leak from senescing wheat tissue (Wadey et al., 1994).

The existing knowledge on cesium chemistry and cycling guided the design of several studies directly aimed at assessing the potential for phytoremediation of ^{137}Cs. Phytotech, Inc., conducted a rhizofiltration demonstration in Ukraine, in collaboration with the International Institute of Cell Biology in Kiev. The feasibility of remediating a pond in the vicinity of the Chernobyl nuclear reactor contaminated with ^{137}Cs and ^{90}Sr was demonstrated both in greenhouse tests and subsequently in the field. Sunflower plants were grown hydroponically in the pond water (80 Bq/L ^{137}Cs) in greenhouse experiments. In 12 days, 90% Cs removal was achieved. A larger selection of plant species was then tested for Cs and Sr removal directly in the contaminated pond in a 4–8-week rhizofiltration study. All bioaccumulation factors reported were above 350 for Cs; sunflower roots accumulated up to 8 times more Cs than roots of timothy or foxtail, yielding a bioaccumulation coefficient of 4900–8600. The highest bioaccumulation factor was found in foxtail. From these data, investigators estimated that 55 kg of dry sunflower biomass would be necessary to remove the entire radioactivity in the pond (9.2×10^6 Bq ^{137}Cs and 1.4×10^8 Bq ^{90}Sr) (Vasudev et al., 1996; Dushenkov et al., 1997b).

These rhizofiltration treatability studies showed that, on the purely technical side, phytoremediation offers a definite, immediate possibility for the treatment of radiocesium-contaminated wastewater or groundwater; however, no detailed cost evaluation is available to date to compare the cost of rhizofiltration with that of currently available technologies. Also, information on costs and technical challenges associated with the disposal of the radiocontaminated biomass is not available, nor is an assessment of the surface area needs and their implications for the practical deployment of this approach.

A treatability study of phytoextraction of ^{137}Cs from soils at Brookhaven National Laboratory (BNL) was also conducted in recent years (Cornish et al., 1997; Lasat et al., 1998). In these investigations, ammonium nitrate was more effective in releasing radiocesium from the BNL soil at concentrations above 0.2 M in bench-scale leaching studies than weak and strong acids; approximately 25% of the soil radiocesium was released. Also, ammonium nitrate and ammonium chloride were more effective than potassium nitrate under the same conditions. Greenhouse studies revealed that whereas adding ammonium stimulated the uptake of radiocesium into cabbage, tepary beans, Indian mustard, and reed canarygrass grown in aged, contaminated BNL soil, it also depressed biomass production. In field trials at BNL, *Amaranthus retroflexus* (redroot pigweed) accumulated the highest levels of ^{137}Cs, yielding a concentration ratio (CR) of about 3. Unlike the pot tests, however, ammonium amendments failed to show any visible effect on ^{137}Cs uptake under field conditions. The authors suggest as possible explanations of this that either the cesium or the ammonium percolated deeper than the root zone. A linear relationship was found between soil ^{137}Cs in the range of 3.7–33 Bq/g (100–900 pCi/g) and shoot ^{137}Cs concentration.

Other field trials at BNL (Cornish et al., 1997) included Indian mustard, corn, sweet corn, and other plant species. Although Indian mustard appeared to suffer from the low pH of the BNL soil, it accumulated between 74 and 148 Bq/g (2000 and 4000 pCi/g) ^{137}Cs in the shoots in approximately 7 weeks. Corn, probably because of its high biomass, exhibited the highest average removal of ^{137}Cs. From these and the previously mentioned trials, the investigators calculated that under the best scenario (using *Amaranthus*, the best performing plant, at maximum aboveground biomass production and CR), the annual removal of ^{137}Cs from the upper 30 cm of soil would be approximately 3% of the total cesium; 6% removal was considered achievable by further optimizing amendment techniques and adopting multiple cropping. Although cost evaluations are premature at this stage, it is also unclear whether the remediation time of less than 15 years hypothesized by Cornish et al. (1997) is acceptable to regulators and stakeholders.

Another bench scale treatability study assessed the possibility of utilizing phytoremediation to remediate a ^{137}Cs-contaminated surface soil at the ANL West site in Idaho. In this case, in which the contamination levels were quite lower than the ones at BNL and close to the remediation goals, phytoremediation was selected as the preferred remediation method (contingent on the bench scale studies results) because of its estimated 50–75% cost savings compared to the other available technologies (excavation and disposal and containment in situ). Bench scale treatability studies indicated potential yearly ^{137}Cs removals of 2–4%, with projected times to achieve the cleanup goals of 4–7 years. These data, quite similar to those previously mentioned, are being verified on a larger scale in a 2-year field test, started in early 1999, which encompasses the whole area to be remediated. The radioactively contaminated biomass is expected to be incinerated at the INEEL licensed radioactive waste incinerator facility (Negri et al. 1988; Idaho Department of Health and Welfare, 1988).

STRONTIUM-90 (^{90}Sr)

Among the radionuclides of environmental concern, ^{90}Sr, a fission product with a half-life of approximately 28 years is probably the most mobile and bioavailable for plant uptake (Nisbet and Shaw, 1994; Essington et al., 1962). Its biological importance lies in its similarities with calcium in that it is deposited in animal bone. Unlike Cs, which has a definite affinity for micaceous clays, Sr binding in soils is more dependent on typical exchange sites commonly included in the cationic exchange complex. Percent base saturation and surface area of clays, together with the presence of organic matter, play a definite role in determining Sr bioavailability to plants. Sr bioavailability to plants thus strongly depends on CEC and soil types. Lysimeter experiments with artificially contaminated soils showed over a number of years that Sr concentration ratios for cabbage and carrot were maximum in sandy soil (values of 2–3), elevated in loam (1.5–2.7), and low in peat (<0.5). Concentration ratios in peat were also consistently low over the 5-year study. Humic acids present in the organic matter of peat soil appeared to be responsible for the decreased Sr uptake by forming organometal complexes that retain Sr and Ca more strongly than the ion-exchange sites in mineral soils (Nisbet and Shaw, 1994). In turn, loamy soils with higher CEC tended to exhibit higher retention of Sr than did sands. Irrigation with water rich in calcium, which supposedly displaced Sr below the root zone, appeared to be the reason for a decrease in Sr uptake over the same 5-year time frame observed in both loam and sand. In other studies, applying Ca to soils low in Ca induced a steady downward movement of ^{90}Sr (Squire, 1966). A very complete, albeit not recent, review of Sr soil/plant relationship is found in Francis (1978).

In Russian and Ukrainian soils, 40–98% of the total radiostrontium was found in the easily extractable fractions of a sequential extraction conducted 5 years after the Chernobyl disaster. Strontium mobility is thought to depend on the physicochemical form of the fallout (condensed particles dispersed at longer distances bearing more soluble Sr than heavier, quickly deposited fuel particles), because the mobility of the Chernobyl-derived ^{90}Sr was found to increase with the distance from the reactor (Oughton et al., 1992). Unlike ^{137}Cs, the rate of fixation of soluble ^{90}Sr by soils is believed to be slower than the rate of mobilization from the fallout particles; therefore, an increase in mobile Sr with time in these areas, as particles undergo weathering processes in the soil, is a serious concern (Salbu et al., 1994). In fact, ammonium acetate-extractable fractions of ^{90}Sr within the 10-km zone around the Chernobyl reactor increased from 7–30% in 1987 to 27–95% in 1991 (Popov, 1992). Various authors also noted that Sr migrated rapidly through the soil profile: vertical migration of radiostrontium was noted in podzolic soils, peat soil, and transitional swamps formed under anaerobic conditions and was considered very limited in sphagnum peat bogs. Sr is considered to be very mobile in reducing-anaerobic conditions (Francis, 1978). Rainfall tends to increase the movement of radiostrontium through the soil profile by mass transfer, independent of the amounts of precipitation and water permeability of the soil (Francis, 1978).

Several studies have examined the relationship between Sr and Ca in soil and plants. The relative availability of Sr and Ca is generally influenced by CEC and by the percent Ca saturation. In studies conducted by Veresoglou et al. (1995), Sr and Ca concentrations of plant species grown in the same soil were linearly and positively related to extractable Ca in soil. Strontium an calcium uptake showed a correlation coefficient of 0.9. As compared with the transfer factor (TF) for Sr, the TF for Sr/Ca concentration in plants exhibited much lower variation among plant species grown in the same soil and was proposed by the authors as an index of Sr extractability. In general, adding calcium carbonate has been shown to reduce Sr uptake in experiments conducted by Andersen (1967b), where Sr concentration in crops decreased with increasing concentrations of exchangeable Ca; Francis (1978) confirms these findings. Nisbet and Shaw (1994) found that up to 20–25% meqCa/100 g soils, exchangeable Ca appeared to be negatively correlated with Sr uptake. Soils that were richer in Ca did not show this correlation.

In what can be considered one of the earliest studies of phytoremediation of radionuclides, the uptake of ^{90}Sr was studied in water hyacinth (*Eichornia crassipes*) as a function of pH of the hydroponic medium. Uptake was highest at pH 9 and lowest at pH 4. More ^{90}Sr than ^{137}Cs was taken up, and the aggregate uptake increased with the activity of the medium almost linearly with time. Approximately 80–90% of the activity was confined to the root system, with some migration to leaves and petioles (Jayaraman and Prabhakar, 1982). In other studies, hydroponically grown sunflowers reduced Sr concentrations from 200 to 35 µg/L within 48 h and to 1 µg/L in 2 days of exposure (Dushenkov et al., 1997b).

In studies of Sr uptake by 44 plant species grown on the same soil, the lowest concentrations of Sr were found in the Gramineae, intermediate amounts in Crucifers, and highest levels were found in legumes and Umbelliferae. The Sr content of various plant parts varied significantly, with root crops having the highest amounts in tops (Andersen, 1967b).

Plants of the Chenopodiaceae have been shown to take up significant amounts of this and other radionuclides. Russian thistle (*Salsola kali*) is a weedy chenopode that is frequently found over disturbed burial land at disposal sites in the western United States. Elevated activity levels and ^{90}Sr concentration ratios averaging 250 were found in Russian thistle growing on disturbed soil, including buried waste disposal areas at the INEEL site in Idaho (Arthur III, 1982; Blanchfield and Hoffman, 1984). Plants in the *Atriplex* genus also proved to be, in a desert environment, the highest accumulators of ^{90}Sr (Wallace and Romney, 1972). Also, very elevated activity levels of ^{90}Sr were found in Russian thistle growing at the Hanford site in Washington (Johnson et al., 1994). Activity levels of radiostrontium in the needles and wood of pine trees growing above defense waste lysimeters at the Savannah River plant were 10 and 15 Bq/g ^{90}Sr

(285 and 410 pCi/g) respectively. Almost four fold-higher levels of ^{90}Sr were found in the biomass of weeds growing in the same lysimeters, which demonstrates the redistribution of Sr from the pine needles to the upper soil layers in a mobile form, which, in turn, was taken up by the weeds (McIntyre and Hawkins, 1988).

Ponderosa pine and Monterey pine seedlings accumulated approximately 1.5 and 4.5% of Sr when grown on an artificially contaminated peat moss–perlite medium. Both duration of exposure and initial medium concentration were curvilinearly related to the accumulation of Sr by the seedlings. Most of the Sr was found in the shoots ($\geq70\%$), in part because the biomass present in aboveground tissue is significantly higher than that in roots (Entry et al., 1993). Haghiri (1969) reports that partial mining by crops of available Sr may have occurred in highly productive agricultural soils. Regarding the mobility of strontium within the plant, grapevine leaves took up only traces of foliarly applied Sr, and essentially no Sr was redistributed within other plant parts (Zehnder et al., 1995). Abundant evidence reveals, however, that once taken up by the roots, ^{90}Sr is translocated to aboveground biomass in degrees that are species-dependent (Francis, 1978).

In a study of the influence of chelating agents on plant uptake of Sr from a calcareous soil, the addition of EDTA, DTPA, CDTA, EDDHA, and other chelating agents did not influence Sr uptake, with the exception of NTA supplied at 100 ppm (Essington et al., 1962). Ectomychorrizae have been shown to influence the amounts of Sr accumulated by trees; this effect is, however, highly dependent on a specific fungus-tree pair (Entry et al., 1996).

Phytotech, Inc., and the International Institute of Cell Biology in Kiev, investigated the phytoremediation (rhizofiltration) potential for the removal of Sr contamination from surface water, both in hydroponic studies and in the field in Ukraine. Batch studies showed that sunflower plants decreased ^{90}Sr from the initial level of 1200 Bq/L to less than 200 (>80% Sr removal) in 12 days. In field trials at a contaminated pond near Chernobyl, sunflowers proved quite effective in removing Sr, with bioaccumulation coefficients in excess of 600 for both shoots and roots. Shoots, which showed the highest CF for Sr, contained 2.5×10^6 Bq ^{90}Sr/kg dry plant (Vasudev et al., 1996; Dushenkov et al., 1997a). No studies were found that tested specifically phytoremediation of ^{90}Sr contaminated soil.

URANIUM (U)

Uranium is a naturally occurring primordial (i.e., associated with the formation of the earth) radionuclide. Naturally occurring uranium consists of ^{234}U, ^{235}U, and ^{238}U (all have half-lives between 10^5 and 10^9 years). Uranium is a key element of the nuclear fuel cycle.

Uranium is not considered essential for plant growth by most authors. Levels of U that are toxic to plants have been reported over a wide range of concentrations, up to values of 3700 Bq (1×10^4 pCi) g^{-1} dry weight (Cornish et al., 1995b). Concentrations as low as 1 μg/g in soil decreased growth parameters in wheat (Aery, 1998); on the other side, no negative effect was found by Meyer and McLendon (1997) on the emergence and growth of three grasses at levels below 25000 mg/kg depleted uranium (administered as weathered schoepite from deployed munitions). Contrasting data on phytotoxic levels of U in soils are believed to be, at least in part, the result of varying U bioavailability among soil types and conditions.

Uranium in soils is generally considered to be relatively immobile. Natural uranium is released in nature by the weathering of rocks or by ore deposits. A critical factor in weathering is the action of bicarbonate ions, which cause readily soluble uranyl complexes to form. Like many other metals, uranium mobility is considered to be limited by its precipitation as insoluble compounds (e.g., oxides) and by its adsorption to soil organic matter and clays. Mobile forms are transported as divalent uranyl (UO_2^{2+}) ions (Schulz, 1965; Nishita et al., 1978) or hexavalent carbonate complexes in the presence of high concentrations of CO_2 [$UO_2(HCO_3)_3^-$ and UO_2 $(CO_3)_3^{4-}$], or they are adsorbed by colloidal humic and fulvic acids and other low-molecular-weight complexing agents naturally found in groundwater (Shanbhag and Choppin, 1981).

Plant uptake of uranium is limited to mobile species and is affected by soil pH. Most researchers agree that plants take up very small amounts of uranium from soils. Typical concentration factors are usually in the 10^{-2}–10^{-4} range (Atomic Energy Control Board, 1988; Schreckhise and Cline, 1980b); legumes have higher values than grasses. Saric et al. (1995) noted that uranium was readily translocated to shoots and seemed to accumulate in older leaves rather than in young ones.

Baumgartner et al. (1996) state that uranium may accumulate in or on roots and precipitates on the root surface or apoplast. The presence of elevated nutrient levels has been correlated with higher levels of uranium uptake, as a consequence of a higher metabolism [Sultanbaev, (1974), cited in Harris (1989)]. Despite considerable interest in the evaluation of uranium uptake from soil and concentration ratios in plants, insufficient data are available that relate such uptake with the fraction of the actinide actually bioavailable to plants. In other words, most uptake rates are referred to total U concentrations in soils and may underestimate or otherwise not correctly reflect the uptake of mobile species of uranium, such as the ones found in groundwater.

In support of the potential for uptake of mobile species, twigs of black spruce (*Picea mariana*) collected in Saskatchewan were found to have U concentrations exceeding 1000 mg/kg, which reflected a localized uranium biogeochemical anomaly. These high levels were related not to the soil and bedrock concentration of uranium, but to uranium-rich formation water moving upward from 60–250 m below the surface, and to deeply penetrating boreal forest tree roots taking up the water (Dunn, 1980). Deep-rooting tree species in Utah (*Quercus, Juniperus*) were found to have 140–1600 mg/kg U in their roots (Cannon, 1952). Levels of 0.89–3.3 mg U/g were found in the ash of piñon pine growing near a waste pit at Los Alamos (Fresquez et al., 1995).

A limited number of trials of phytoremediation of uranium-contaminated soil and groundwater have been conducted with varying degrees of success. Cornish et al. (1995a) conducted preliminary investigations of the application of phytoremediation for the removal of U from soils at the Fernald site in Ohio and at an uranium waste dump in Montana. In these tests, which did not involve any soil treatment, only modest

AVRIL ROBARTS LRC

uranium uptake was noted, with higher concentrations in roots than in tops, even though the fraction of U that was considered bioavailable (extracted in ammonium acetate) was 8–20% of the total uranium.

Chelating agents and organic acids, such as citric and other organic acids that are present in root exudates, have been shown to participate in the process of plant uptake and may act as slow-release sources of actinides. Citric acid extractions have been used in conjunction with soil-washing-based remediation technologies because of the well-known ability of citrate to form strong complexes with uranium. Because of the harsher extracting conditions mandated by the need to keep reaction times short in batch reactor extractions, using citric acid to "wash" soil from the Fernald, OH, site presented a series of inconveniences (such as the dissolution of carbonate minerals and iron and aluminum sesquioxide coatings on soil particles, with substantial loadings of the secondary waste) (Soil Decon Task Group, 1993). These problems, however, appear to be much less important when citrate is used in situ for soil conditioning in phytoremediation applications. In this case, efficient dissolution rates are achievable in relatively longer times and over multiple croppings, thus decreasing the rate of application and the impact of the acidic chemical on soil components. Because mobilization must be synchronized with plant uptake as much as possible, phytoremediation is not confined to the quick reaction times of soil-washing procedures, so removal rates of comparable effectiveness can be achieved with less chemically harsh conditions. Citrate–U complexes have been shown to remain undegraded over an extended period under neutral pH conditions, but to be degraded by light and at alkaline pH (Huang et al., 1998a; Dodge and Francis, 1994).

The effectiveness of using a chemical treatment to increase uranium solubility appears, however, to depend on the nature of the contamination. Studies conducted on different locations from the Fernald site suggest that some of these soils may contain uranium in discrete particles (emitted from an incinerator) that are more resistant to chemical leaching than uranium chemically bound to soil components (Soil Decon Task Group, 1993).

Other innovative leaching procedures proposed for the extraction of U from low-grade ore rely on microbial activity (bacteria of the genera *Thiobacillus* and *Leptospirillium*) that does not attack the uranium itself, but instead, creates the conditions necessary for its dissolution by producing ferric sulfate from pyrite (Muñoz et al., 1995). To date, these technologies have not been studied in conjunction with phytoremediation.

Dushenkov et al. (1997a) conducted hydroponic treatability studies (from small batch to commercial scale) to assess the potential of treating uranium-contaminated wastewater (20–870 µg/L U) at the RMI site in Ashtabula, OH. Of four different plant species tested (sunflower CV SF-187 and Mammoth Giant, beans, and *Brassica juncea*), the two sunflower cultivars proved to be the most effective in decreasing the uranium concentration in the water, with more than 95% removal in 24 h. Subsequent experiments showed that U removal depended on the medium's pH (higher uptake at pH 5 than 7), but it was not affected by the initial U concentration over the range

10–2400 µg/L. Although sunflower shoots did not accumulate significant amounts of uranium, their roots had average bioaccumulation coefficients in excess of 6000.

A commercial-scale pilot rhizofiltration system was then set up at the Ashtabula site and operated at varying flow rates and influent uranium concentrations. The system flow diagram and performance are shown in Figures 8-2 and 8-3 (Dushenkov

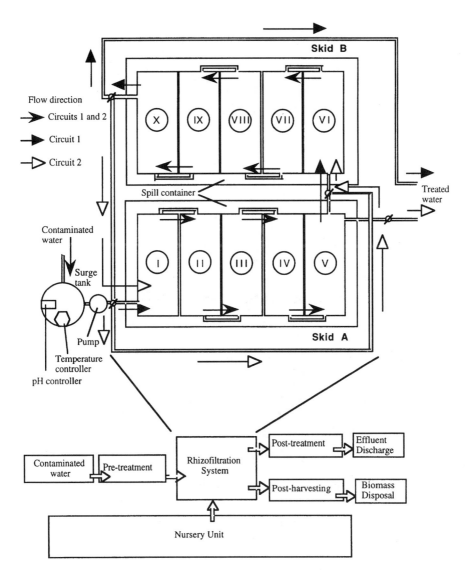

Figure 8-2 Scheme of the commercial rhizofiltration unit with a flow diagram of rhizofiltration system at the Ashtabula site. Six-week old sunflower plants were placed in the system for 2 weeks. During the operation, plants in skids A and B were alternatively harvested and replaced with a new set of plants from the nursery. Flow circuits were changed weekly to ensure effective removal of U at low concentrations with fresh sunflower roots. [Reproduced with permission from Dushenkov et al. (1997a).]

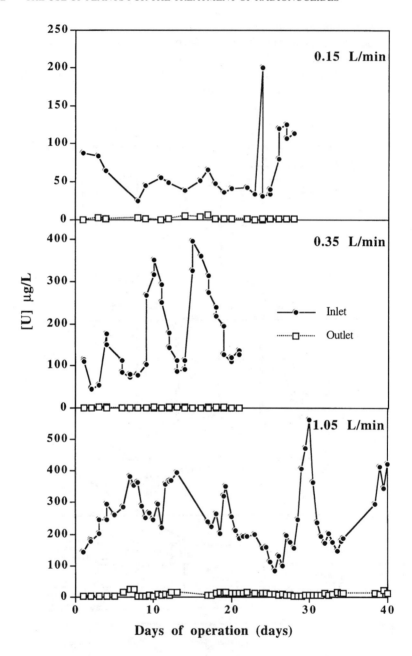

Figure 8-3 Pilot rhizofiltration system performance at the Ashtabula, OH, site. Sunflower plants were grown hydroponically for 6 weeks and then were transferred to the rhizofiltration system for 2 weeks. Plants were periodically harvested and replaced with fresh plants from the nursery. The pilot rhizofiltration system was assembled in the greenhouse. A surge tank was used for the pretreatment of the source water. The water temperature was maintained at 24°C, and the pH was automatically adjusted to 5.5 ± 0.3. System performance at three different flow rates is shown in this figure. [Reproduced with permission from Dushenkov et al. (1997a).]

et al., 1997a). Although a flow rate of 1 L/min was found to be the highest rate that would give the desired performance under normal contamination levels, the same flow rate was not able to reduce the U concentration below the 20-µg/L drinking-water standard when the influent concentration was artificially raised to >1000 µg/L. While limitations to the system were found, the Ashtabula studies showed that treating 2000 million L/yr of contaminated water would generate approximately 25 kg/day of fresh biomass (4.56 g biomass/m³ of water) at concentrations exceeding 1% uranium in the roots. Cost evaluations are not available to compare the cost-effectiveness of this approach with that of baseline technologies.

Huang et al. (1998b) conducted phytoextraction experiments using soil also from the Ashtabula site. In this case, applying citric acid at 20 mmol/kg increased uranium concentrations in the soil solution from 1.2 to 240 mg/L (a 200-fold increase), while temporarily decreasing soil pH of 0.5–1 unit. Shoot accumulation of uranium was found to be concentration-dependent, with dramatic increases in uranium at applications of citric acid above 10 mmol/kg. Uranium accumulation was noted at 24 h after the application and reached a maximum in the plants (>4500 mg/kg) 3 days after treatment. *Brassica juncea*, *B. chinensis*, *B. narinosa*, and amaranth showed a more than 1000-fold increase in uranium accumulation compared with the untreated control. The total phytoextraction of uranium was, however, higher in the *Brassicas* because amaranth produced about 30% of their biomass.

PLUTONIUM (Pu)

Although no direct work has been found describing the phytoremediation of Pu contamination, a significant amount of literature has been published on plutonium behavior in soil and uptake by plants. Plutonium isotopes are distributed in the environment as a consequence of nuclear weapons testing, fuel reprocessing facilities, production facilities, and accidental releases. The isotopes of most concern are $^{239-240}$Pu, ^{241}Pu, and ^{238}Pu.

Plutonium in aqueous solutions is commonly found in the III, IV, V, and VI oxidation states as Pu^{3+}, Pu^{4+}, PuO_2^+, and PuO_2^{2+}, respectively; the oxidation IV state is the most stable. In soil, under oxidizing conditions, PuO_2 is considered the most stable form at all pH. Under reducing conditions, PuO_2 would be the most stable form at pH above 4 (Nishita, 1979). While the physicochemical states of Pu in soil are not fully understood, most authors agree that it is tightly bound to the soil particles. At physiological pH for plants, significant water hydrolysis and subsequent sorption or precipitation onto soil particles were reported to occur in the absence of complexing ligands (Garland et al., 1987). Results from a sequential extraction conducted on contaminated soil samples from the Mound site in Ohio suggested that, more than 20 years after a plutonium nitrate spill, about 0.2% of the Pu still found in the soil was readily available/exchangeable, more than 50% was bound to the organic matter, 40–43% was associated with the oxides, and 3% was insoluble (Argonne National Laboratory, 1994).

AVRIL ROBARTS LRC

Plutonium mobility has been evaluated in desorption studies. Data in the literature show a wide range of K_d (distribution coefficient, concentration in sediment/ concentration in water) values in soil solutions (10^3–10^6 mL/g, depending on soil characteristics and leaching system) (Sanchez et al., 1982). Because of its relatively low mobility, Pu is typically found in the upper layers of soil, unless the soil is disturbed (Francis, 1973).

Plutonium is not considered to be very available to plants. It has not been reported as homologous to any useful cation, but data in the literature on this subject are almost nonexistent. Concentration factors (CFs) in plant tissue are also very wide-ranging. CFs reported in the literature are 10^{-8}–10^{-2}; average values are around 10^{-4}–10^{-5}. (Wallace et al., 1979; Price, 1973a; Adriano et al., 1980). Plants that were studied in the literature for Pu uptake typically consist of species grown in controlled experiments (mostly crops) and plant material (native or invading flora) analyzed in environmental studies. The strong potential for the aerial deposition of radionuclides on plant surfaces and the relatively low levels of contamination involved make it difficult to extrapolate the phytoremediation potential on the basis of these studies. Although no real plutonium hyperaccumulator was found, Noshkin (1972) reported that North Atlantic Sargassum had a high affinity for plutonium, with a CF of 21,000 over the marine water donor.

Several researchers have reported that Pu-contaminated soils in the kBq/g (μCi/g) soil range allowed the growth of various plant species without showing phytotoxic effects (Price, 1973a; Schreckhise and Cline, 1980a; Wallace et al., 1980). Vaughan et al. (1975) noted toxic effects in soybean and barley plants grown in 3.7 MBq (100 μCi/g) soil. However, both CF studies and evaluations of phytotoxicity rarely relate the findings with "bioavailable" quota; rather, they intend to determine this quota by measuring plant uptake. Livens et al. (1994) found that plant/soil concentration ratios of both Am and Pu expressed as ratio of concentration in plant tissue to concentration in soil solution were consistently many orders of magnitude higher than when calculated on soil concentration. This finding suggests that the critical, limiting factor in determining uptake potential may indeed be contaminant mobility in the substrate and, secondarily, plant-specific discrimination factors.

Factors contributing to increased dissolution of plutonium from soil suggested by various authors include pH and Eh changes; the direct presence of natural organic matter (such as decaying roots), which has complexing characteristics that prevent the readsorption of Pu onto soil particles [Romney et al. (1970), cited in Francis (1973)]; and the production by the microflora of extracellular metabolites able to complex Pu (Wildung and Garland, 1987). Nelson et al. (1987), in a study of natural waters/ sediment K_d values, describe natural dissolved organic compounds (DOCs), such as humic materials, as important complexing agents for many metals in surface waters, which could be responsible for a decrease in adsorption of Pu found in sediments associated with waters rich in DOCs.

Some of the effects of varying soil conditions have been reported in terms of how Pu mobility relates to its accumulation in plant tissues, as is the case in the evaluation of the effects of soil amendments and fertilization. Rediske et al. (1955) [cited in

Francis (1973)] reported that plant uptake of Pu was more than 3 times greater from acidic soil than from alkaline soil, and Romney et al. (1976) [cited in Harris (1989)] found that adding sulfur resulted in a significant increase in the uptake of americium and $^{239-240}$Pu. Plant uptake of plutonium was found to be greatest in alkaline soil. Bicarbonate appears to form very stable complexes with Pu^{IV}, and it is possible that these complexes could lead to increased plant uptake and translocation from calcareous soils (Schulz et al., 1976).

Organic chelated radionuclides have been reported to be more bioavailable for plant uptake than inorganic forms (Essington et al., 1962, 1963; Price, 1973b; Romney et al., 1970). Romney et al. (1970) found that DTPA increased plant uptake and translocation of americium and plutonium. Price (1973b) found that glycolate, oxalate, and citrate complexes enhanced plutonium uptake by tumbleweed and cheatgrass. Studies on chelators and plutonium bioavailability (mostly, however, conducted on artificially contaminated soil) are summarized by Harris (1989). Glyphosate, a broad-spectrum herbicide of known metal-chelating properties, was found to increase crop uptake in loam soil of ^{241}Am but not of ^{239}Pu. Desorption of Am occurred probably because stable, more bioavailable Am–glyphosate complexes formed; this effect was not noted in peat and sand soil, probably because of their low pH values. The authors attribute the general lack of response in the case of Pu to the fact that glyphosate typically creates complexes with di- and trivalent metals, but not with tetravalent ones, such as Pu (Nisbet and Shaw, 1990).

Plutonium uptake and translocation in plant tissues appears to vary with plant species, tissues, and age (Hakonson, 1975; Garland et al., 1987) and with soil physical, chemical, and microbiological characteristics. Garland et al. (1974) reported that plant uptake of Pu increased following the incubation of a Ritzville soil (pH 6.8) with carbon and nitrogen sources to provide maximum microbial activity. Changes in Pu bioavailability may occur as microbial activity increases (Romney and Davis, 1972; Francis, 1973; Nishita et al., 1978). Beckert and Au (1976) have shown *Aspergillus niger*, a common soil fungus, to be able to attack plutonium dioxide in a culture medium, take it up, and transfer part of it to its spores. They hypothesized that high levels of microbial activity in the rhizosphere may be increasing the fraction of Pu that is available to plants.

There is some evidence that Pu uptake by plants (in more heavily contaminated substrates) increases over time (Francis, 1973). Some authors recorded this phenomenon at the second or fifth year of the experiment, although others debate it. Possible causes of the reported increase are a chelating effect of decaying roots or simply the fact that, over time, more perennial roots come in contact with the Pu. Measurements of the distribution of Pu in shoots and roots of barley showed that Pu, once in the plant, was rather mobile; older leaves seemed to be accumulating Pu more than younger ones (Vaughan et al., 1975). As concentrations of Pu were much higher in leaves than stems, Garland et al. (1987) suggested, after conducting thin-layer electrophoretic analyses, that Pu is transported within the plant as an anionic organic Pu complex.

TRITIUM

Large green plants are capable of moving large amounts of soil solution into the plant body through the roots and evaporate the water out of the leaves as pure water vapor during transpiration. For example, data show that a single willow tree can, on a hot summer day, transpire more than 19 m^3 of water (5000 gal) (Miami Conservancy District, 1991). Plants transpire water to move nutrients from the soil solution through the roots (which function as a highly dispersed, fibrous uptake system) to leaves and stems, the sites of photosynthesis, and for cooling. During this process, contaminants present in the soil water are also taken up and sequestered, metabolized, or vaporized out of the leaves, along with the transpired water.

Tritium is known to be directly incorporated in water and biological tissues; moreover, plants transpire tritiated water vapor (IAEA, 1981). Studies conducted by the University of Heidelberg in natural ecosystems suggested that heavy plant growth may pull water from the soil at a rate so fast that tritium diffusion is no longer the controlling factor in mixing, and that considerably less isotopic mixing may take place (IAEA, 1987). Work conducted at the Maxey Flats Disposal Site suggested that trees could also be bioindicators of tritium contamination (Rickard and Kirby, 1987).

An innovative application of engineered phytoremediation is at the deployment stage at the Argonne National Laboratory site in Illinois. At this site tritium is present as a cocontaminant with volatile organic compounds (VOCs) in the groundwater, approximately 10 m deep in the glacial subsoil. The phytoremediation installation, which involves the use of high-transpiring, deep-rooted phreatophytes to provide hydraulic control of the contaminated plume (Nyer and Gatliff, 1996; Argonne National Laboratory, 1998), is deployed to address principally the VOCs contamination, and will allow the collection of data on the phytoevaporation of the tritiated water. A condition for the approval of the project by the regulatory agencies was the determination that no significant increase in radiation exposure to the most exposed individual would arise, even in the worst case scenario of complete release in the air of tritiated vapor using the highest tritium concentrations ever found in the groundwater to be remediated.

Although modeling studies would always be necessary to assess the potential hazards involved in the aerial release of tritiated water vapor for every application, in many cases (such as the one mentioned above) a phytoevaporation system could be more efficient and cost effective than traditional pump-and-treat systems.

ACKNOWLEDGMENT

This work was supported by the U.S. Department of Energy (DOE), Energy Research, Laboratory Technology Research Program, under CRADA with Applied Natural Sciences, Inc., and by the DOE Argonne Group West.

REFERENCES

Adriano, D. C. et al., 1980. Uptake of transuranic nuclides from soil by plants grown under controlled environmental conditions. In *Transuranic Elements in the Environment*, W. Hanson, ed., DOE/TIC-22800.

Aery, N. C., 1998. Influence of uranium on the growth of wheat. *J. Environ. Biol.*, **19**(1):15–23.

Andersen, A. J., 1967a. *Investigations on the Plant Uptake of Fission Products from Contaminated Soils. I. Influence of Plant Species and Soil Types on the Uptake of Radioactive Strontium and Cesium*, Risö Report No. 170, Roschilde, Denmark.

Andersen, A. J., 1967b. *Investigations on the Plant Uptake of Fission Products from Contaminated Soils. II. The Uptake of Radioactive Strontium Placed at Different Depths in the Soils*, Risö Report 174, Roschilde, Denmark.

Argonne National Laboratory, 1994, *Mound-ACT*DE*CON Feasibility Study, Phase II: Final Report*, Dec. 1994 (unpublished).

Argonne National Laboratory, 1998. Conceptual Design Report for the 317 Area French Drain (SWMU No. 11), Rev. 1. Prepared for the U.S. Department of Energy, Argonne, IL.

Arthur, W. J., III, 1982. Radionuclide concentrations in vegetation at a solid radioactive waste-disposal area in southeastern Idaho. *J. Environ. Qual.* **11**(3):394–399.

Atomic Energy Control Board (1988), *Effect of Soil Type on Radionuclides in Plants: Field Study*, Environmental Research Branch, Whiteshell Nuclear Research Establishment, Atomic Energy of Canada Ltd. INFO-0290, Dec. 1998.

Baumgartner, D. J. et. al., 1996. *Plant Responses to Simulated and Actual Uranium Mill Tailings Contaminated Groundwater*. A report to UMTRA Project, U.S. DOE, Environmental Research Laboratory, Univ. Arizona.

Beckert, W. F., and F. H. F. Au, 1976. Plutonium uptake by a soil fungus and transport to its spores. In *Transuranium Nuclides in the Environment, Proc. Symp.* San Francisco, CA, Nov. 17–21, 1975, jointly organized by USERDA and IAEA, IAEA-SM199/72.

Blanchfield, L. A., and L. G. Hoffman, 1984. *Environmental Surveillance for the INEL Radioactive Waste Management Complex and Other Areas*, Annual Report 1983, EG&G2312, INEL, Aug. 1984.

Broadley, M. R., and N. J. Willey, 1997. Differences in root uptake of radiocaesium by 30 plant taxa. *Environ. Pollut.* **97**(1):2.

Bruckmann, A., and V. Wolters, 1994. Microbial immobilization and recycling of ^{137}Cs in the organic layers of forest ecosystems: Relationship to environmental conditions, humification, and invertebrate activity. *Sci. Total Environ.* **157**:249–256.

Cannon, H. L., 1952. The effect of uranium-vanadium deposits on vegetation of the Colorado Plateau. *Am. J. Sci.* **250**:735–770.

Cornish, J. E. et al., 1995a. Evaluation of in situ phytoremediation of uranium-contaminated soils in Ohio and Montana. Paper presented at the Hazardous Management '95 Conf., Tucson, AZ, March 1995.

Cornish, J. E. et al., 1995b. Phytoremediation of soils contaminated with toxic elements and radionuclides. Paper presented at the In-Situ and On-Site Bioreclamation Symp., San Diego, CA, April 27, 1995.

Cornish, J. E. et al., 1997. *Phytoextraction Treatability Study: Removal of ^{137}Cs from Soils at BNL's Hazardous Waste Management Facility Site*, MSE Technology Applications, Butte, MT (Western Environmental Technology Office).

Dodge, C. J., and A. J. Francis, 1994. Photodegradation of Ukranian-citrate complex with uranium recovery. *Environ. Sci. Technol.* **28:**1300–1308.

Dunn, C. E., 1980. In *Mineral Exploration: Biological Systems and Organic Matter*, D. Carlisle et al., eds., J. R. Rubey Series, Vol. 5, Prentice-Hall, Englewood Cliffs, NJ, pp. 139–149.

Dushenkov, S. et al., 1997a. Removal of uranium from water using terrestrial plants. *Environ. Sci. Technol.* **31:**3468–3474.

Dushenkov, S. et al., 1997b. In *Phytoremediation: A Novel Approach to an Old Problem—Global Environmental Biotechnology*, D. L. Wise, ed., Elsevier Science B.V., pp. 563–572.

Entry, J. A. et al., 1993. Accumulation of cesium[137] and strontium-[90] in Ponderosa pine and Monterey pine seedlings. *J. Environ. Qual.* **22:**742–746.

Entry, J. A. et al., 1996. Phytoremediation of soil contaminated with low concentrations of radionuclides. *Water, Air Soil Pollut.* **88:**167–176.

Essington, E. et al., 1962. Influence of chelates on availability of fission products to plants grown in a contaminated soil. *Soil Sci.* **94**(2):96–105.

Essington, E. et al., 1963. Effect of chelating agents on the uptake of Y91, Ru106, Ce144, and Pm147 by beans grown in a Calcareous Soil. *Soil Sci.* **95**(5):331–337.

Francis, C. W., 1973. Plutonium mobility in soil and uptake by plants: A review. *J. Environ. Qual.* **2**(1):67–70.

Francis, C. W., 1978. *Radiostrontium Movement in Soils and Uptake in Plants*, Technical Information Center, TID-27564, U.S. DOE, Oak Ridge, TN.

Fresquez, P. R. et al., 1995. *Radionuclide Concentrations in Vegetation at Radioactive-Waste Disposal Area G during the 1994 Growing Season*, Los Alamos National Laboratory Report LA-12954-MS, Los Alamos, NM 87545.

Garland, T. R. et al., 1974. Factors affecting uptake and distribution of plutonium in barley and soybean plants. In *Pacific Northwest Laboratory Annual Report for 1974 to the U.S. Atomic Energy Commission*, Division of Biomedical and Environmental Sciences, Part 2, Ecological Sciences, B. E. Vaughan et al., BNWL-1950 PT2-UC48.

Garland, T. R. et al., 1987. Factors affecting adsorption, transport, and form of plutonium in plants, In *Environmental Research on Actinide Elements*, CONF-841142 (DE86008713).

Grauby, A. et al., 1990. Study of the possibility of alternating soil-plant transfer after an accident by application of moisture to the soil and by *Foliar* Spraying. In *Transfer of Radionuclides in Natural and Semi Natural Environments*, G. Desmet et al., eds., Elsevier Applied Science, London, pp. 403–410.

Haghiri, F., 1969. Effects of soil and crop management practices on the removal of strontium-90 by plant uptake, leaching runoff and erosion. *Agron. J.* **61:**793–796.

Hakonson, T. E., 1975. Environmental pathways of plutonium into terrestrial plants and animals. *Health Phys.* **29:**583–588.

Harris, G. A., 1989. *Radionuclide Uptake in Plants*, Statistics and Reliability Engineering, EG&G Idaho Inc., EGG-SARE-8769, DE90 007290, prepared for the U.S. DOE, Idaho Operations Office.

Horrill, A. D., and G. Clint, 1994. Caesium cycling in Heather Moorland ecosystems. In *Toxic Metals in Soil-Plant Systems*, S. M. Ross, ed., Wiley, New York.

Huang, F. Y. C. et al., 1998a. Biodegradation of uranium-citrate complexes: Implications for extraction of uranium from soils. *Environ. Sci. Technol.* **32:**379–382.

Huang, J. W. et al., 1998b. Phytoremediation of uranium-contaminated soils: Role of organic acids in triggering uranium hyperaccumulation in plants. *Environ. Sci. Technol.*, **32**(13):2004–2008.

IAEA, 1981. *Tritium in Some Typical Ecosystems*, Technical Reports Series no. 207, Vienna.

Idaho Department of Health and Welfare, Division of Environmental Quality (1998), *Proposed Plan for Waste Area Group 9, Argonne National Laboratory-West*, Idaho National Engineering and Environmental Laboratory, Jan. 1998.

International Atomic Energy Agency (IAEA), 1987. *Tritium and Other Environmental Isotopes in the Hydrological Cycle*, Technical Reports Series no. 73, Vienna.

Jayaraman, A. P., and S. Prabhakar, 1982. The water hyacinth's uptake of ^{137}Cs and ^{90}Sr and its decontamination potential as an approach to the zero-release concept. In *Proc. Internatl. Symp. Migration in the Terrestrial Environment of Long-Lived Radionuclides from the Nuclear Fuel Cycle*, Knoxville, TN, July 27–31, 1981, International Atomic Energy Agency, Vienna.

Johnson, A. R. et al., 1994. *Historical Records of Radioactive Contamination in Biota at the 200 Areas of the Hanford Site*, WHC-MR-0418, Westinghouse Hanford Company, Richland, WA.

Kirchman, R. et al., 1993. Terrestrial pathways. In F. Warner and R. M. Harrison, eds., *Radioecology After Chernobyl—Biogeochemical Pathways of Artificial Radionuclides, Scope 50*, Wiley, New York, pp. 101–176.

Kirton, J. A. et al., 1990. Derivation of soil to plant concentration ratios for upland plant species in the glasshouse and comparison with field and experimental measurements. In *Transfer of Radionuclides in Natural and Semi Natural Environments*, G. Desmet et al. eds., Elsevier Applied Science, London.

Lasat, M. M. et al., 1997. Potential for phytoextraction of ^{137}Cs from a contaminated soil. *Plant Soil* **195:**99–106.

Lasat, M. M. et al., 1998. Phytoremediation of radiocesium-contaminated soil: Evaluation of cesium-[137] bioaccumulation in the shoots of three plant species. *J. Environ. Qual.* **27:**165–169.

Livens, F. R. et al., 1994. The relationship between concentrations of Pu and Au in soil intensive waters and their uptake by plants. *Sci. Total Environ.* **155**(2):151–159.

McIntyre, P. F., and R. H. Hawkins, 1988. *Radionuclide Uptake by Pine Trees Growing on Lysimeters*, Technical Division, Savannah River Laboratory DPST-88-910, Aiken, SC.

Menzel, R. G., 1965. Soil-plant relationships of radioactive elements. *Health Phys.* **11:**1325–1332.

Meyer, M. C., and T. McLendon, 1997. Phytotoxicity of depleted uranium on three grasses characteristic of different successional stages. *J. Environ. Qual.* **26:**748–752.

Miami Conservancy District, 1991. *Aquifer Update*, no. 1:1.

Muñoz, J. A. et al., 1995. A study of the bioleaching of a Spanish uranium ore. Part 1: A review of the bacterial leaching in the treatment of uranium ores. *Hydrometallurgy* **38:**39–57.

Negri, M. C., R. R. Hinchman, and G. T. Ulbricht, 1988. Operable Unit 9-04 Phytoremediation Bench Scale Testing. A: Interceptor Canal Mound Soil. Final Report to ANL-W, Argonne Natinal Laboratory, IL, Nov. 1998.

Nelson, D. M. et al., 1987. Chemical speciation of plutonium in natural waters. In *Environmental Research on Actinide Elements*, J. E. Pinder III et al., eds., U.S. DOE Conf. 841142, DE86008713.

Nisbet, A. F., and S. Shaw, 1990. Effects of the herbicide glyphosate on the uptake of ^{239}Pu and ^{241}Am to vegetation. In *Transfer of Radionuclides in Natural and Semi-Natural Environments*, G. Desmet et al., eds., Proc. of a Workshop organized by the Commission of the European Communities, in Passariano, Italy, Sept. 11–15, 1989, Elsevier Applied Science, London and New York.

Nisbet A. F., and S. Shaw, 1994. Summary of a five-year lysimeter study on the time dependent transfer of ^{137}Cs, ^{90}Sr, $^{239-240}$Pu, and ^{241}Am to crops from three contrasting soil types, 2: Distribution between different plant parts. *J. Environ. Radioact.* **23:**171–187.

Nishita H. et al., 1960. Influence of K and Cs on release of ^{137}Cs from three soils. *Soil Sci.* **89:**167–176.

Nishita, H., 1979. *A Review of Behavior of Plutonium in Soils and Other Geologic Materials*, prepared for U.S. Nuclear Regulatory Commission, NUREG/CR-1056, UCLA12-1193.

Nishita, H. et al., 1978. *Radionuclide Uptake by Plants*, U.S. Nuclear Regulatory Commission, NUREG/CR0336, UCLA 12-1158.

Noshkin, V. E., 1972. Ecological aspects of plutonium dissemination in aquatic environments. *Health Phys.* **22:**537–549.

Nyer, E. K., and E. G. Gatliff, 1996. Phytoremediation. *Groundwater Monit. Remed.* **16**(1):58–62.

Øhlenschlaeger, M., and G. Gissel-Nielsen, 1989. *Transfer of Radiocaesium to Barley, Rye Grass, and Pea*, Riso-M-2831, Riso National Laboratory, DK-4000, Roskilde, Denmark, pp. 16–20.

Oughton, D. H. et al., 1992. Radionuclide mobility and bioavailability in Norwegian and Soviet soils. *Analyst* **117:**481–486.

Popov, V. E., 1992. Chemical forms of radionuclides in alluvial soils of Chernobyl 10-km zone: Depth distribution, transformation, and mechanisms of migration. *Proc. Internatl. Symp. Radioecology, Chemical Speciation-Hot Particles.* Commission of the European Communities/International Union of Radioecologists, Znojmo.

Price, K. R., 1973a. A review of transuranic elements in soils, plants, and animals. *J. Environ. Qual.* **2**(1).

Price, K. R., 1973b. *Tumbleweed and Cheatgrass Uptake of Transuranium Elements Applied to Soil as Organic Acid Complexes*, Battelle Pacific Northwest Laboratories, BNWL-1755, sponsored by the Atlantic Richfield Hanford Company under United States Atomic Energy Commission Contract AT(45-1)-2130.

Prister, B. et al., 1992. Efficiency of measures aimed at decreasing the contamination of agricultural products in areas contaminated by the Chernobyl NPP accident. *Sci. Total Environ.* **112:**79–87.

Prozorov, L. B. et al., 1997. Leaching of cesium-137 from contaminated soils with various reagents. *Radiochemistry* **39**(3):283–287.

Rickard, W. H., and L. J. Kirby, 1987. Trees as indicators of subterranean water flow from a retired radioactive waste disposal site. *Health Phys.* **52**(2):201–206.

Robison, W. L., and E. L. Stone, 1992. The effect of potassium on the uptake of ^{137}Cs in food crops grown on coral soils: Coconut at Bikini Atoll. *Health Phys.* **62**:496–511.

Rogers, R. D., and S. E. Williams, 1986. Vesicular-arbuscular mycorrhiza: Influence on plant uptake of cesium and cobalt. *Soil Biol. Biochm.* **18**(4):371–376.

Romney, E. M., and J. J. Davis, 1972. Ecological aspects of plutonium dissemination in terrestrial environments. *Health Phys.* **22**:551–557.

Romney, E. M. et al., 1970. Persistence of plutonium in soil, plants and small mammals. *Health Phys.* **19**:487–491, Oct. 1970.

Rühm, W. et al., 1996. Migration of ^{137}Cs and ^{134}Cs in different forest soil layers. *J. Environ. Radioact.* **33**(1):63–75.

Salbu, B. et al., 1994. The mobility of ^{137}Cs and ^{90}Sr in agricultural soils in the Ukraine, Belarus, and Russia, 1991. *Health Phys.* **67**(5):518–528.

Sanchez, A. L. et al., 1982. Distribution coefficients for plutonium and americium on particulates in aquatic environments. In *Environmental Migration of Long-Lived Radionuclides*, IAEA-SM-257/90P, 188–203.

Saric, M. R. et al., 1995. Uranium in plant species grown on natural barren soil. *J. Plant Nutr.* **18**(7):1509–1518.

Schreckhise, R. G., and J. F. Cline, 1980a. Comparative uptake and distribution of plutonium, americium, curium, and neptunium in four plant species. *Health Phys.* **38**:817–824.

Schreckhise, R. G., and J. F. Cline, 1980b. Uptake and distribution of ^{232}U in peas and barley. *Health Phys.* **38**:341–343.

Schultz, R. K. et al., 1960. On the soil chemistry of cesium-137. *Soil Sci.* **89**:16–27.

Schulz, R. K., 1965. Soil chemistry of radionuclides. *Health Phys.* **11**:1317–1324.

Schulz, R. K. et al., 1976. Uptake of plutonium and americium by plants from soils. In *Proc. Symp. Transuranium Nuclides in the Environment*, San Francisco, Nov. 1975, USERDA-IAEA, IAEA, Vienna 1976, IAEA-SM-199/92.

Seel, J. F. et al., 1995. Uptake of ^{137}Cs in vegetable crops grown on a contaminated lakebed. *Health Phys.* **68**:793–799.

Shanbhag, P. M., and G. R. Choppin, 1981. Binding of *Uranyl* by humic acid. *J. Inorg. Nucl. Chem.* **43**(12):3369–3372.

Soil Decon Task Group, 1993. Removal of uranium from uranium-contaminated soils. Phase 1: Bench scale testing. In *Uranium in Soils Integrated Demonstration*, ORNL-6762, Oak Ridge National Laboratory, TN, Sept. 1993.

Squire, H. M., 1966. Long term studies of ^{90}Sr in soils and pastures. *Radiat. Bot.* **6**:413–423.

Sweeck, L., et al., 1990. The specific interception potential of soils for radiocesium. In *Transfer of Radionuclides in Natural and Seminatural Environments*, G. Desmet et al., eds., Elsevier Applied Science, London, pp. 249–258.

Tegen, I., and H. Dorr, 1996. Mobilization of cesium in organic rich soils: Correlation with production of dissolved organic carbon. *Water, Air, Soil Pollut.* **88**:133–144.

United Nations, 1988. Sources, Effects and Risks of Ionizing Radiation. United Nations Scientific Committee on the Effects of Atomic Radiation. 1988 Report to the General Assembly, with annexes. U.N., New York.

Valcke, E., and A. Cremers, 1994. Sorption-desorption dynamics of radiocaesium in organic matter soils. *Sci. Total Environ.* **157**:275–283.

Vasudev, D. et al., 1996. Removal of radionuclide contamination from water by metal accumulating terrestrial plants. Paper presented at the Spring National Meeting, "In-Situ and Sediment Remediation," New Orleans, LA, Feb. 25–29, AIChE.

Vaughan, B. E. et al., 1975. *Pacific Northwest Laboratory Annual Report for 1974 to the USAEC Division of Biomedical and Environmental Research, Part 2: Ecological,* BNWL-1950, PT2, UC-489.

Veresoglou, D. S., et al., 1995. Transfer factors for Sr as influenced by species Ca uptake and soil on availability. *Plant Soil* **175:**225–232.

Wadey, P. et al., 1994. Radionuclide transport above a near-surface water table: II. Vertical distribution of gamma activities within soil profiles in relation to wheat rooting density and soil-to-plant transfers. *J. Environ. Qual.* **23:**1330–1337.

Wallace, A., and E. M. Romney, 1972. *Radioecology and Ecophysiology of Desert Plants at the Nevada Test Site.* Environmental Radiation Division, Los Angeles Soil Science and Agricultural Engineering Univ. Calif., Riverside, TID-25954.

Wallace, A. et al., 1979. *Biological Transport of Radionuclides at Low Level Waste Storage Sites,* U.S. Nuclear Regulatory Commission, Washington, DC, 1979-NUREG/CR-0701, RW, RE.

Wallace, A. et al., 1980. *Vegetational Cover in Monitoring and Stabilization of Shallow Land Burial Sites,* NUREG/CR-1358, UCLA 12-1235.

Wauters, J. et al., 1994. Availability of radiocaesium in soils: A new methodology. *Sci. Total Environ.* **157:**239–248.

Wildung, R. E., and T. R. Garland, 1987. Plutonium interactions with soil microbial metabolites: Effects on plutonium sorption by soil. In *Environmental Research on Actinide Elements,* J. E. Pinder, III et al., eds., U.S. DOE Conf. 841142, Aug. 1987 (NTIS no. DE860087-13).

Willey, N. J., and M. H. Martin, 1997. A comparison of stable cesium uptake by six grass species of contrasting growth strategy. *Environ. Pollut.* **95:**311–317.

Zehnder, H. J. et al., 1995. Uptake and transport of radioactive cesium and strontium into grapevines after leaf contamination. *Radiat. Phys. Chem.* **46**(1):61–69.

9

PHYTOSTABILIZATION OF METALS USING HYBRID POPLAR TREES

JERALD L. SCHNOOR

INTRODUCTION

Phytostabilization refers to stabilizing contaminated soils and sediments in place using vegetation, and to mitigating the migration of toxic contaminants in soils. Hydraulic control is possible, in some cases, because of the large volume of water that is transpired through plants, which prevents migration of leachate toward groundwater or receiving waters. Phytostabilization is especially applicable for metal contaminants at waste sites where the best alternative is to hold contaminants in place. Metals do not degrade, so immobilizing them in situ is often the best alternative at sites with modest contamination levels (below risk thresholds) or vastly contaminated areas where a large-scale removal action or soil washing is not feasible. Low-level radionuclide contaminants can also be held in place by phytostabilization, and if their half-lives are not too long, the alternative can result in significant risk reduction.

Fast-growing hybrid poplar trees have been used in a variety of climate zones in riparian zone applications to stabilize soils, decrease windblown dust, and decrease vertical migration of pollutants. Most risk assessments at metals waste sites indicate that the greatest cancer risk is from groundwater contamination and windblown dust. Direct ingestion of windblown dust (waste materials) by children on playgrounds is an important exposure route, as are the pathways of windblown dust from pesticides to vegetables to humans, or to cattle and then to humans. Phytostabilization can serve to diminish these exposure pathways dramatically. It can be considered as a sole

Phytoremediation of Toxic Metals: Using Plants to Clean Up the Environment, edited by Ilya Raskin and Burt D. Ensley.
ISBN: 0-471-19254-6 ©2000 John Wiley & Sons, Inc.

remediation method or used in tandem with other technologies for stabilizing soils. It can also be used as a cap and closure at landfill sites in lieu of typical designs that utilize impermeable liners and compacted clay (Licht et al., 1994).

Especially if contamination is in the upper 2–3 m of soil, deep-rooted phreatophytic trees (poplar, cottonwood, willow) can decrease the downward migration of leachate via evapotranspiration. The Salicacae (Salix) family of trees is ubiquitous in temperate climates. They grow well from Alaska to the subtropics. Cultivation of hybrid poplar trees can begin from 2-m "whips," cuttings that have preformed root initials. When planted at a depth of 2 m, they form a dense root mass that will take up large quantities of moisture, increase soil suction, and decrease downward migration of pollutants. In the winter dormant season, there may be some leakage of water through the system but, normally, precipitation is not great during this period. The trees can grow more than 2 m in the first growing season, and they reach a height of 6–8 m after 3 years when planted at a density of 4000 trees per hectare with an annual carbon fixation of approximately 2.5 kg/m^2 (Licht, 1990). Various maintenance practices have been adopted, and, in some settings, the trees can remain with very little attention after the second season for 20 years or more.

Phytostabilization has the following advantages:

- Low cost relative to soil removal, soil washing, or solidification
- Decreased human exposure due to mitigation of windblown dust and leachate
- Increased organic matter content of the soil, which binds metals and organics
- Aesthetically pleasing alternative that begins ecological restoration

Disadvantages of phytostabilization include concerns about leaf litter fall and whether toxic residues might be blown off site. This concern may be tested in a greenhouse, small plot, or field study to determine whether uptake and translocation of toxic metals into the leaves of trees exceed standards. In general, cadmium, zinc, strontium, cesium, and arsenic (arsenate) are the most bioavailable metals because of their chemical similarity to nutrients (calcium, potassium, and phosphate). Zinc is itself a nutrient that can be accumulated to toxic levels in plants from waste materials. Lead, Cr(III), and uranium are of lesser concern because of their lower translocation rates, and they can bind tightly to and thus be immobilized by root tissues. Other problems associated with phytostabilization include toxicity and the difficulty of establishing a vigorously growing stand of trees. Also, there is some concern regarding the mobilization of metals by soluble root exudates, but this problem can be minimized if hydraulic control is achieved at the site.

The purpose of in situ phytostabilization is to (1) stabilize wastes and prevent exposure pathways via wind and water erosion of waste metals, (2) prevent the vertical migration of contaminants to groundwater (hydraulic control), and (3) physically and chemically immobilize contaminants by binding to roots and organic matter in soils, and by chemical fixation with soil amendments including lime, phosphate rock, aged manure, and other materials.

HYDRAULIC CONTROL

Hydraulic control refers to the "solar pump" that is established when trees are deeply and densely rooted and growing at a site. Large quantities of soil moisture can be reliably removed from the soil profile, and a vertical upward migration of soil moisture throughout much of the year can be established. Groundwater is protected because vertical migration of leachate through the soil profile is minimized. Hydraulic control depends on the rooting depth and density, hydraulic conductivity of the waste and soils, growth rate of the trees (and cover grasses), and climatological factors such as humidity, sunlight, wind speed, and rainfall. To perform a proper estimate of the hydraulic regime, a mathematical model is needed to calculate the soil moisture balance on a daily basis. For the purpose of this chapter, the HELP3 (Hydrologic Evaluation of Landfill Performance, phase 3) water balance model, developed by the Waterways Experiment Station of the U.S. Army Corps of Engineers, was utilized (Schroeder et al., 1994).

Figure 9-1 is a schematic of a water balance in a phytostabilization system. Precipitation comes into the site and results in evapotranspiration, runoff, percolation to groundwater, and/or storage in the soil profile. For hydraulic control to be viable, the system must act as a "sponge and pump." When precipitation is great, the system relies on the field capacity of the soil/waste to absorb the water. Later, the trees will

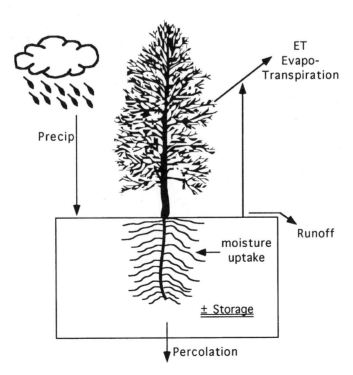

Figure 9-1   Water balance for the root zone of a phytostabilization effort using hybrid poplar trees.

effectively "dewater" the soil by transpiring the soil moisture from roots through the leaves and to the atmosphere. Runoff and percolation are the two greatest risk factors, which must be evaluated and controlled within an acceptable level. When deciduous trees are dormant in the winter, phytostabilization depends on the soil moisture holding capacity to the absorb precipitation that will be transpired out of the system the following spring. If the soil is frozen during the winter, presumably snowmelt runoff would not have time to contact the waste materials and should not be of detrimental water quality.

The hydrologic balance can be expressed by the following equation:

$$\text{Storage} = \text{precipitation} - \text{evapotranspiration} - \text{percolation} - \text{runoff} \qquad (9\text{-}1)$$

Overland runoff is a problem on steeply sloped terrain with relatively impermeable soils. These are not typical conditions at most waste sites. Results of the HELP3 model for a site at Burlington, Iowa, receiving 36 in. of precipitation each year are shown in Figure 9-2. Daily water balances were performed by the model, and the results are monthly summaries of storage, precipitation, evapotranspiration, and percolation (Kinney, 1996). Soils at the site were silt–loam with a saturated hydraulic conductivity of 0.0001 cm/s. Three-year-old hybrid poplar trees were simulated with a planting density of 1500 trees per acre, an evapotranspiration rate of 600 gal per tree per season, a leaf area index (LAI) of 3.0, and a rooting depth of 6 ft. Runoff was negligible in this case.

In Iowa, rainfall is greatest in the summer months when evapotranspiration is greatest. This is a common (and fortuitous) climatological circumstance that must be

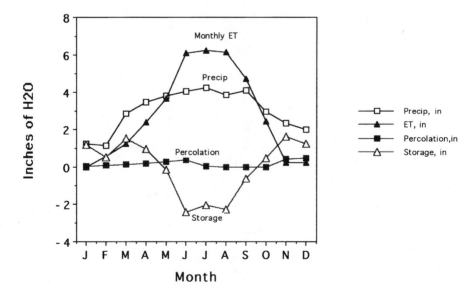

Figure 9-2   Monthly water balance terms for average climatological conditions at Burlington, IA (all units in inches of water per month).

determined for each site. During the warm months of May through September, when the trees are actively transpiring, the *storage* term in Eq. (9-1) is negative because the soils are dewatered by evapotranspiration (Fig. 9-2). Then, during the cooler months of October through May, the *storage* term is positive and the soils absorb water because they have a large storage capacity remaining from the previous summer when they were "pumped" quite dry. The cumulative water balance is shown in Figure 9-3. Cumulative precipitation is balanced by cumulative evapotranspiration and percolation. Percolation to groundwater must be kept to a minimum. For this case, the total percoation was only 2.1 in. per year (Kull, 1995). This is surprisingly low and compares favorably with an 18-in. clay cap at the site with a saturated hydraulic conductivity of 10^{-7} cm/s (Fig. 9-4). In fact, the poplar tree system allowed less percolation than all the alternatives simulated in Figure 9-4, including no cover (barren soil), a good grass cover, and a clay cap. The effectiveness of the poplar phytostabilization system at this site is due to vigorously growing trees (the solar pump) and the soil moisture holding capacity that is characteristic of this hydrogeologic setting. Each site must be evaluated independently.

Rooting depth is a key parameter in the HELP3 model that determines the percolation rate through the root zone. For the site at Burlington, Iowa, with soils of 0.0001 cm/s saturated hydraulic conductivity, there was a large difference in percolation rate between trees rooted only 36 in. deep and those rooted at 72 in. depth (Fig. 9-5). This nonlinear response is the result of percolation rates that become excessive when the soil moisture holding capacity is exceeded (in the case of a 36-in. root depth). In addition, evapotranspiration decreases when trees do not have a sufficient soil profile from which to draw water. In this simulation, at least 4 ft (48 in.)

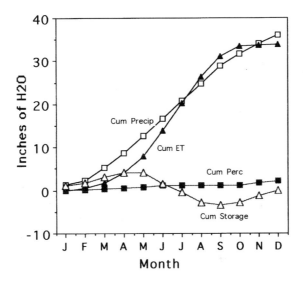

Figure 9-3   Cumulative water balance for average climatological conditions at Burlington, IA (inches of water).

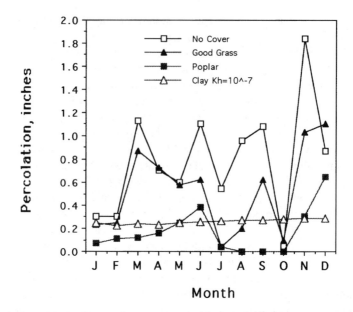

Figure 9-4   Percolation through various caps at Burlington, Iowa, in inches of water per month. (*No cover* refers to a barren plot; *Good grass* refers to a well-managed grass crop on top of the waste; *Poplar* refers to a phytostabilization cap with 4 ft of rooting depth in native soil on top of waste; and *Clay* refers to an 18-in. compacted clay cap with a saturated hydraulic conductivity of 10^{-7} cm/s on top of waste).

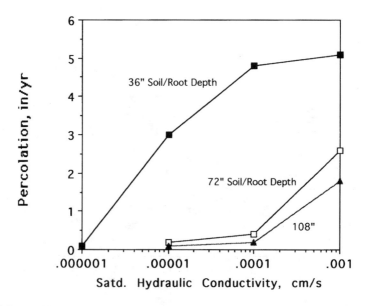

Figure 9-5   Annual percolation through a poplar tree cap at Burlington, IA, with soils of various saturated hydraulic conductivity rates and rooting depths.

of rooting depth was needed to prevent more than 3 in. of percolation to groundwater. This may require the addition of clean topsoil at the site to ensure a deep root zone. Also, if soils and wastes are too permeable, it is difficult to prevent percolation (leachate formation) through the site (Fig. 9-5). Phytostabilization and hydraulic control are possible at many sites with the proper seasonal rainfall distribution, deep rooting zones, and relatively impermeable soils.

IMMOBILIZATION AND CHEMICAL CONTROL

Metal ions can bind to soil particles at waste sites and be essentially immobilized. This phenomenon is variously named in scientific literature as surface complexation, ion exchange, (ad)sorption, and/or metal-particle binding. Surface complexing materials in soils are many and varied, including hydrous oxides, clay particles, organic matter, roots, and microorganisms.

Soils and waste materials provide surfaces for complexation of metal ions, acid–base reactions, and anions in the case of metalloids such as arsenate. In the following equations, reactions that may affect metal mobility at the surface of a soil particle are illustrated.

Metal ion binding:
$$SOH + M^{2+} \rightleftharpoons SOM^+ + H^+ \qquad (9\text{-}2)$$

$$(SOH)_2 + M^{2+} \rightleftharpoons (SO)_2M + 2H^+ \qquad (9\text{-}3)$$

Acid–base reactions:
$$SOH \rightleftharpoons SO^- + H^+ \qquad (9\text{-}4)$$

$$SOH + H^+ \rightleftharpoons SOH_2^+ \qquad (9\text{-}5)$$

Organic acids sorption:
$$SOH + H_2A \rightleftharpoons SOAH^{2-} + 2H^+ \qquad (9\text{-}6)$$

Anion sorption:
$$SOH + A^{2-} \rightleftharpoons SA^- + OH^- \qquad (9\text{-}7)$$

$$(SOH)_2 + A^{2-} \rightleftharpoons S_2A + 2OH^- \qquad (9\text{-}8)$$

where SOH represents hydrous oxide surface sites, M^{2+} is the metal ion, and A^{2-} is an oxyanion such as arsenate ($HAsO_4^{2-}$).

Figure 9-6 shows the "sorption edges" for three hypothetical metal cations forming surface complexes on hydrous oxides [e.g., amorphous iron oxides $FeOOH_{(s)}$ or aluminum hydroxide $Al(OH)_{3(s)}$]. Equations (9-2) and (9-3) indicate that the higher is

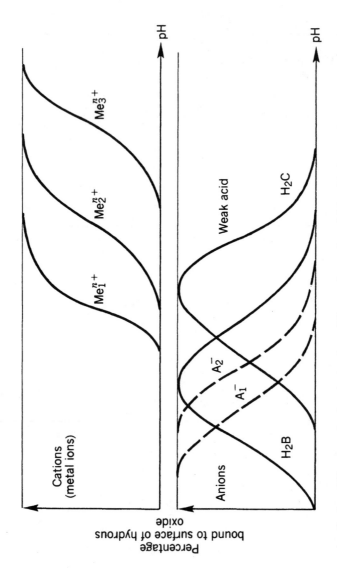

Figure 9-6 Sorption edges for three different metal cations (top) and two different anions and two fulvic acids (bottom) onto hydrous oxide surfaces in soils. [(Adapted from Stumm (1992)].

the pH of the solution, the greater the reactions will proceed to the right. The adsorption edge culminates in nearly 100% of the metal ion bound to the surface of the particles at high pH. At exactly what pH that occurs depends on the acidity and basicity constants for the surface sites [Eqs. (9-4) and (9-5)] and the strength of the surface complexation reaction [Eqs. (9-2) and (9-3)].

Similarly, anions such as arsenate and chromate can form surface complexes on hydrous oxides (Fig. 9-6). Most particles in soils are negatively charged at neutral pH. At low pH, the surfaces become positively charged and anion sorption is facilitated [Eqs. (9-7) and (9-8)]. Weak acids (such as organic acids, fulvic acids) typically have a maximum sorption near the pH of their first acidity constant (pH $= pK_a$), a consequence of the equilibrium equations [Eqs. (9-4)–(9-6)].

Metal cations are most tightly bound by those ions that form strong complexes with $-OH$ groups on the surface of minerals and hydrous oxides in waste materials. The strength of the binding usually occurs in the order $Cr(III) > Pb^{2+} > Cu^{2+} > Cd^{2+} > Zn^{2+} > Ni^{2+}$ (Stumm, 1992). Hydrous ferric oxides, $FeOOH_{(s)}$, are ubiquitous in the environment, and they can form coatings on the surface of other particles creating many surface sites for metals binding, disproportionate to the fraction of elemental iron that is present in the material.

Similar to metal binding on hydrous oxides at waste sites, metals can also bind to organic materials. One purpose of planting trees is to establish a dense, deep root mass that can sorb and immobilize contaminants:

Metal binding to organics: $ROH + M^{2+} \rightleftharpoons ROM^+ + H^+$ (9-9)

$$(ROH)_2 + M^{2+} \rightleftharpoons (RO)_2M + 2H^+ \qquad (9\text{-}10)$$

where ROH represents the binding site on an organic material. Here, metal cations are bound to particulate organic matter by bonding to phenolic, carboxylic, and hydroxy bridging groups in soil humus, wastes, or roots. The typical hierarchy of binding strength of soil organic matter for divalent cations is $Hg^{2+} > Cu^{2+} > Pb^{2+} > Cd^{2+} > Ni^{2+} > Co^{2+} > Zn^{2+}$ (Schnoor, 1996).

Addition of aged manure, straw, digested sewage sludge, or compost can be an efficient method of adding organic matter to inorganic waste sites such as mine tailings, abandoned smelters, plating wastes, and other industrial wastes. Thus, the water holding capacity of the waste is enhanced, which helps with establishing trees or grasses, and the binding capacity of the soils is improved for immobilizing metals [Eqs. (9-9) and (9-10)].

Addition of lime (CaO) and limestone ($CaCO_3$) is an effective amendment to neutralize acid soils and wastes in order to establish vegetation and to bind cationic metals on inorganic wastes (Fig. 9-6). Phosphate rock (such as calcium hydroxyapatite) can be an effective amendment for lead wastes. Lead (Pb^{2+}) forms strong surface complexes with such minerals, and it may even lead to precipitation of insoluble minerals such as lead phosphate $Pb_3(PO_4)_2$.

DESIGN CONSIDERATIONS

Plant Density

Planting density depends on the application, but a few typical designs will be reported here. For hybrid poplar trees, 1000–2000 trees per acre are typically planted with a conventional tree planter at 12–18 in. depth or in trenched rows 1–6 ft deep. Poplars have the ability to root along the entire buried depth. If a row conformation is used, the trees may be spaced with 2 ft between trees and 10 ft between rows. The poplars are planted simply as "sticks," long cuttings that will root and grow rapidly in the first season. Several phreatophytes in the Salix family, such as willow and cottonwood, can be planted in a similar manner. Hardwood trees and evergreens may require a lower planting density initially. A high initial planting density assures a significant amount of evapotranspiration in the first year which is normally desirable, but the trees will naturally thin themselves by competition to 600–800 trees per acre over the first six years. If desirable, hybrid poplars can be harvested on a 6-year rotation and sold for fuelwood or pulp and paper, and the trees will grow back from the cut stump (coppicing trait). The dense, deep root system stays in place to sustain growth for the next year. The lifetime of hybrid poplars such as *Populus deltoides x nigra* DN-34 (Imperial Carolina) is on the order of 30 years, which is usually sufficient as the design life of the project.

Irrigation and Maintenance

For terrestrial phytoremediation applications, it is often desirable to include irrigation costs in the design, on the order of 10–20 in. of water per year. Irrigation of the plants ensures a vigorous start to the system even in a drought. Irrigation may be necessary in dry climates, but too much water is not desirable either because of the threat of leachate generation. Hydrologic modeling may be required to estimate the rate of percolation to groundwater under irrigation conditions. Over time, irrigation should be withdrawn from the site, provided the area receives sufficient rainfall to sustain the plants. Maintenance costs should be considered in the design of phytoremediation systems. Mowing, replanting, pruning, harvesting, monitoring vegetation for contaminants, and fertilizer costs should all be included in the initial design costs if they are needed [operation and maintenance (O & M)].

Amendments

Agronomic inputs may include the nutrients necessary for vigorous growth of the vegetation. These include N/P/K from commercial fertilizer mixes, and carbon addition and soil conditioners such as aged manure, digested sewage sludge, compost, straw, or mulch (Hsu et al., 1993). It is critical that the site soils have sufficient water holding capacity to sustain vegetation. This is often not the case at mine tailings sites, abandoned smelters, and metals wastes from industry. In these cases, soil amendments are necessary to improve soil tilth and to allow for water to be absorbed for plants. Sometimes it is desirable to neutralize pH by lime addition. A standard agronomic analysis of site soils will allow assessment of soil amendments that are necessary (Jordahl, 1997). Phosphate rock or phosphate fertilizers are effective in binding lead and zinc. They can be added to trenches or disked into the soil prior to planting.

Nutrient Stoichiometry

Biomass production can be estimated at 7 tons dry matter/acre per year for fast-growing hybrid poplar trees (Licht, 1990). More than this can be achieved from a well-managed site with sufficient rainfall. In some cases, wood can be harvested from the site on a 6-year rotation for pulp or paper, and an income stream can be generated to cover the operation and maintenance costs. Poplar trees regenerate from the cut stump, so replanting is not necessary, but pruning may be needed to keep the trees from becoming too bushy. The amount of nitrogen stored in woody tissue is typically 0.5–1.0%, and nitrogen requirements can be calculated. Typical nitrogen fertilizer additions of 50–100 lb per acre are added 1–3 times per year. Stoichiometries of woody tissue and leaf tissue are available in the literature to estimate major nutrient uptake requirements (Jordahl, 1997).

CASE STUDY

An 18-m stretch of Whitewood Creek was a U.S. Superfund site due to contamination of surface and groundwater with arsenic and cadmium (arsenopyrite is the major mineral in the tailings) from 130 years of gold mining activity. It is located in the Black Hills of western South Dakota below the town of Whitewood. The University of Iowa planted 3100 hybrid poplar trees on one acre of the site to a depth of 1.6 m in old mine tailings along Whitewood Creek in April 1991 (Aoki, 1992). A commercial N/P/K fertilizer was used at recommended rates to ensure vigorous early growth of the cuttings. Roots formed along the entire length of the cutting in the soil, and a root mass was established to prevent leachate generation and percolation to groundwater.

Metals concentrations at the site were confined to the upper 6 ft of soil and tailings, which makes phytostabilization more viable. Arsenic, cadmium, lead, and zinc concentrations are shown in Figure 9-7 (Hse, 1996). Arsenic was the element of concern according to risk assessments at the site, but cadmium was monitored also because of its potential for uptake and translocation to leaves, thus creating a new pathway of exposure at the site.

At the end of the first growing season, the trees had grown to four feet in height. There was some evidence of phytotoxicity of the tailings in the field and in laboratory pot studies, but vegetation was growing. Leaves, stems and roots were collected from the field and lab to compare arsenic and cadmium uptake and translocation. Poplar leaves in the field did not take up very much cadmium or arsenic, averaging 1.1 and

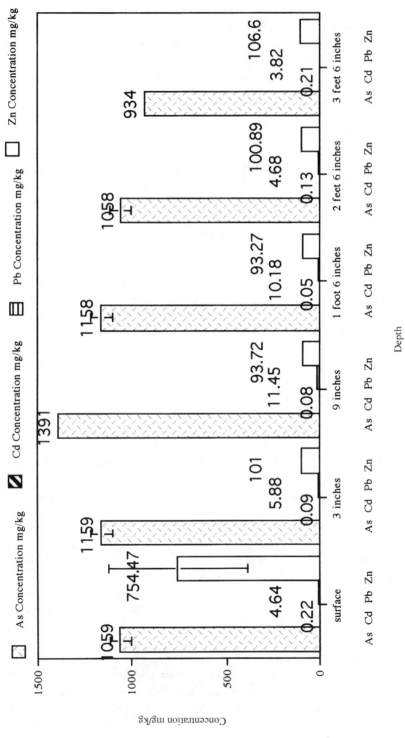

Figure 9-7 Concentrations of arsenic, cadmium, lead, and zinc in subsurface soil at various soil depths, Whitewood Creek, SD, mine tailings site (all concentrations are in mg/kg dry weight). [Adapted from Hse (1996)].

27 mg/kg dry weight, respectively (Hse, 1996). These concentrations are below most limits (such as Environmental Protection Agency 503 guidelines for biosolids) established for field application of municipal sewage sludge or compost. Results indicate that uptake of metals by the poplar trees was not a serious concern relative to published guidelines and native vegetation at the site.

Predicting the uptake of metals based on the bulk soil concentration is difficult. Bioavailability is a complex issue. However, there was a weak relationship between the total digestible arsenic concentration in soils at the site and the concentration that was monitored in the stem tissue of young poplar trees (Fig. 9-8). There was also a correlation between the zinc and the cadmium concentration in leaves of poplar trees (Fig. 9-9). However, as zinc concentrations increased, the cadmium uptake leveled off at two sites that were studied; there seems to be a competitive relationship between zinc and cadmium due to the similarity in their chemistry. If surplus zinc is available to plants, they will not take up much cadmium. In general, relationships between total digestible metal in soil (or chelate extractable fractions) and the concentration of metals found in leaves were not statistically significant.

At the Whitewood Creek site, South Dakota, approximately 150 poplar trees were still living in 1995 from a total of 6000 trees that were planted at the site between 1991 and 1993. Overall survival rates were very low due to possible phytotoxicity (Aoki, 1992), harsh climate, and animal browse. One difficulty was that the site is rather remote, and there was no one to maintain it on a daily basis. Fencing was needed to prevent animal browse (deer and cattle) that was very heavy at times. Severe ice

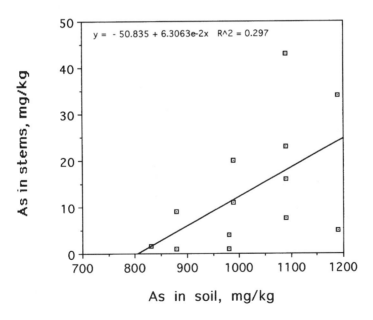

Figure 9-8 Concentrations of arsenic in stems of hybrid poplar trees and in soil in the root zone at Whitewood Creek, SD, 1995 (all concentrations are in mg/kg dry weight).

AVRIL ROBARTS LRC

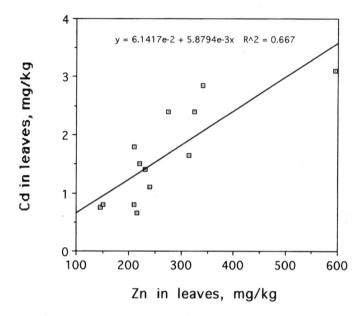

$$y = 6.1417e\text{-}2 + 5.8794e\text{-}3x \quad R^2 = 0.667$$

Figure 9-9 Concentrations of cadmium and zinc in leaves of hybrid poplar trees grown in mine tailings at Whitewood Creek, SD, 1995 (all concentrations are in mg/kg dry weight).

storms in October 1992 and in 1994 killed many of the trees. It was difficult to establish vegetation at the Whitewood Creek site given multiple stresses on the trees. At a second site established in 1994 at an abandoned smelter in Dearing, Kansas, the addition of aged manure and straw allowed success in establishing a phytostabilization system with greater than 50% survival of the trees.

EQUILIBRIUM CHEMICAL MODELING

Conceptually, the uptake of free metal ions by plants depends on the chemistry of the soil solution, which, in turn, depends on the sorption and surface precipitation of metals on soil particles (Fig. 9-10). A simple chemical equilibrium model was developed to simulate the binding of lead to soils at the Whitewood Creek site. Lead was bound to both inorganic and organic sites, and the model indicated that Pb^{2+} would not become mobile until $pH_{soil} < 5$. Hse (1996) noted that 99.9% of Pb(II) was bound by inorganic and organic complexes at surface sites and was not bioavailable for plant uptake and mobilization.

The model includes aqueous speciation and adsorption to two types of binding sites: >SOH, an inorganic oxide sorbent, and >ROH, an organic binding site. Equilibrium constants were obtained from the literature (Schnoor, 1996) and by model calibration. Two variables—the total concentration of organic binding sites [>ROH] and the equilibrium stability constant for the formation of a surface complex between Pb^{2+} and >ROH—were chosen in order to calibrate the model. It was

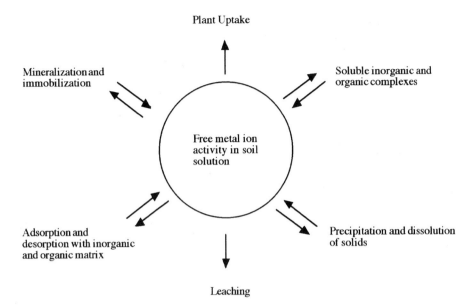

Figure 9-10 Schematic of processes influencing free metal ion activities in soil solution of a phytostabilization system.

determined that [>ROH] = 0.005 M and log K_{ROPb^+} = *0.0 at the Whitewood Creek, South Dakota site.*

$$Pb^{2+} + >ROH \rightleftharpoons >ROPb^+ + H^+ \qquad (9\text{-}11)$$

where >ROH is the concentration of organic binding sites and >ROPb$^+$ is the concentration of lead bound on the sites.

The chemical equilibrium model included the following information and predominant reactions:

Reactions	Equilibrium Constant
$[Pb(OH)^+] = K_{PbOH^+}\,[Pb^{2+}]\,[H^+]^-$	$\log K_{PbOH^+} = -7.7$
$[Pb(OH)_2] = K_{Pb(OH)_2}\,[Pb^{2+}]\,[H^+]^{2-}$	$\log K_{Pb(OH)_2} = -17.1$
$[Pb(OH)_3^-] = K_{Pb(OH)_3^-}\,[Pb^{2+}]\,[H^+]^{3-}$	$\log K_{Pb(OH)_3^-} = -28.1$
$[Pb_2(OH)^{3+}] = K_{Pb_2(OH)_3^+}\,[Pb^{2+}]^2\,[H^+]^-$	$\log K_{Pb_2(OH)_3^+} = -6.4$
$[PbFA] = K_{PbFA}\,[Pb^{2+}]\,[FA^{2-}]$	$\log K_{PbFA} = 5.5$
$[PbCO_3] = K_{PbCO_3}\,[Pb^{2+}]\,[CO_{2(g)}]\,[H^+]^{2-}$	$\log K_{PbCO_3} = -11.18$
$[Pb(CO_3)_2^{2-}] = K_{Pb(CO_3)_2^{2-}}\,[Pb^{2+}]\,[CO_{2(g)}]^2\,[H^+]^{4-}$	$\log K_{Pb(CO_3)_2^{2-}} = -28.16$
$[PbNO_3^+] = K_{PbNO_3^+}\,[Pb^{2+}]\,[NO_3^-]$	$\log K_{PbNO_{3+}} = 1.17$
$[Pb(NO_3)_2] = K_{Pb(NO_3)_2}\,[Pb^{2+}]\,[NO_3^-]^2$	$\log K_{Pb(NO_3)_2} = 1.40$
$[PbSO_4] = K_{PbSO_4}\,[Pb^{2+}]\,[SO_4^{2-}]$	$\log K_{PbSO_4} = 2.8$
$[>SOH_2^+] = K_{>SOH_2^+}\,[>SOH]\,[H^+]$	$\log K_{>SOH_2^+} = 4.0$, surface charge = 1

$[>SO^-] = K_{>SO^-} [>SOH][H^+]^-$ $\log K_{>SO^-} = -9.0$, surface charge $= -1$

$[>SOPb^+] = K_{>SOPb^+} [Pb^{2+}] [>SOH][H^+]^-$ $\log K_{>SOPb^+} = 4.0$, surface charge $= 1$

$[>ROPb^+] = K_{>ROPb^+} [Pb^{2+}] [>ROH][H^+]^-$ $\log K_{>ROPb^+} = 0.0$

$[H_2CO_3] = K_{H_2CO_3} [CO_{2(g)}]$ $\log K_{H_2CO_3} = -1.5$

$[HCO_3^-] = K_{HCO_3^-} [CO_{2(g)}] [H^+]^-$ $\log K_{HCO_3^-} = -7.85$

$[CO_3^{2-}] = K_{CO_3^{2-}} [CO_{2(g)}] [H^+]^{2-}$ $\log K_{CO_3^{2-}} = -18.18$

$[H_2FA] = K_{H_2FA} [FA^{2-}] [H^+]^2$ $\log K_{H_2FA} = 11.8$

$[HFA^-] = K_{HFA^-} [FA^{2-}] [H^+]$ $\log K_{HFA^-} = 7.0$

A constant capacitance model was utilized to obtain the intrinsic equilibrium constants for binding of lead at the surface of hydrous oxides (Schnoor, 1996). The following parameters were utilized:

Specific surface area = 4 m^2/g

Number of active surface sites = 0.0000032 mol/g

Concentration of adsorbent = 750 g/L

Ionic strength = 0.005 mol/L

Capacitance of inner layer = 2.9 F/m^2

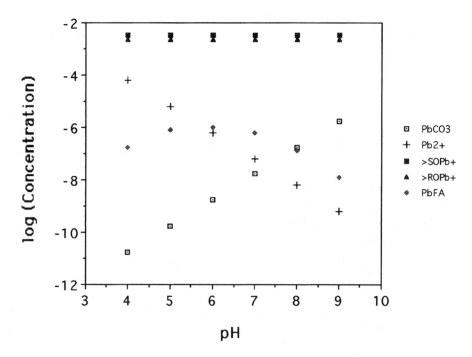

Figure 9-11 Log concentration of Pb^{2+} species versus pH at Whitewood Creek mine tailings site, MacμQL model (actual soil pH is 5.0–6.5, 1:1 in H_2O).

A matrix was constructed of all the pertinent equilibrium reactions, and it was solved using a chemical equilibrium program MacµQL (Müller, 1993). Results are shown in Figure 9-11. There was an increase in free Pb^{2+} ions predicted between soil pH values of 4.0 and 9.0. This is within the range of soil pH values that could be generated at the Whitewood Creek site as a result of oxidation of reduced sulfur in some areas (acid mine drainage). Above pH 5.5, the free Pb^{2+} ion concentration becomes very small, which is normally the case. Where low-pH soils are encountered, lime and limestone should be added to neutralize the wastes. Almost all of the Pb^{2+} in Figure 9-11, is absorbed to inorganic complex surface sites (>SOPb$^+$) and organic surface sites (>ROPb$^+$) where the pH is greater than 5.5.

The model provided a framework for analysis of the chemical contamination problem at the Whitewood Creek site. It helped set research priorities, and it established the need to measure binding site capacities at hazardous waste sites.

SUMMARY

Phytostabilization is a viable remedy at some metals-contaminated sites where it is desirable to immobilize the contaminants in situ. There are distinct cost advantages in such an approach if vegetation can be established at the site. Phytotoxicity of wastes and uptake of metals into tree leaves are concerns that need to be considered in such a remediation effort. Modeling of the water balance is necessary to estimate the potential for percolation of leachate to groundwater and surface runoff. In many circumstances, phytostabilization can successfully achieve hydraulic control and chemical immobilization at hazardous waste sites.

ACKNOWLEDGMENTS

The author is grateful to several students and colleagues who have been leaders in applying phytostabilization technology: Jay Brady, Randy Kinney, Daniel Aoki, Warren Hse, Jim Jordahl, and Craig Just, University of Iowa; Louis Licht, Ecolotree; and Larry Erickson, Kansas State University. The Great Plains and Rocky Mountain Hazardous Substance Research Center (EPA) has promoted phytoremediation for the past 7 years and has been a leader in getting the technology transferred to the field. At the University of Iowa, the Centers for Global and Regional Environmental Research, Biocatalysis and Bioprocessing, and the Environmental Health Sciences Research Center have assisted in this goal.

REFERENCES

Aoki, D. F., 1992. *The Uptake of Arsenic and Cadmium Mine Tailings by Poplar Trees*. M.S. thesis, Univ. Iowa, Iowa City.

Hse, W., 1996. *Metals Soil Pollution and Vegetative Remediation by Using Poplar Trees at Two Heavy Metal Contaminated Sites*. M.S. thesis, Univ. Iowa, Iowa City.

Hsu, M. S. et al., 1993. *Compost Sci. Util.* **1**(4):36–48.

Jordahl, J. L., 1997. *Fate, Transport and Toxicity of Soluble Salts Applied to Hybrid Poplar (Populus spp.) Trees*. Ph.D. dissertation, Univ. Iowa, Iowa City.

Kinney, R. S., May 1996. *Applications of the Hydrologic Evaluation of Landfill Performance Model at the Iowa Army Ammunition Plant, Middletown, Iowa*. Report submitted in partial fulfillment of the M.S. degree, Univ. Iowa, Iowa City.

Kull, D., 1995. *Landfill Leachate and Poplar Tree Perimeter Buffer*. M.S. thesis, Univ. Iowa, Iowa City.

Licht, L. A., 1990. *Poplar Trees for Nonpoint Source Pollution Control and Biomass Production*. Ph.D. dissertation, Univ. Iowa.

Licht, L. A. et al., 1994. Ecolotree™ cap at the Barje landfill in Ljubljana, Slovenia. *Proc. 2nd Internatl. Conf. Energy and Environment: Transitions in East Central Europe*, Prague, Czech Republic, Nov. 1994.

Madison, M. F., and L. A. Licht, 1991. Landfill cap closure utilizing a tree ecosystem. *Proc. American Society of Agricultural Engineers* (ASAE), Albuquerque, NM, Dec. 1991.

Müller, B., 1993. *MacμQL—A Chemical Equilibrium Model*, Swiss Federal Institute for Environmental Science and Technology (EAWAG-ETH), Dübendorf, Switzerland, 39 pp.

Schnoor, J. L., 1996. *Environmental Modeling*. J. Wiley, New York.

Schroeder, P. R. et al., 1994. *The Hydrologic Evaluation of Landfill Performance (HELP) Model—Version 3*, EPA/600/R-94/168 a,b, 1994, U.S. Environmental Protection Agency, Cincinnati, OH.

Stumm, W., 1992. *Chemistry of the Solid-Water Interface*. J. Wiley, New York.

10

PHYTOREDUCTION OF ENVIRONMENTAL MERCURY POLLUTION

CLAYTON L. RUGH, SCOTT P. BIZILY, AND RICHARD B. MEAGHER

INTRODUCTION

The primary objective of this research is to examine a possible alternative to standard and current practices for remediating soils contaminated with mercury. Although mercury is typically found at trace levels in the Earth's crust, anthropogenic activities have created areas with elevated and potentially dangerous concentrations (Nriagu, 1979; Porcella et al., 1995). Mercury and mercurial compounds are nonnutritive, heavy metals, which are hazardous to all biological organisms. Bacteria have evolved at least one mechanism for colonizing mercury-contaminated environments, an operon of mercury resistance (*mer*) genes encoding transporters and enzymes for biochemical detoxification. *Mer*[+] bacteria convert organic and ionic mercury compounds to the volatile and less toxic elemental form, Hg(0), which rapidly evaporates through cell membranes and/or walls. By genetically engineering plants with two of the *mer* genes (*merA, merB*), we hoped to express the biochemical detoxification pathway and enable the plants to remove and detoxify mercury. Presently, we have tested these genes in three plant species, a laboratory model plant, *Arabidopsis thaliana* L., a widely researched crop plant, tobacco (*Nicotiana tabacum* L.), and a forest tree, yellow-poplar (*Liriodendron tulipifera* L.) and have demonstrated that transgenic plants grow in the presence of otherwise lethal levels of ionic and organic mercury. Furthermore, these plants enzymatically and quantitatively

Phytoremediation of Toxic Metals: Using Plants to Clean Up the Environment, edited by Ilya Raskin and Burt D. Ensley.
ISBN: 0-471-19254-6 ©2000 John Wiley & Sons, Inc.

151

AVRIL ROBARTS LRC

reduce mercurial compounds from spiked growth media (Rugh et al., 1996; Bizily et al., 1999).

These studies demonstrate that genetic engineering offers a useful approach for adapting plants to new roles in environmental restoration. In particular, it will be feasible to engineer a variety of species that are native to specific terrestrial, wetland, and aquatic habitats for phytoremediation. By combining sets of genes that confer activity toward different pollutants, it may also be possible to develop plants that detoxify a broad range of chemicals.

MERCURY POLLUTION

Atmospheric Emission

Mercury emissions have elevated the global atmospheric pool from ~2 to ~4 × 10^6 kg of mercury, primarily as Hg(0), over the twentieth century (Slemr and Langer, 1992). Smokestack emissions of mercury occur in three principal forms: free Hg(II), particle-bound Hg(II) (Hg_p), and Hg(0). Hg(II) and Hg_p have atmospheric residence times on the order of weeks, unlike Hg(0), which can be retained ~0.5–2.0 years in the global atmospheric pool (Lindqvist and Rodhe, 1985; Lindqvist, 1991). Although mercury deposition rates have declined in parallel with patterns of mercury use over the last three decades (Benoit et al., 1994; Swain et al., 1992), the global atmospheric Hg(0) pool has increased in magnitude. This apparent inconsistency may be explained by significant declines in overall particulate matter pollution and a concordant reduction in Hg_p formation and deposition (Hudson et al.,1995).

The principal sources of atmospheric mercury emission in the United States are fossil fuel and medical waste incineration, which collectively account for >80% of all anthropogenic sources (Keating et al., 1997). In order to achieve a 90–95% reduction in U.S. emissions from these sources, the U.S. Environmental Protection Agency is issuing recommendations for removing mercury-containing materials for separate disposal prior to incinerating medical and municipal wastes (Keating et al., 1997).

Terrestrial and Aquatic Release

Mercury and mercurial compounds have been used in a variety of consumer and agricultural products such as paints, crop sprays, and dental amalgams. Several industrial processes use mercury or mercurial compounds: chloralkali synthesis (used in electrodes), wood pulping (for bleaching and as a fungicide), and precious-metal extraction (to form amalgams). One of mankind's most notorious environmental mishaps was the contamination of Minamata Bay, Japan in the 1950s and 1960s by runoff from an acetylaldehyde plant (Takizawa, 1979; Harada, 1995). This prolonged, incidental release of organic mercury claimed more than 1000 lives and caused irreparable neurological damage in 5000–6000 people, all of whom were exposed by consuming contaminated seafood (Harada, 1995). The Minamata tragedy helped curtail the extensive industrial use of mercury and motivated governmental agencies to enact strict controls on its disposal. Because of the increased documentation of the health hazards of mercury exposure, most consumer use also has been discontinued and manufacturers have replaced mercurial compounds with alternative agents or processes (Nriagu, 1979).

The use of mercury to extract gold from ore remains a significant source of environmental contamination in economically and technologically underdeveloped countries and regions (Pfeiffer et al., 1993; Akagi et al., 1995). The gold mining industry in the Brazilian Amazon alone consumed an average of 137 tons of mercury per year during the 1980s (Ferreira and Appel, 1990). On the basis of gold production records and emissions factors of 1.3–2.0 kg Hg released per kilogram of Au produced, Pfeiffer et al. (1993) estimate that 130 tons Hg/yr were released into the Amazon forests. Most of this sum (65–83%) was emitted into the atmosphere during the roasting of the Hg/Au amalgam with the remainder spilling into soils and waterways (Pfeiffer et al., 1993). Soils and sediments near many of the Brazilian mining areas have been monitored and the reported levels of mercury contamination are as high as 19.8 ppm (Pfeiffer et al., 1993). This is a relatively low concentration when compared to data for many American sites contaminated by runoff from nuclear and chloralkali plants, although it is still alarming considering that the pollution in the Amazon is dispersed over an enormous area. As anthropogenic mercury deposits in and around waterways increase in magnitude, mercury may eventually reach hazardous concentrations in fish and other wildlife.

Mercury Biogeochemistry

Ionic mercury(II) is the most prevalent terrestrial species and is found associated with a number of different anions, such as chloride, sulfides, hydroxides, and organics (Anderson, 1979). Hg(II) compounds may undergo various chemical transformations by abiotic (Allard and Arsenie, 1991) or bacterial processes (Robinson and Tuovinen, 1984) (Fig. 10-1a). Abiotic and microbial reduction of Hg(II) from Hg-containing sites generates 25–50% of global atmospheric Hg(0) emissions (Fitzgerald, 1995; Keating et al., 1997). In vitro and in situ analyses have demonstrated that abiotic reduction of ionic mercury by humic and fulvic acids releases Hg(0) vapor (Allard and Arsenie, 1991). Sunlight appears to promote humic acid-associated Hg(II) reduction (Xiao et al., 1995).

Some bacteria colonize mercury-contaminated habitats and are capable of enzymatically detoxifying Hg(II) and organomercurial compounds (Summers and Lewis, 1973; Robinson and Tuovinen, 1984). These characteristics are conferred by the *mer* operon, a series of genes that encodes two regulatory proteins, two mercury transport peptides, and, for broad-spectrum mercury resistance, two catalytic peptides (Fig. 10-1a) (Summers, 1986). Broad-spectrum resistant bacteria express the enzyme organomercurial lyase (MerB) (Tezuka and Tonomura, 1976; Begley et al., 1986), which catalyzes the protonolytic cleavage of the carbon–mercury bond of several forms of organic mercury (Fig. 10-1b). This reaction provides the inorganic substrate

AVRIL ROBARTS LRC

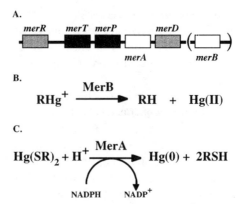

Figure 10-1   The genetics and biochemistry of bacterial mercury resistance. (*a*) The *mer* operon contains a series of five or more genes that collectively confer mercury resistance to bacteria. *MerA* and *merB* encode the enzymes mercuric reductase and organomercurial lyase, respectively. *MerT* and *merP* code for Hg(II) transport proteins, used by bacteria to shuttle mercury to the cytoplasm for enzymatic mercury detoxification. *MerD* and *merR* encode proteins that regulate gene expression of the *mer* operon. (*b*) Organomercurial lyase (MerB) detoxifies organic mercury (RHg) by catalyzing the cleavage of the carbon–mercury bond. (*c*) Mercuric reductase (MerA) reduces Hg(II), in the form of a thiointermediate enzymatic substrate, to volatile Hg(0).

[Hg(II)] for MerA (mercuric reductase), an NADP-requiring, FAD-containing mercuric ion oxidoreductase that catalyzes the electrochemical reduction of Hg(II) to Hg(0) (Fox and Walsh, 1982) (Fig. 10-1*c*). Since elemental mercury is relatively inert and is volatile at ambient temperatures, it diffuses out of bacterial cells without binding to cellular components.

Anaerobic bacteria such as the sulfate-reducing strains methylate Hg(II), accounting for at least 90% of environmental methylmercury (MeHg) formation (Compeau and Bartha, 1985). Bacterial methylmercury synthesis has been observed to occur more efficiently under acidic, saline, and anoxic conditions (Ridley et al., 1977; Beijer and Jernelov, 1979). Aerobic bacteria also produce methylmercury, although at much lower efficiencies than sulfate reducers (Olson and Cooper, 1976).

Mercury Toxicity

Mercury and mercury compounds are nonnutrient poisons that cause severe physiological effects when ingested in relatively small doses. The primary mechanism of cellular toxicity is the reaction of charged forms of mercury with reduced sulfhydryl groups on critical proteins. The binding of mercury to these macromolecules inhibits their normal biological functions and disrupts cellular metabolism. Common symptoms of mercury poisoning include tremors, sensory disturbances, ataxia, and paresthesia. Histological studies have confirmed that mercury poisoning can lead to the atrophy of the central nervous system.

The toxicokinetics of the three principal forms of mercury, Hg(0), Hg(II), and R-Hg$^+$ are substantially different. Elemental mercury is readily absorbed through the lungs and passively diffuses through other biological tissues yet has a relatively short residence time. It neither accumulates nor reacts with most biochemicals unless first oxidized to Hg(II), and this may be catalytically driven by peroxidase or catalase enzymes (Ogata and Aikoh, 1984). Victims who experience chronic exposure may suffer neurological dysfunction among other symptoms. Intense, short-term Hg(0) exposure can cause rapid death due to respiratory failure as reported in a study involving members of a Michigan family (Kanleun and Gottlieb, 1991).

Despite its high chemical reactivity, Hg(II) does not freely diffuse through cellular membranes and is inefficiently absorbed through the intestines. Studies have reported that ~80–85% of ingested doses are excreted in feces (Miettinen, 1973) with the rest being eliminated over a period of days or months through urine (Rahola et al., 1973). As Hg(II) is processed and filtered by the body, it is likely to cause damage to the kidneys, liver, and gastrointestinal tract.

Methylmercury is lethal at much lower doses than Hg(0) or Hg(II) because it is reactive yet highly mobile. It is efficiently absorbed from the gastrointestinal tract with an estimated uptake of 95% in humans (Aberg et al., 1969). Once inside the circulatory system, MeHg is rapidly distributed throughout the body. Unlike Hg(II), it is readily transported across blood vessel walls and into tissues. The membrane permeability of MeHg has been variously attributed to its basic hydrophobic character, to its tendency to associate with counterions (e.g., Cl^-) to form neutral salts (Mason et al., 1996), and to the recognition of a cysteine-bound MeHg complex by a transport protein (Aschner and Aschner, 1990; Kerper et al., 1992). Although it may have a widespread impact on many basic cellular processes, MeHg poisoning is typically identified by aberrant neurological responses. MeHg has been shown to interfere with acetylcholine neurotransmitter release by disrupting the flow of Ca^{2+} through ion channels (Atchison and Hare, 1994). Initial symptoms of poisoning may include shaking, difficulty in walking, and other sensory impairments. At greater durations and/or magnitudes of exposure, victims experience a rapid central nervous system (CNS) deterioration. The syndrome caused by MeHg poisoning has been termed Minamata disease following the outbreak at Minamata Bay, Japan described in the sections on terrestrial and aquatic release and on mercury pollution remediation. Because MeHg has a long residence time and is efficiently transferred between trophic levels it is biomagnified through the food web (Jernelov and Lann, 1971). Bioconcentration factors ($B_f = \log [CH_3Hg^+ \text{ biota}]/[CH_3Hg^+ H_2O]$) have been estimated at 5–6 units for fish indicating that parts per trillion levels in waters can translate into parts per million (ppm) levels in fish (Watras and Bloom, 1992). For habitual consumers of fish, these tissue concentrations are sufficient to cause neurotoxicity.

STANDARD METHODS FOR MERCURY POLLUTION REMEDIATION

As with most xenobiotic pollution, mercury contamination is difficult and expensive to remediate. The restoration of heavy-metal-contaminated areas typically involves the relocation or intensive on-site treatment of large quantities of soils and sediments. These techniques are often extremely disruptive to the indigenous wildlife and may delay the full ecological recovery of polluted sites.

Bulk relocation of mercury-contaminated soils and sediments is the most common method for the remediation of polluted sites. High cost, modest efficiency, and incidental Hg mobilization are drawbacks to this approach (Schultz et al., 1995). In recent years new technologies have minimized the loss of mercury-containing sediments during dredging operations and simultaneously raised the efficiency of pollutant recovery (Herbich, 1995). Improvements include the use of hydraulic suction dredgers to avoid violent mechanical "cutting" into sediment floors and the addition of camera monitoring systems to measure turbidity and permit the control of suction force. Both of these dredging strategies were employed to reclaim the contaminated Minamata Bay area (Yoshinaga, 1995). During this operation, sediment resuspension was carefully controlled as greater than 1.5 million m^3 of mercury-containing sediments were moved onto the innermost 580,000 m^2 of the site over a 10-year period. The sediment mounds were enclosed in a constructed watertight revetment, overlaid with plastic, and then capped with clean soil. Using these advanced techniques, the hazards of dredging can be mitigated, although the cost remains prohibitively high for most sites.

Dredged, excavated waste is often relocated a large distance, incurring the additional expenses of transportation and landfill disposal. Once buried, organic leachates and anaerobic conditions found in landfills have been demonstrated to promote Hg(II) mobilization and methylation (Niebla et al., 1976). To prevent the escape of mercury and other heavy-metal toxins, cements or resin polymers may be added to solidify contaminated sludge material. Waste solidification reduces volume and water content and prevents the leaching of most hazardous materials. However, mercury is more prone to leaching than other chemicals and may not be adequately contained by thermal solidification treatments or low-temperature stabilization. Complexing of the ionic mercury with sulfides prior to treatment may help minimize this problem (Hamilton and Bowers, 1997). Such intensive treatments can be expensive and disrupt the utility of the polluted site. However, for severely contaminated environments, excavation and stabilized burial will offer the most rapid method for bulk containment of the mercury hazard.

MERCURY BIOREMEDIATION

Bioremediation is the use of living bacteria or immobilized biomass for the treatment and cleanup of hazardous wastes. Toxin-degrading organisms may be added to waste sites (bioaugmentation) or amplified by enriching the substrate with essential nutrients (biostimulation) (Bandyopadhyay et al., 1994). Alternatively, mats of living cells or dead biological material may be used to absorb and filter pollutants from effluent streams. Some bacteria enzymatically reduce particular heavy metals to insoluble precipitates as a means of detoxifying them (Silver and Phung, 1996). When added to wastes these bacterial strains may help make heavy metals more recoverable (Rawlings and Silver, 1995).

In light of their ability to detoxify and volatilize mercury, mer^+ bacteria are obvious candidates for bioremediation. *Pseudomonas* strain K-62 was shown to reduce mercurial compounds to Hg(0) from industrial wastes (Suzuki et al., 1968), although its rate of volatilization was sharply affected by suboptimal salt concentrations. Similarly, Hansen et al. (1984) treated mercury-contaminated sewage with a continuous culture of mercury-resistant bacteria and observed that the bacteria removed 98% of Hg(II) by converting it to Hg(0). Given that bacteria are extremely sensitive to environmental conditions and are difficult to manipulate in the field, their most promising role as bioremediators will probably be in controlled industrial applications.

Biosorption is the use of immobilized cells or biomaterials to nonspecifically bind and remove reactive and particulate heavy metals from wastestreams (White et al., 1997). The strategy has been applied to mercury and other contaminants. For example, Wilkinson et al. (1989) used immobilized algal cultures to accumulate Hg(II) and found that algae filtered out more than 95% of the total mercury treatment from the medium. Unfortunately, biosorption reactor flow rates and volumes are usually very limited and will probably restrict this technology to small-scale applications (Gadd and White, 1993). In addition, biosorption is sensitive to contaminants such as oils or particulates, making it unsuitable for many field-site cleanup applications (Brierly et al., 1989).

PLANT INTERACTIONS WITH MERCURY

Plants are capable of extracting a variety of metal ions from their growth substrates, including mercury. Field and laboratory experiments on plant uptake and phytotoxicity have demonstrated that plants accumulate mercury from both soil and atmospheric sources (Godbold, 1991; Godbold and Hutterman, 1986; Lindberg et al., 1979). Mercury phytoavailability and uptake depend on both the physiological state of the plant (Hanson et al., 1995) and the chemical characteristics of the soil (Adriano, 1986).

Plant transpiration has been observed to significantly elevate Hg(0) levels at the canopy–atmosphere interface above mercury-containing soils (Kozuchowski and Johnson, 1978; Hanson et al., 1995). This observation suggests that plants naturally reduce Hg(II) at a low level, and provide a conduit for rhizosphere-generated Hg(0). By absorbing mercury from both the ground and the atmosphere, plants play a dynamic role in the cycling of Hg(0) at the soil–air interface (Siegel and Siegel, 1979; Lindberg et al., 1986). Plants will acquire Hg(II) from soil and aqueous media, though it is largely retained in the root tissues (Suszcynsky and Shann, 1995). At sites with high atmospheric mercury loads, mercury is detectable in the foliage of local vegetation (Huckabee et al., 1983). Although some of the foliar content may be due to root uptake and subsequent transport, most of it is directly deposited into leaves as Hg(0) (Lindberg et al., 1979; Suszcynsky and Shann, 1995). The oxidation of Hg(0) vapor by leaf catalases and peroxidases causes the less mobile mercurial Hg(II) to accumulate in these tissues (Du and Feng, 1983).

Plants tolerate moderate levels of inorganic mercury in the soil and atmosphere but may grow more slowly than in "clean" environments. When challenged with toxic levels of mercury, they display severely stunted growth, leaf curling, and chlorosis. Mercury is known to affect photosynthesis and oxidative metabolism by interfering with electron transport by the membrane-rich mitochondria and chloroplasts (Siegel et al., 1974). Mercury diminishes the uptake of water (Beauford et al., 1977) and the homeostasis of nutrients and electrolytes by altering membrane potentials and inhibiting ion channels and pumps (Godbold, 1991; Kennedy and Gonsalves, 1987, 1989). Whereas organic and inorganic mercury seem to damage plants via the same basic mechanisms, compounds such as phenylmercuric acetate and methylmercury chloride appear to be two to three orders of magnitude more toxic than Hg(II) (Godbold, 1991; Bizily et al., 1998). Because organomercurials are more lipophilic, they may have higher affinity for subcellular organelles, accounting for their severe effects at lower exposures.

MERCURY PHYTOREMEDIATION BY TRANSGENIC PLANTS

As stated previously, the ultimate goal of this work is to develop plants that provide an alternative to standard practices for mercury cleanup. Basic principles describing the biogeochemical transformations of mercury and its subsequent interaction with aquatic food webs in the environment are schematically depicted in Fig. 10-2a. The transfer of the bacterial organomercurial lyase (merB) and mercuric reductase (merA) genes into plants may enable direct management of mercury cycling at contaminated sites, thereby protecting endangered ecosystems (Fig. 10-2b). This experimental approach holds great promise for reducing mercury-related hazards and stands as a model for the future application of transgenic plants to beneficial environmental goals.

Introduction of Bacterial Genes into Plants for Mercury Detoxification

To improve the expression of mer genes in plants, we modified the bacterial merA DNA sequence by reducing the GC content in a 9% block of the protein coding region and adding plant regulatory elements (Rugh et al., 1996). When transferred to the laboratory model plant (Arabidopsis thaliana L.), and to tobacco (Nicotiana tabacum L.), the new gene construct (merA9) conferred resistance to 50 µm Hg(II) (Figs. 10-3a–10-3d), suggesting that merA plants enzymatically reduce Hg(II) and evaporate away Hg(0). We also examined the ability of yellow-poplar (Liriodendron tulipifera L.) tissue cultures and plantlets to express modified merA gene constructs (Rugh et al., 1998). Three reconstructed genes were used for the transformation of yellow-poplar proembryogenic masses (PEMs), each having different amounts (0%, 9%, and 18% blocks, respectively) of altered coding sequence. Each of these constructs was shown to confer Hg(II) resistance to transformed tissue cultures. Regenerated yellow-poplar plantlets containing the merA18 gene germinated and grew vigorously in media containing 50 µM Hg(II) (Figs. 10-3e and 10-3f). Given the distant evolutionary relationship of yellow-poplar, tobacco, and Arabidopsis, these results indicate that most, if not all, plants should be capable of merA transgene expression.

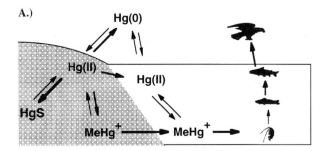

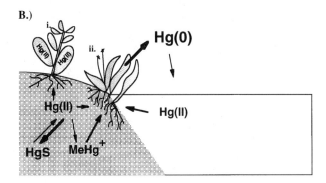

Figure 10-2   Biogeochemical cycling of mercury in the environment. (*a*) Hg (II) is the predominant form of mercury in soils, usually as stable HgS. Hg(II) may be reduced to volatile Hg(0) abiotically or by *mer*[+] bacteria, or methylated, primarily by anaerobic bacteria. MeHg biomagnifies in the aquatic food web, accumulating in sediment-dwelling invertebrates and increasing in concentration at higher trophic levels. MeHg can be demethylated by both abiotic or microbial processes. (*b*) In the future, plants transformed with *mer* genes may help extract mercurials from polluted sites, either (i) translocating Hg(II) to their shoots (phytosequestration) or (ii) evaporating mercurials as reduced Hg(0) (phytovolatilization).

Arabidopsis was also transformed with constructs containing the bacterial *merB* gene. Plants that expressed MerB were able to germinate and survive on growth medium treated with the organomercurial compounds, phenylmercuric acetate (PMA) or methylmercury (MeHg) (Figs. 10-4*a* and 10-4*b*). Note that the levels of organomercurials in these assays were one to two orders of magnitude lower than the concentrations of Hg(II) required for wild-type seed mortality in *merA* assays. This clearly indicates that organomercurial compounds are more toxic to plants than ionic mercury.

Mercury Phytovolatilization

Closed-system assays were performed to compare the rates at which *merA* and wild-type plants reduce Hg(II) and evaporate Hg(0) from spiked media. *Arabidopsis* seedlings were transferred to reaction tubes containing 25 µM HgCl$_2$ in buffered medium (Rugh et al., 1996). The medium and head space were evacuated each minute

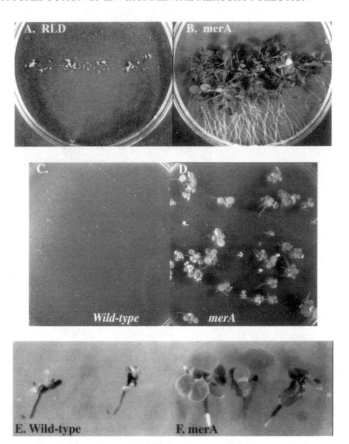

Figure 10-3 Toxicity assays for germination of *merA* and wild-type seeds and embryos on growth media spiked with 50 μM Hg(II): (*a*) wild-type *Arabidopsis thaliana* (ecotype RLD) seeds, (*b*) merA9 *A. thaliana* plants grown from seeds ~3 weeks; (*c*) wild-type tobacco seeds (no germination or survival); (*d*) *merA* tobacco from seeds ~2 weeks; (*e*) wild-type yellow-poplar regenerants; (*f*) *merA* yellow-poplar from tissue culture-derived embryos, germinated on selective medium ~2 weeks.

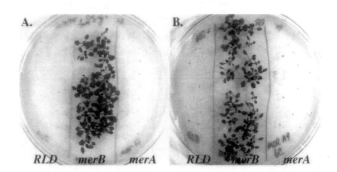

Figure 10-4 Toxicity assays for germination of (L to R) wild-type, *merB*-transformed, and *merA*-transformed *Arabidopsis* seeds ~4 weeks on medium containing organomercurial salts: (*a*) 1 μM methylmercury; (*b*) 1 μM phenylmercuric acetate (PMA).

over a 10-min period and sampled for Hg(0) using a portable mercury vapor analyzer (Arizona Instruments; Phoenix, AZ). *MerA9* plants volatilized 3–4 times as much Hg(0) as did wild-type plants (Fig. 10-5). In an analogous experiment, *merA18* yellow-poplar plantlets were placed in test tubes containing gel medium spiked with 10 μM Hg(II). The tubes were evacuated every 12 h over a period of 6 days (Rugh et al., 1998a). Hg(0) readings indicated that *merA18* plantlets released elemental mercury at approximately 10 times the rate of nontransformed control plantlets (Fig. 10-6). These experiments independently verify that a functional mercuric reductase is expressed in transgenic plant lines of both *Arabidopsis* and yellow-poplar.

Hydroponic assays were used to quantify the rates at which *merA* tobacco removed Hg(II) from spiked medium (Heaton et al., 1998). Approximately 10-cm-tall plants, from independent *merA9* and wild-type lines, were removed from soil and placed in an aqueous nutrient medium for a one-week acclimation period. After a week, the old medium was replaced with fresh hydroponic solution containing 5μM Hg(II). Both the medium and the plants were sampled or harvested at several time points over 7 days. For all samples, the entire dose of Hg(II) was bound to or absorbed by the plant's root system in fewer than 18 h (data not shown). The *merA9* tobacco plants chemically reduced nearly 40% of their initial Hg(II) load after one day and eliminated it from the system (Fig. 10-7). Over the course of the assay, they did not exhibit symptoms of stress and showed declining mercury tissue concentrations at each sample point, reducing ~70% of the Hg(II) dosage from the hydroponic system after 7 days. Wild-type plants wilted on the first day, then recovered during the following 2 days and reduced ~20% of the mercury from the system by the end of the experiment. Although the single dose of Hg(II) caused an initial shock to the wild-type plants, they were able to recover, presumably by growing new root tissue and reducing mercury at low, background levels. It is likely that repeated doses of Hg(II) into the medium would prevent wild-type plants from recovering and eventually cause death.

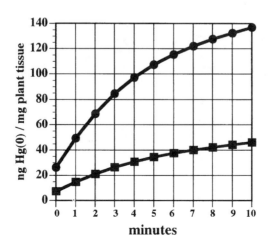

Figure 10-5 Hg(0) volatilization assays for wild-type (■) and *merA* (●) *Arabidopsis* plantlets.

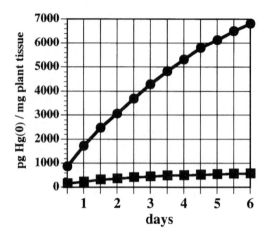

Figure 10-6   Hg(0) volatilization assays for wild-type (■) and *merA* (●) *Liriodendron* plantlets.

merA plants will presumably elevate rates of environmental Hg(0) volatilization well beyond background levels. In certain situations this may not be a desirable result as some areas may have a "funneling" topography or unique meteorological conditions that promote the rapid redeposition of Hg(0) and make the goal of atmospheric dilution unachievable. At appropriate locations, the extreme toxicity of methylmercury justifies the accelerated volatilization and atmospheric dilution of Hg(0) back to normal, low environmental levels. Should our expectations hold true, the establishment of *merA* communities of plants will continuously prevent the accumulation of mercurial compounds in soils and sediments, avoiding the need for maintenance or harvesting.

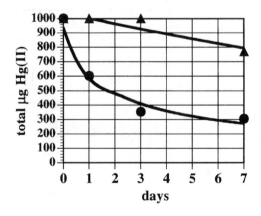

Figure 10-7   Tobacco hydroponic assays for Hg(II) reduction and removal with wild-type (▲) and *merA*-transformed (●) plants.

Mercury Phytosequestration

For areas where phytovolatilization is unsuitable due to the hazards of releasing Hg(0), an alternative strategy must be considered. One option is to deploy plants that sequester high mercury loads in harvestable tissues. The ability of wild plants to accumulate mercury is largely dependent on their natural physiochemical and structural characteristics (Hanson et al., 1995). The overexpression of peroxidases and catalases in specific tissues in the plant shoot may allow Hg(0) to be captured from the evapotranspiration stream (Du and Feng, 1983). This strategy for mercury sequestration may be further enhanced by root-specific expression of the *merA* and *merB* genes to detoxify charged mercurials prior to transport to shoots. Plants have not been found to hyperaccumulate mercury to the extent that some do for other heavy metals, such as copper, nickel, and zinc (Baker and Brooks, 1989). Plants, in general, may require special genes to enhance their uptake, retention, and tolerance of Hg(II). Specifically, plants will have to be engineered to load Hg(II) into tissues (e.g., cell wall) or organelles (e.g., vacuole) in the shoot, where it can be safely accumulated. Plant expression of modified bacterial mercury-transport genes, *merP* and *merT* (Fig. 10-1a), may provide a means for improving mercury uptake and organelle- and tissue-specific targeting. Plants naturally use a variety of chelators for metal homeostasis and tolerance, such as metalothioneins (Yeargan et al., 1992), phytochelatins (Zenk, 1996), and organic acids (Lee et al., 1977). Modification and regulated expression of genes coding for these or their biosynthetic enzymes may assist in transporting Hg(II) to storage targets and for protecting cellular functions during uptake.

Although the major benefit of phytosequestration is the minimal release of gaseous mercury, potential problems do exist. For one, herbivores may consume shoot-bound Hg(II). Methods to avoid this might include the frequent harvesting of Hg(II)-loaded plant shoots or the sowing of edible, non-accumulating crops at the periphery of sites to attract herbivores away from hyperaccumulators. A second disadvantage of phytosequestration is the added cost of managing, retrieving, and disposing of Hg-containing tissues. These expenses increase proportionally with the inaccessability of a site, but are still likely to be far less than those associated with engineering-based approaches.

Mercury Phytoremediation Crop Design and Selection

In our initial efforts to develop mercury-detoxifying plants, we used *Arabidopsis*, tobacco, and yellow-poplar principally because they are amenable to genetic transformation and plant regeneration. These species are not necessarily adapted to the conditions at most contaminated field sites. In areas for which mercury phytovolatilization is most suitable, the "harvestability" of the plant will not be an issue, and site compatibility will be the primary criterion for selecting plants. Mercury pollution is frequently deposited in lowland regions, including riparian and estuarine sites, salt marshes, and freshwater wetlands. Facultative and obligate wetland grasses such as cattail (*Typha* spp.), rush (*Juncus* spp.), bulrush (*Scirpus* spp.), and cordgrass (*Spartina* spp.) have been used in constructed wetlands for sewage and wastewater treatment. Of these species, *Spartina* is best adapted to polluted saline and brackish ecosystems. Emergent aquatic plants help to stabilize sediments, contribute organic matter, and aerate sediments (Reddy et al., 1990; Stengel, 1993) making conditions less conducive to mercury methylation.

Trees also possess physiological and structural characteristics that make them good candidates for mercury phytoremediation (Stomp et al., 1993). Their abundance of nonliving, woody tissues enables them to load heavy metals away from sensitive, vital plant organs. They were shown to be effective for extracting heavy metals from sewage, in which they were grown, cut, and resprouted in continuous cycle (Riddell-Black, 1994). Furthermore, their long lives allow for prolonged service without replanting. Since trees have long periods of nonreproductive juvenility, they may be harvested prior to the onset of flowering and seed production in order to prevent the release of transgenic offspring.

Some trees, such as sweetgum (*Liquidambar styraciflua* L.), willow (*Salix* spp.), and poplar (*Populus* spp.), are particularly water-tolerant and are valuable for treating inundated bottom lands. These species could be used in combination with herbaceous plants to construct a dynamic community for the phytoremediation of freshwater wetlands. Aquatic and woody macrophytes possess many physiological attributes to complement novel detoxifying abilities conferred by transgenes. However, many of these species have not been used for bioengineering, and new protocols will have to be developed for their genetic transformation and regeneration from tissue cultures.

CONCLUSION

Since the Minamata Bay disaster, industrialized nations have looked for alternatives to mercury consumption and have sharply curtailed its use. Some developing countries continue to use mercury in a number of processes without controlling its release into the environment. This practice may eventually affect indigent, local communities that utilize unmonitored waterways and floodplains for food and water supplies. These populations are at a particular risk of methylmercury exposure if their diet largely consists of fish from contaminated waters.

Few polluted areas are eligible for high-cost, intensively engineered cleanup strategies, such as those that were employed at Minamata Bay. Less disruptive and more affordable alternatives must be devised. The use of biological organisms may provide a means to greatly reduce costs, minimize incidental mobilization, and help stabilize sites for ecological recovery as they are decontaminated. Bacteria may be effective for the treatment of industrial or municipal wastes, though their specialized growth requirements cause them to be impractical for many environmental applications. Furthermore, the recovery of microbe-bound toxins, such as heavy metals, is problematic and may not be feasible in the field.

Plants possess a natural physiochemical apparatus that is well adapted to infiltrate and purify contaminated land and water. Plants are easier to contain than environmentally released microorganisms and centuries of agricultural practices have made their harvest a routine and efficient practice. Many studies have demonstrated that plants have great potential for extracting and detoxifying pollutants. However, there are no known plant species that naturally degrade or hyperaccumulate mercury and mercurial compounds. Bacterial pathways for detoxifying mercury and other poisons are genetically and biochemically well characterized. The tools of genetic engineering have enabled us to combine microbial capabilities with the physiological and structural advantages offered by plants. We have demonstrated that bacterial mercury-resistance genes can be used to confer broad-range mercury detoxifying and reducing capabilities to a variety of plant species. One limitation to the use of *mer* plants will be for Hg-polluted sites that contain toxins in addition to mercury. As the field of environmental biotechnology develops, however, it is likely that researchers will discover or engineer genes to provide resistance to other compounds. By adding appropriate sets of genes, we may be able to customize plants to the specific demands that each polluted site presents.

REFERENCES

Aberg, B. et al., 1969. Metabolism of methylmercury compounds in man: Excretion and distribution. *Arch. Environ. Health* **19**:478–484.

Adriano, D. C., 1986. *Trace Elements in the Terrestrial Environment.* Springer-Verlag, pp. 299–328.

Akagi, H. et al., 1995. Human exposure to mercury to goldmining in the Tapajos River Basin, Amazon, Brazil: Speciation of mercury in human hair, blood and urine. *Water, Air, Soil Pollut.* **80**:85–94.

Allard, B., and I. Arsenie, 1991. Abiotic reduction of mercury by humic substances in aquatic systems: An important process for the mercury cycle. *Water, Air, Soil Pollut.* **56**:457–464.

Anderson, A., 1979. Mercury in soils. In *The Biogeochemistry of Mercury in the Environment*, J. O. Nriagu, ed., Elsevier/North-Holland Biomedical Press, pp. 79–113.

Aschner, M., and J. L. Aschner, 1990. Mercury neurotoxicity: Mechanisms of blood-brain barrier transport. *Neurosci. Biobehav. Rev.* **14**(2):169–176.

Atchison, W. D., and M. F. Hare, 1994. Mechanisms of methylmercury-induced neurotoxicity. *FASEB J.* **8**:622–629.

Baker, A. J. M., and R. R. Brooks, 1989. Terrestrial higher plants which hyperaccumulate metallic elements—a review of their distribution, ecology and phytochemistry. *Biorecovery* **1**:81–126.

Baker, A. J. M. et al., 1994. The possibility of in situ heavy metal decontamination of polluted soils using crops of metal-accumulating plants. *Resources Conserv. Recycl.* **11**:41–49.

Bandyopadhyay, S. et al., 1994. Engineering aspects of bioremediation. In *Remediation of Hazardous Waste Contaminated Soils*, D. L. Wise and D. J. Trantolo, eds., Marcel-Decker, New York, pp. 55–77.

Beauford, W. et al., 1977. Uptake and distribution of mercury within higher plants. *Physiol. Plant.* **39**:261–265.

Begley, T. P. et al., 1986. Bacterial organomercurial lyase: Overproduction, isolation and characterization. *Biochemistry* **25:**7186–7192.

Beijer, K., and A. Jernelov, 1979. Methylation of mercury in aquatic environments. In *The Biogeochemistry of Mercury in the Environment*, J. O. Nriagu, ed., Elsevier/North-Holland Biomedical Press, pp. 203–230.

Benoit, J. M. et al., 1994. Modeling the biochemical cycle of mercury in lakes: The mercury cycling model (MCM) and its application to the MTL study lakes. In *Mercury Pollution: Integration and Synthesis*, C. J. Watras and J. H. Huckabee, eds., Lewis Publishers, pp. 187–202.

Bizily, S. P. et al., 1999. Phytoremediation of methylmercury pollution: *MerB*-expression in *Arabidopsis thaliana* confers resistance to organomercurials. *Proc. Natl. Acad. Sci.* (USA).

Brierly, C. L. et al., 1989. Applied microbial processes for metals recovery and removal from freshwater. In *Metal Ions and Bacteria*, T. J. Beveridge and R. J. Doyle, eds., J Wiley, New York, pp. 359–382.

Compeau, G. C., and R. Bartha, 1985. Sulfate-reducing bacteria: Principal methylators of mercury in anoxic estuarine sediment. *Appl. Environ. Microbiol.* **50:**498–502.

Du, S. -H., and S. C. Feng, 1983. Catalase activity of C_3 and C_4 species and its relationship to mercury vapor uptake. *Environ. Exper. Bot.* **23:**347–353.

Ferreira, R. C. H., and L. E. Appel, 1990. *Estudo Detalhado de Fontes e Usos de Mercurio. Relatorio Preliminar.* Centro de Technologia Mineral, Rio de Janeiro.

Fitzgerald, 1995. Is mercury increasing in the atmosphere? The need for an atmospheric mercury network (AMNET). In *Mercury As a Global Pollutant*, D. B. Porella, J. W. Huckabee, and B. Wheatley, eds., Kluwer, Boston, pp. 245–254.

Fox, B., and C. T. Walsh, 1982. Mercuric reductase. Purification and characterization of a transposon-encoded flavoprotein containing an oxidation-reduction-active disulfide. *J. Biol. Chem.* **10:**2498–2503.

Gadd, G. M., and C. White, 1993. Microbial treatment of heavy metal pollution—a working biotechnology. *Trends Biotechnol.* **11:**353–359.

Godbold, D. L., 1991. Mercury-induced root damage in spruce seedlings. *Water, Air, Soil Pollut.* **56:**823–831.

Godbold, D. L., and A. Huttermann, 1986. The uptake and toxicity of mercury and lead to spruce (*Picea abies karst*) seedlings. *Water, Air, Soil Pollut.* **31:**509–515.

Hamilton, W. P., and A. R. Bowers, 1997. Determination of acute Hg emissions from solidified/stabilized cement waste forms. *Waste Manage.* **17:**25–32.

Hansen, C. L. et al., 1984. Bacterial removal of mercury from sewage sludge. *Biotechnol. Bioeng.* **26:**1330–1333.

Hanson, P. J. et al., 1995. Foliar exchange of mercury vapor: Evidence for a compensation point. *Water, Air, Soil Pollut.* **80:**373–382.

Harada, M., 1995. Minamata disease: Methylmercury poisoning in Japan caused by environmental pollution. *Crit. Rev. Toxicol.* **25:**1–24.

Heaton, A. C. P. et al., 1998. Phytoremediation of mercury and methylmercury polluted soils using genetically engineered plants. *J. Soil Contam.* **7:**497–509.

Herbich, J. B., 1995. Removal of contaminated sediments: Equipment and recent field studies. In *Dredging, Remediation, and Containment of Contaminated Sediments*, K. R. Demars et al., eds., ASTM Press, pp. 77–111.

Huckabee, J. W. et al., 1983. Distribution of mercury in vegetation at Almaden, Spain. *Environ. Pollut., Series A* **30:**211–224.

Hudson, R. J. M. et al., 1995. Anthropogenic influences on the global mercury cycle: A model-based analysis. *Water, Air, Soil Pollut.* **80:**265–272.

Jernelov, A., and H. Lann, 1971. Mercury accumulation in food chains. *Oikos* **22:**403–406.

Kanleun, R. B., and C. A. Gottlieb, 1991. A clinical pathological study of four adult cases of acute mercury inhalation toxicity. *Arch. Pathol. Lab. Med.* **115:**56–60.

Keating, M. H. et al., 1997. *Mercury Study Report to Congress.* EPA-452/R-9. U.S. Environmental Protection Agency, Washington, DC, Vol. I, Section 3, pp. 6–7.

Kennedy, C. D., and F. A. N. Gonsalves, 1987. The action of divalent zinc, cadmium, mercury, copper and lead on the trans-root potential and H^+ efflux of excised roots. *J. Exper. Bot.* **38:**800–817.

Kennedy, C. D., and F. A. N. Gonsalves, 1989. The action of divalent Zn, Cd, Hg, Cu and Pb ions on the ATPase activity of a plasma membrane fraction isolated from roots of *Zea mays. Plant Soil* **117:**167–175.

Kerper, L. E. et al., 1992. Methylmercury transport across the blood-brain barrier by an amino acid carrier. *Am. J. Physiol.* **262**(5):R761–R765.

Kozuchowski, J., and D. L. Johnson, 1978. Gaseous emissions of mercury from an aquatic water plant. *Nature* **274:**468–469.

Lee, J. et al., 1977. Isolation and identification of a citrate-complex of nickel from nickel-accumulating plants. *Phytochemistry* **16:**1503–1505.

Lindberg, S. E. et al., 1979. Atmospheric emission and plant uptake of mercury from agricultural soils near the Almaden mercury mine. *J. Environ. Qual.* **8:**572–578.

Lindberg, S. E. et al., 1986. Atmospheric deposition and canopy interactions of major ions in a forest. *Science* **231:**141–145.

Lindqvist, O., and H. Rodhe, 1985. Atmospheric mercury: A review. *Tellus* **37B:**136–159.

Lindqvist, O., 1991. *Mercury in the Swedish Environment.* Kluwer Academic Publishers, p. 261.

Mason, R. P. et al., 1996. Uptake, toxicity, and trophic transfer of mercury in a coastal diatom. *Environ. Sci. Technol.* **30:**1835–1845.

Niebla, E. E. et al., 1976. Effect of municipal landfill leachate on mercury movement through soils. *Air, Water, Soil Pollut.* **5:**399–401.

Nriagu, J. O., ed., Production and uses of mercury. In *The Biogeochemistry of Mercury in the Environment,* Elsevier/North-Holland Biomedical Press, 1979, pp. 23–41.

Ogata, M., and H. Aikoh, 1984. Mechanism of metallic mercury oxidation *in vitro* by catalase and peroxidase. *Biochem. Pharmacol.* **33:**490–493.

Olson, B. H., and R. C. Cooper, 1976. Comparison of aerobic and anaerobic methylation of mercuric chloride by San Francisco Bay sediments. *Water Res.* **10:**113–116.

Pfeiffer, W. C. et al., 1993. Environmental fate of mercury from gold mining in the Brazilian Amazon. *Environ. Rev.* **1:**26–37.

Porcella et al., eds., *Mercury As a Global Pollutant,* Kluwer, Boston, 1995, 1336 pages.

Rahola, T. et al., 1973. Elimination of free and protein-bound ionic mercury in man. *Ann. Clin. Res.* **5:**214–219.

Rawlings, D. E., and S. Silver, 1995. Mining with microbes. *Bio/Technology* **13:**773–778.

Reddy, K. R. et al., 1990. Oxygen transport through aquatic macrophytes: The role in wastewater treatment. *J. Environ. Qual.* **19**:261–267.

Riddell-Black, D. M., 1994. Heavy metal uptake by fast growing willow species. In *Willow Vegetation Filters for Municipal Wastewaters and Sludges*, P. Aronsson and K. Perttu, eds., Swedish Univ. Agriculture, Section of Energy Forestry Report 50, pp. 145–151.

Ridley, W. P. et al., 1977. Biomethylation of toxic elements in the environment. *Science* **197**:329–333.

Robinson, J. B., and O. H. Tuovinen, 1984. Mechanisms of microbial resistance and detoxification of mercury and organomercurial compounds: Physiological, biochemical and genetic analyses. *Microbiol. Rev.* **48**:95–124.

Rugh, C. L. et al, 1996. Mercuric ion reduction and resistance in transgenic *Arabidopsis thaliana* plants expressing a modified merA gene. *Proc. Natl. Acad. Sci.* (USA) **93**:3182–3187.

Rugh, C. L. et al., 1998. Development of transgenic yellow-poplar for mercury phytoremediation. *Nature Biotechnol.* **16**:925–929.

Schultz, T. et al., 1995. A mercury model used for assessment of dredging impacts. *Water, Air, Soil Pollut.* **80**:1170–1180.

Siegel, B. Z., and S. M. Siegel, 1979. Biological indicators of atmospheric mercury. In *The Biogeochemistry of Mercury in the Environment*, J. O. Nriagu, ed., Elsevier/North-Holland Biomedical, pp. 131–159.

Siegel, S. M. et al., 1974. Release of volatile mercury from vascular plants. *Physiol. Plant.* **32**:174–176.

Silver, S., and L. T. Phung, 1996. Bacterial heavy metal resistance: New surprises. *Annu. Rev. Microbiol.* **50**:753–789.

Slemr, F., and E. Langer, 1992. Increase in global atmospheric concentrations of mercury inferred from measurements over the Atlantic Ocean. *Nature* **355**:434–437.

Stengel, E., 1993. Species-specific aeration of water by different vegetation types in constructed wetlands. In *Constructed Wetlands for Water Quality Improvement*, G. A. Moshiri, ed., Lewis Publishers, Ann Arbor, MI, pp. 427–434.

Stomp, A. -M. et al., 1993. Genetic improvement of tree species for remediation of hazardous wastes. *In Vitro Cell. Devel. Biol.* **29P**:227–232.

Summers, A. O., 1986. Organization, expression, and evolution of genes for mercury resistance. *Annu. Rev. Microbiol.* **40**:607–634.

Summers, A. O., and E. Lewis, 1973. Volatilization of mercuric chloride by mercury-resistant plasmid-bearing strains of *Escherichia coli*, *Staphylococcus aureas*, and *Psuedomonas aeroginosa*. *J. Bacteriol.* **113**:1070–1072.

Suszcynsky, E. M., and J. R. Shann, 1995. Phytotoxicity and accumulation of mercury subjected to different exposure routes. *Environ. Toxicol. Chem.* **14**:61–67.

Suzuki, T. et al., 1968. Studies on the removal of inorganic mercurial compounds in waste by the cell-reused method of mercury-resistant bacterium. *J. Ferment. Technol.* **12**:1048–1055.

Swain, E. B. et al., 1992. Increasing rates of atmospheric mercury deposition in midcontinental North America. *Science* **257**:784–787.

Takizawa, Y., 1979. Epidemiology of mercury poisoning. In *The Biogeochemistry of Mercury in the Environment*, J. O. Nriagu, ed., Elsevier/North-Holland Biomedical Press, pp. 325–367.

Tezuka, T., and K. Tonomura, 1976. Purification and properties of an enzyme catalyzing the splitting of carbon-mercury linkages from mercury-resistant *Pseudomonas* K-62 strain. *J. Biochem.* **80:**79–87.

Watras, C. J., and N. S. Bloom, 1992. Mercury and methylmercury in individual zooplankton: Implications for bioaccumulation. *Limnol. Oceanogr.* **37:**1313–1318.

White, C. et al., 1997. The role of microorganisms in biosorption of toxic metals and radionuclides. *Int. Biodeter. Biodegrad.* **35:**17–40.

Wilkinson, S. C. et al., 1989. Mercury accumulation and volatilization in immobilized algal cell systems. *Biotechnol. Lett.* **11:**861–864.

Xiao, Z. F. et al., 1995. Influence of humic substances on photolysis of divalent mercury in aqueous solution. *Water, Air, Soil Pollut.* **80:**789–798.

Yeargan, R. et al., 1992. Tissue partitioning of Cd in transgenic tobacco seedlings and field grown plants expressing the mouse metallothionein I gene. *Transgen. Res.* **1:**261–267.

Yoshinaga, K., 1995. Mercury-contaminated sludge treatment by dredging in Minamata Bay. In *Dredging, Remediation, and Containment of Contaminated Sediments*, K. R. Demars et al., eds., ASTM Press, pp. 182–191.

Zenk, M. H., 1996. Heavy metal detoxification in higher plants: A review. *Gene* **179:**21–30.

11

THE PHYSIOLOGY AND BIOCHEMISTRY OF SELENIUM VOLATILIZATION BY PLANTS

M. P. DE SOUZA, E. A. H. PILON-SMITS, AND N. TERRY

INTRODUCTION

Selenium is a major environmental pollutant in California and other areas worldwide. It is a contaminant of agricultural irrigation drainage water derived from irrigated seleniferous soils, and it is found in wastewater produced by oil refineries, electric power plants and other industries. Oil refinery effluent mainly contains selenite (Se^{4+}) (Hansen et al., 1998), whereas agricultural drainage water mainly contains selenate (Se^{6+}) (McNeal and Balisteri, 1989). Other chemical forms of Se found in the environment include colloidal or elemental selenium (Se^0), selenide (Se^{2-}), organic forms such as selenoamino acids, and volatile organic forms such as dimethyl selenide (DMSe), dimethyldiselenide, dimethylselenone, dimethylselenylsulfide, and methaneselenol (Rosenfeld and Beath, 1964; Reamer and Zoller, 1980; Fan et al., 1997).

Although Se is a micronutrient for humans and animals, slightly larger doses above the Se requirement can result in toxic effects (Wilber, 1983). The environmental catastrophe at the Kesterson reservoir in California, where Se accumulated to high levels in fish and birds, causing reproductive and developmental defects and death, clearly demonstrates that soil and water contaminated with Se can be toxic to wildlife (Ohlendorf et al., 1986; Saiki and Lowe, 1987). The wildlife deformities observed at the Kesterson reservoir and in agricultural drainage basins currently in use have been

Phytoremediation of Toxic Metals: Using Plants to Clean Up the Environment, edited by Ilya Raskin and Burt D. Ensley.
ISBN: 0-471-19254-6 ©2000 John Wiley & Sons, Inc.

thought to be due to the bioaccumulation of Se in the food chain (Skorupa and Ohlendorf, 1991; Maier and Knight, 1994). The cleanup of areas contaminated with Se is, therefore, of primary importance.

Phytoremediation offers a cheap and environmentally friendly technology for the cleanup of soil and water contaminated with Se (Banuelos and Schrale, 1989; Banuelos and Meek, 1990; Terry and Zayed, 1994; Wu et al., 1996). The fact that both wetland and upland plants can bioconcentrate Se in their tissues to levels that are orders of magnitude higher than those found in contaminated water and soil is utilized in phytoremediation, (Banuelos and Schrale, 1989; Banuelos and Meek, 1990; Terry et al., 1992; Wu et al., 1996; Hansen et al., 1998). The removal of Se from polluted soil and water into plant tissues is heavily dependent on the chemical species of the element (Mikkelsen et al., 1989; Blaylock and James, 1994). In general, more oxidized forms of Se, such as selenate or selenite, are highly soluble and therefore more easily removed by plants, whereas reduced inorganic forms, such as selenide or elemental Se, are much less bioavailable.

Vegetated constructed wetlands remove Se from selenite-contaminated wastewater very effectively. For example, our research has shown that the 36-ha Chevron-constructed wetland in Richmond, CA removed ~90% of the Se entering the wetland as selenite-contaminated oil refinery effluent (Hansen et al., 1998). Much of the Se extracted was removed in the first 12-ha section of the wetland, suggesting that relatively small areas of wetland are capable of removing significant amounts of Se. The Chevron wetland is populated mainly by saltmarsh bulrush (*Scirpus robustus*) and cattail (*Typha latifolia*), with some stands of brass button (*Polygonum* sp.), rabbitfoot grass (*Polypogon monspeliensis*), and other grass species. Part of the Se was removed to the sediments, and some was accumulated in plants.

A significant proportion of the Se removed by the Chevron wetland was volatilized to the atmosphere by wetland plants and microbes; specifically, selenite was metabolized by plants and microbes to volatile forms such as dimethylselenide (DMSe). The extent of Se removal by volatilization varied over the year, with as much as 10–30% of the Se removed by the wetland lost through volatilization. Removal of Se by volatilization has several advantages; most importantly, it minimizes Se buildup in sediments and plant tissues where it may enter the food chain. Besides removing Se from the site, volatilization has the advantage of detoxifying inorganic Se; this is because the major form of volatile Se, DMSe, is 500–600 times less toxic than its inorganic forms (Wilber, 1980; Ganther et al., 1966, McConnell and Portman, 1952). DMSe gas has a short lifetime (9.6 days), and studies on the fate of DMSe in the troposphere reveal that any DMSe released into the atmosphere will most likely be diluted, dispersed away from the polluted site, and deposited in Se-deficient areas by air currents (Atkinson et al., 1990).

Selenium volatilization has been measured from other wetlands and aquatic ecosystems (Chau et al., 1976; Cooke and Bruland, 1987; Karlson and Frankenberger, 1990; Thompson-Eagle and Frankenberger, 1990; 1991; Allen, 1991; Velinsky and Cutter, 1991; Azaizeh et al., 1997; Hansen et al., 1998; Zhang and Moore, 1997), and from vegetated and bare soils (Abu-Erreish et al., 1968; Francis et al., 1974; Doran and

Alexander, 1977; Reamer and Zoller, 1980; Zieve and Peterson, 1981; Frankenberger and Karlson, 1988; Karlson and Frankenberger, 1989; Biggar and Jayaweera, 1993; Lin and Terry, unpublished). These studies showed that both plants and microbes can volatilize Se in situ, and that the rate of volatilization is dependent on temperature, Se concentration, Se speciation, and other environmental factors. Furthermore, the availability of carbon and protein sources limit microbial volatilization.

Selenium volatilization has been measured from unvegetated soil and water at rates comparable to vegetated areas (Lin and Terry, unpublished). This suggests that bacteria, fungi, and algae can volatilize Se at rates similar to those of vascular plants or plant–microbe associations. Indeed, pure cultures of bacteria, fungi, and algae tested under laboratory conditions produce volatile Se at rates as high or higher than plants when expressed per gram dry weight of tissue (Fleming and Alexander, 1972; Thompson-Eagle et al., 1989; Brady et al., 1996; Rael and Frankenberger, 1996; Fan et al., 1997; de Souza and Terry, 1997). However, other investigators have shown that plants added to bare soil significantly increased the rate of Se volatilization (Zieve and Peterson, 1984; Biggar and Jayaweera, 1993; Duckart et al., 1992). Plants contribute fixed carbon, which can support large populations of microbes, especially in the rhizosphere. Furthermore, plants accumulate larger amounts of Se in their tissues than bacteria. Rhizosphere microbes facilitate Se uptake and volatilization by plants (Zayed and Terry, 1994; de Souza and Terry, 1997). It is clear that both plants and rhizosphere microbes are necessary for the optimum remediation of Se-contaminated sites.

Phytoremediation of Se through plant accumulation and volatilization accounts for a significant portion of the Se removed from contaminated wastewater, soils, and sediments (Cooke and Bruland, 1987; Velinsky and Cutter, 1991; Allen, 1991; Zhang and Moore, 1997). Many plant species have been shown to volatilize and accumulate Se (Lewis et al., 1966; Asher et. al., 1967; Zieve and Peterson, 1984; Duckart et al., 1992; Terry et al., 1992). Screening experiments with different wetland plants and crop species to identify the best Se-accumulating and volatilizing species revealed that there was substantial variation in the ability to phytoextract and volatilize Se (Duckart et al., 1992; Terry et al., 1992; Pilon-Smits et al., 1999b). Cattails and saltmarsh bulrush were determined to be good wetland plants for Se phytoremediation because they produce a lot of biomass, and accumulate and volatilize Se at reasonably high rates (Pilon-Smits et al., 1999b). With respect to crop plants, members of the Brassicaceae, known to accumulate and volatilize sulfur, were shown to be excellent Se volatilizers (Terry et al., 1992). Thus, it is very likely that S and Se are taken up and volatilized by the same pathway. Indeed, the rate of Se volatilization was shown to be inversely correlated with the sulfate concentration in the nutrient solution (Zayed and Terry, 1992).

Since the successful Chevron study, we have built a constructed wetland in Corcoran, CA, to treat selenate-contaminated agricultural drainage water from the Tulare Lake Drainage District (TLDD). This project is being conducted in collaboration with scientists from the University of California Salinity/Drainage Task Force, and the TLDD. The TLDD wetland consists of 10 flow-through wetland cells, seven of which contain monotypic stands of rabbitfoot grass, saltmarsh bulrush, Nypa forage saltgrass (*Distichlis spicata*), cattail (2 cells), smooth cordgrass (*Spartina alterniflora*), or baltic rush (*Juncus balticus*). Two cells have combinations of the different plant species, and one cell is an unplanted control that contains naturally occurring algae and bacteria. Constructed in May 1996, this wetland appears to be successful at removing Se from the agricultural drainage water, although much more work needs to be done to confirm the efficacy of this approach. The wetland plants accumulate Se and volatilize it to different extents.

A major goal of our studies at Berkeley is to improve the efficiency of Se phytoremediation by plants and their associated microbes using a multidisciplinary research approach. First, we perform physiological experiments to determine which biochemical steps in the Se volatilization pathway are rate limiting. Second, we speciate the different chemical forms of Se in plant tissue, using X-ray absorption spectroscopy, to determine which intermediates are being accumulated in vivo. Third, we use genetic engineering to overexpress putative rate limiting enzymes, and then determine the physiological effects of these genetic changes. Finally, we investigate the role of microbes in the various phytoremediation processes. In this chapter we examine the state of our knowledge with respect to the various steps involved in Se assimilation and volatilization by plants [for other reviews, see Anderson (1993) and Lauchli (1993)]. For each step we give an overview of what is known in the literature, followed by the most recent results from our work in Se physiology, molecular biology, and microbiology. This is followed by a discussion of the step(s) that may be rate-limiting for Se assimilation and volatilization.

THE ASSIMILATION OF SELENIUM BY PLANTS

The Uptake and Translocation of Selenate and Selenite

In many plant species, selenate is generally thought to be taken up actively by the same mechanism as sulfate is absorbed, that is, by a sulfate/H^+ cotransporter protein located in the root plasma membrane (Legget and Epstein, 1956) (Fig. 11-1). Selenite uptake, on the other hand, appears to be passive (Bange, 1973; Arvy, 1993). The evidence that selenate is taken up via a sulfate permease is based on physiological studies showing that sulfate ions were antagonistic to the uptake of selenate ions into plants (Hurd-Karrer, 1938; Mikkelsen et al., 1988; Zayed and Terry, 1992). Additional support for this view has been put forward by Smith et al. (1995), who cloned two sulfate permease genes from the plant *Stylosanthes hamata*. When expressed in the yeast deletion mutant YSD1, these genes were shown to transport *selenate* over the plasma membrane (Frank Smith, personal communication). Two types of plant sulfate transporter protein have been identified, with different expression patterns and different affinities for sulfate. The SH1 protein is thought to be involved in sulfate uptake over the cell membrane, while the SH3 protein may be involved in internal sulfate transport between (sub)cellular compartments (Smith et al., 1995).

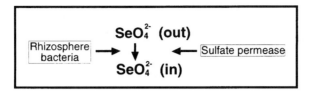

Figure 11-1   Selenate uptake from the external medium into roots is mediated by a protein transporter in the root-cell membrane, sulfate permease. Selenate uptake into roots is enhanced by rhizosphere bacteria that interact with the plant to produce a heat-labile proteinaceous compound that stimulates selenate uptake, possibly by a novel pathway.

Some plant species are able to accumulate *very* high concentrations of Se in their tissues, including *Astragalus bisulcatus* (Ulrich and Schrift, 1968); these are termed *Se hyperaccumulators*. There is some evidence that Se hyperaccumulators have Se-specific transporters that preferentially take up selenate over sulfate: the Se:S ratio in Se hyperaccumulating plants is much higher than in non-accumulators (Rosenfeld and Beath, 1964; Bell et al., 1992). The test plant used in our research is Indian mustard (*B. juncea*). Indian mustard is not a Se hyperaccumulator, but it does accumulate Se to high concentrations. It is an excellent phytoremediating plant species because it grows rapidly, producing a large biomass, and it can tolerate fairly toxic environments. It is also relatively easy to genetically manipulate.

In order to determine whether the uptake of selenate is rate-limiting with respect to Se assimilation in Indian mustard, we overexpressed the two sulfate permease clones for SH1 and SH3 from *Stylosanthes hamata* (generously supplied by Frank Smith) in *B. juncea* using a modification of the method described by Barfield and Pua (1991). The SH1, but not SH3, transgenics accumulated more Se in their tissues than wild type (Hwang and Terry, unpublished results) showing that the SH1 sulfate permease is rate-limiting selenate uptake in plants. This research is the first to show that the sulfate transporter, SH1, is responsible for selenate uptake in vivo in plants. It is possible that the SH3 transgenic plants do not accumulate more selenate than wild type because the SH3 transporter has kinetic properties different from those for SH1 for selenate uptake or that SH1 and SH3 are located on different membranes (cell or vacuole).

In addition to the variation in the uptake process for selenate and selenite (i.e., active versus passive transport, respectively), the translocation patterns within the plant differ. When we supplied Indian mustard plants with selenate or selenite, the concentration of Se in roots was comparable, but the Se concentration in shoots of selenate-treated plants was about 10-fold higher than in selenite-supplied plants (de Souza et al., 1998). Thus, a large proportion of selenate is translocated to the shoot, whereas selenite remains mainly in the root. Other studies have shown similar results (Smith and Watkinson, 1984; Arvy, 1993; Zayed et. al., 1998).

The Role of Microbes in the Plant Uptake of Selenate and Selenite

Microbes also play a role in the uptake of selenate by plants (deSouza et. al., 1999) (Fig. 11-1b). In one approach, we inhibited microbial activity by supplying antibiotics to Indian mustard plants, and in the other approach we inoculated sterile plants with isolated microbes. Both types of experiment gave the same conclusion: bacteria contribute to 77–88% of the selenate uptake by Indian mustard roots and to 16–21% of selenite uptake (Table 11-1).

To investigate the mechanism of the enhanced accumulation of selenate by plant–microbe combinations compared to sterile plants, we added filtered growth medium from nonsterile plants to sterile plants amended with selenate and found that the bacteria-free medium contained soluble compounds produced by the plant–microbe interaction, which increased Se uptake by sterile plants (de Souza et. al., 1999). However, when the bacteria-free medium was boiled, the effect was lost. It was also observed that bacteria-inoculated plants had greater levels of amino acids (especially serine) in their rhizosphere than did sterile plants (de Souza et. al., 1999). Thus, it appears that the rhizosphere–bacteria increase plant uptake of Se through the excretion of heat-labile proteinaceous compounds by the bacteria–plant combination.

Assimilation of Selenate and Selenite to Selenocysteine

The enzymatic activities involved in SeCys biosynthesis are associated mainly (77–100%) with the chloroplast (Ng and Anderson, 1979), which is the presumed location of the sulfur/selenium assimilation pathway. After selenate is taken up by the sulfate transporter, it is activated to adenosine phosphoselenate (APSe) (Fig. 11-2). This step is catalyzed by the enzyme ATP sulfurylase (APS), which can activate both sulfate and selenate in vitro (Shaw and Anderson, 1972; Dilworth and Bandurski, 1977). APSe can then react nonenzymatically with glutathione (or other thiols) to produce glutathione-conjugated selenite ($GSSeO_3$), which can subsequently be reduced to selenodiglutathione (GSSeSG) (Dilworth and Bandurski, 1977). GSSeSG is reduced to the corresponding selenol (GSSeH), a process most likely involving the enzyme glutathione reductase. Alternatively, this reaction may proceed nonenzymatically, involving glutathione as a reductant (Anderson and Scarf, 1983).

TABLE 11-1   Role of Microbes in Se Volatilization and Accumulation in Plant Tissues[a]

Se Species Supplied	Se Volatilization (μg Se/g DW/per day)		Se Accumulation into Roots (μg Se/g DW)	
	Antibiotics	Sterile Plants	Antibiotics	Sterile Plants
Selenate	35 (S)	50 (S)	77 (S)	87 (S)
Selenite	4 (NS)	50 (S)	21 (NS)	16 (NS)
SeMet	0 (NS)	25 (NS)	22 (NS)	0 (NS)

[a]The percent inhibition of Se volatilization or accumulation by Indian mustard plants when the bacterial contribution is eliminated (by antibiotics or sterile conditions) compared to plants with bacteria in the rhizosphere (no antibiotics or nonsterile conditions) (S—significant; NS—nonsignificant differences, where $p < .05$ for differences in means between plants with and without bacteria).

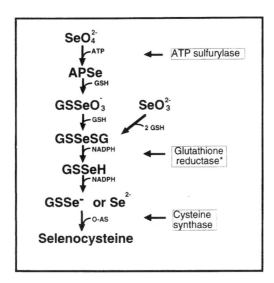

Figure 11-2   Reduction of selenate to selenite, uptake of selenite, and incorporation of selenite into SeCys involve a series of reductive steps. Selenite uptake does not involve bacteria. The asterisk indicates that, on the basis of our glutathione reductase overexpression studies, this enzyme may not be involved in the production of selenide. Instead, it is possible that enzymes responsible for sulfide production may be involved, such as APS reductase.

Other enzymes besides glutathione reductase could also be involved in Se metabolism: the production of selenide could be mediated by either thiosulfonate reductase, APS reductase, or APS kinase plus phosphoadenosine phosposulfate (PAPS) reductase, as proposed for sulfide production in a recent review on sulfur metabolism (Hell, 1997). The selenide produced by these processes can serve as a substrate for cysteine synthase, which couples it to O-acetylserine (OAS), leading to the formation of selenocysteine. Alternatively, by analogy to the sulfate assimilation pathway, GSSe⁻ may be the physiological substrate of cysteine synthase rather than free Se^{2-} (Tsang and Schiff, 1978). If selenite is taken up instead of selenate, it is believed to react nonenzymatically with reduced glutathione to form GSSeSG, which then follows the pathway described above (Fig. 11-2) (Anderson, 1993).

   To investigate which enzymatic step is rate-limiting for Se assimilation, we supplied Indian mustard plants with selenate or selenite and used X-ray absorption spectroscopy to analyze which metabolites accumulated. When selenate was supplied, the main form of Se found inside roots and shoots was selenate. However, when selenite was supplied, the form of Se found in the plants was an organoSe compound like selenomethionine (Table 11-2). Since selenate accumulated in selenate-supplied plants, it appears that the activation of selenate (by ATP sulfurylase) is very slow, and possibly rate-limiting for Se assimilation from selenate. Selenite, however, appears to be rapidly metabolized to organic forms, so that one of the later steps in the pathway is likely to be rate limiting for Se assimilation from selenite.

TABLE 11-2   Se Speciation in Roots and Shoots of Indian Mustard Supplied with 20 μM Selenate, Selenite, or SeMet

Se Species Supplied	Se Species in Shoot	Se Species in Root
Selenate	Selenate	Selenate
Selenite	SeMet	SeMet
SeMet	SeMet	SeMet

The genes encoding the enzymes thought to be involved in the assimilation of selenate into selenocysteine, namely, ATP sulfurylase, glutathione reductase, and cysteine synthase (Anderson, 1993), have been cloned from various sources. An ATP sulfurylase gene (*APS*) was cloned from *Arabidopsis thaliana* (Leustek et al., 1994); a glutathione reductase gene (*gr*) was cloned from *Escherichia coli* and adapted for expression in plants by Foyer et al. (1995); and a cysteine synthase gene was cloned from spinach (Saito et al., 1994). These three genes were overexpressed in Indian mustard.

That ATP sulfurylase is a rate limiting enzyme for the assimilation of Se from selenate is confirmed by the fact that overexpression of ATP sulfurylase led to an increased conversion of selenate to organic forms of Se. X-ray absorption spectroscopic (XAS) analysis showed that, unlike the wild-type plants, which accumulated selenate inside their roots and shoots, the form of Se accumulating in the *APS* plants, (i.e., plants that overexpress ATP sulfurylase) was mainly selenomethionine (50–90%); the remainder was selenate (Pilon-Smits et. al., 1999a). The APS transgenic plants had four-fold higher APS enzymatic activity, and accumulated 3 times as much Se per plant, than the wild type. Furthermore, the APS transgenics were more tolerant to high levels of Se, growing at higher rates than the wildtype (Table 11-3; Pilon-Smits et. al., 1999a). In addition, these results show for the first time that the ATP sulfurylase enzyme is responsible for selenate reduction in vivo.

We have transformed Indian mustard plants to overexpress a bacterial glutathione reductase in the cytoplasm (cytGR) and also in the chloroplast (cpGR); both cytGR

TABLE 11-3   Overexpression of Genes Involved in the S Assimilation Pathway: Effects on Se Metabolism in Indian Mustard

Enzyme Overexpressed	Se Tolerance	Se Accumulation
Sulfate permease	Lower or same	+ 100–200%
ATP sulfurylase	Higher	+ 100–200%
Glutathione reductase	Higher	Same
Cysteine synthase	Same	Same
SAM synthetase	Same	Same

and cpGR seedlings grew better than wild-type seedlings on an agar medium spiked with toxic concentrations of selenate or selenite (Table 11-3, Pilon-Smits et. al., unpublished). However, mature plants did not show any significant difference in Se accumulation or tolerance to high levels of Se compared to wild-type plants. Thus, there is little evidence yet to suggest that glutathione reductase is a rate-limiting enzyme for Se assimilation in Indian mustard. Also, overexpression of cysteine synthase did not increase Se assimilation compared to wild-type plants. Transgenic Indian mustard plants that overexpressed spinach cysteine synthase in the chloroplast did not show any differences with respect to Se assimilation (Table 11-3; Pilon-Smits et. al., unpublished). The conclusion that glutathione reductase and cysteine synthase are probably not rate-limiting for Se assimilation is in agreement with our XAS speciation data, which showed that the conversion of selenite to SeMet is a very rapid process.

Conversion of Se-cysteine to Se-methionine

The enzymes responsible for SeMet synthesis from SeCys are cystathionine γ-synthase, cystathionine β-lyase, and homocysteine methyltransferase (Fig. 11-3). Cystathionine γ-synthase catalyzes the condensation of phosphohomoserine with SeCys to form Se-cystathionine (Dawson and Anderson, 1988). Cystathionine γ-synthase from spinach exhibited a preference for SeCys: it had a higher affinity for SeCys ($K_m = 70 \mu M$) than for Cys ($K_m = 240 \mu M$), and the V_{max} values from these two substrates were similar. Therefore, cystathionine γ-synthase appears to be an important enzyme for Se assimilation because it is more likely to use SeCys as a substrate compared to Cys.

Cystathionine β-lyase cleaves selenocystathionine to selenohomocysteine with the concomitant removal of pyruvate and ammonia (Hell, 1997). Cystathionine β-lyase from *Pisum sativum* does not differentiate between the Se and S forms of

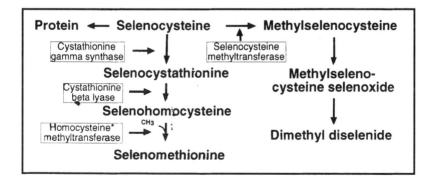

Figure 11-3 Incorporation of selenocysteine into SeMet. SeCys is at a branchpoint where it can proceed along the Se volatilization pathway, or it can be incorporated into protein, or stored as methylSeCys. The asterisk indicates that the conversion of homoselenocysteine to selenomethionine may also be mediated by Met synthase, which uses methyltetrahydrofolate as a methyl donor.

cystathionine since the enzyme had a similar affinity for cystathionine (K_m = 0.31 mM) and selenocystathionine (K_m = 0.35 mM) (McCluskey et al., 1986).

Selenohomocysteine is probably methylated to SeMet by homocysteine methyltransferase using S-adenosylmethionine as a methyl donor (Hell, 1997). Alternatively, Met synthase, which uses methyltetrahydrofolate as a donor in sulfur metabolism, may also be involved in SeMet production (Cossins and Chen, 1997).

To investigate the role of cystathionine β-lyase in Se assimilation in plants, we overexpressed a cDNA from *Arabidopsis thaliana* encoding cystathionine β-lyase (Ravanel et al., 1995) in Indian mustard. The cystathionine β-lyase overexpressing transgenic plants are presently being analyzed with respect to Se tolerance, accumulation, and volatilization.

Selenocysteine may also be incorporated into other compounds, for example, proteins. Incorporation of Se into proteins alters their structure and this may lead to the toxic effects on plant function (Lauchli, 1993). Selenium hyperaccumulating plants are apparently able to tolerate high concentrations of Se by incorporating SeCys into compounds other than proteins, which are less toxic (Brown and Shrift, 1982). For example, Se hyperaccumulating plants can methylate SeCys by SeCys methyltransferase (Neuhierl and Bock, 1996), resulting in methylSeCys, which is stored (Fig. 11-3). MethylSeCys can also be further methylated to produce dimethyldiselenide (DMDSe), which is volatile (Evans et al., 1968; Lewis, 1971) (Fig. 11-3).

Se-methionine to dimethylselenide

Lewis et al. (1974) proposed that selenomethionine was assimilated to a selenonium salt that was then cleaved enzymatically to dimethyl selenide (Fig. 11-4*a*). On the basis of our current research with wetland plants, we propose that SeMet is converted to the selenonium compound, dimethylseleniopropionate (DMSeP) (Fig. 11-4*b*), in analogy to the sulfonium assimilation pathway elucidated by Hanson et al. (1994). Thus, selenomethionine could be methylated to methylselenomethionine by a methyl transferase, using S-adenosylmethionine (SAM) as a methyl donor, which, in turn, is produced by SAM synthetase (SAMS). In this proposed scheme, the methyl-selenomethionine would be decarboxylated, deaminated, and oxidized to DMSeP in a manner analogous to the conversion of S-methylMet to the sulfonium salt dimethylsulfoniopropionate (DMSP) by *Spartina alterniflora* (Hanson et al., personal communication). This view is supported by the fact that we were able to detect DMSeP in wetland plants collected from a seleniferous environment. DMSeP was detected by measuring base-hydrolyzable DMSe from plant tissues digested with NaOH, as described for sulfonium compounds by White (1982). In addition to accumulating DMSeP, we have shown that plants can metabolize and volatilize supplied DMSeP (which we have synthesized) (Table 11-4). Therefore, it is likely that selenomethionine is converted to DMSeP before being volatilized.

In bacteria, DMSeP can be volatilized to DMSe by the enzyme DMSP lyase (Ansede and Yoch, 1997). This S-assimilating enzyme is thought to exist in plants

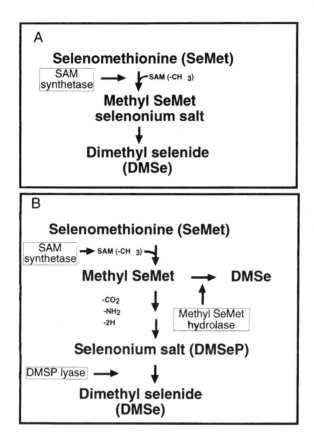

Figure 11-4   Panel *a*, derived from the literature, shows the prevailing view of SeMet conversion to DMSe. Panel *b* is a modified version of Panel *a* that includes our recent results, indicating that DMSeP does exist, and showing additional steps involved in DMSeP synthesis and degradation. Methylselenomethionine may also directly be cleaved to produce DMSe as shown.

(Dacey et al., 1987), and, as in bacteria, may also be responsible for DMSe production from DMSeP (Fig. 11-4*b*). Alternatively, DMSe may be produced directly from methyl selenomethionine via the enzyme *S*-methylmethionine hydrolase (Fig. 11-4*b*) (Mudd and Datko, 1990). In support of this view, phytoplankton supplied with selenomethionine produced DMSe presumably by methylation (Bottino et al., 1984). The genes encoding DMSP lyase, *S*-methylmethionine hydrolase, and the enzymes involved in sulfonium synthesis have not yet been cloned. We have obtained transgenic SAMS plants using a construct that contains the *A. thaliana sams* gene under the control of the 35S promoter (Boerjan et al., 1994). When the transgenic SAMS plants were supplied with selenate, they showed no differences in Se accumulation (Table 11-3); we have not yet tested these plants for their ability to assimilate selenomethionine.

TABLE 11-4 Se Uptake and Volatilization from Indian Mustard Supplied with 20 μM Selenate, Selenite, Selenocysteine, SeMet, or DMSeP for 8 days[a]

Se Species Supplied	Rate of Se Volatilization (μg Se/g DW/d)	Se Accumulation into (μg Se/g DW after 8 days)	
		Shoot	Root
Selenate	0.32 (0.02)	130 (18)	143 (21)
Selenite	0.95 (0.09)	31 (14)	294 (12)
SeCys	1.41 (1.19)	44 (3)	375 (23)
SeMet	6.67 (2.79)	166 (142)	341 (87)
DMSeP	36.16 (20.36)	230 (47)	39 (2)

[a]Numbers in parentheses are standard deviations. (*Note:* SeCys was an impure preparation that also contained some secystine and dithiothreitol).

RATE-LIMITING STEPS FOR SE VOLATILIZATION

In the preceding discussion, we reviewed the various biochemical steps whereby Se is taken up and assimilated by plants to volatile forms of Se. To genetically manipulate plants in order to increase their rate of Se volatilization, it is necessary to identify the rate-limiting enzmes in the pathway to enable us to overcome the rate limitation by overexpression of the enzymes involved. Below, we review the various steps that could be rate limiting to Se volatilization and the physiological activities of the transgenic plants developed in an attempt to overcome the rate limitation.

Se Uptake

The rate of volatilization is highly dependent on the form of Se supplied: rates are highest when Se is supplied as DMSeP, and are progressively lower when Se is supplied (in descending order) as SeMet, SeCys, selenite, and selenate (Table 11-4). With respect to selenate and selenite supplied over an increasing concentration range (DMSeP and SeMet have not been tested), our research has shown that the rate of Se volatilization is linearly correlated with the external Se concentration and with the internal plant tissue Se concentration (de Souza et al., 1998). This suggests that the rate of volatilization is dependent on the concentration as well as the form of Se supplied. In an attempt to overcome the rate limitation imposed by selenate uptake, we developed transgenic Indian mustard plants with higher levels of sulfate permease activity in roots (SH1 transgenics). Despite the increased uptake of selenate by the SH1 transgenics, we have not yet obtained reproducibly higher rates of Se volatilization in the transgenic plants. On the other hand, volatilization rates of Indian mustard plants were increased when selenate uptake into plants was enhanced by the presence of bacteria; this lends support to the idea that Se volatilization is rate-limited by selenate uptake.

Reduction of Selenate to Selenite

Our XAS speciation studies with the transgenic plants that overexpressed ATP sulfurylase showed that the rate limitation imposed by the slow conversion of selenate to selenite had been overcome (Pilon-Smits, et. al., 1999a). The *APS* plants converted selenate to selenite at faster rates because they had four-fold higher levels of *APS* enzymatic activity. Selenite, in turn, was rapidly converted to organoSe forms in plants. The *APS* transgenics also accumulated more Se (especially in shoots), than did the wild type.

Conversion of Selenocysteine to Selenomethionine

Selenocysteine was taken up into the roots at rates comparable to SeMet, but it was volatilized very poorly compared to SeMet (Table 11-4); the rates of volatilization of SeCys were similar to those for selenate and selenite. Selenocysteine is the intermediate at the branchpoint in the Se volatilization pathway outlined in Figure 11-3. At this branchpoint, SeCys can be (1) assimilated to SeMet, (2) incorporated into proteins, or (3) in Se hyperaccumulators, methylated to organoSe forms. Since selenocysteine may move readily into other pathways, one might speculate that the availability of SeCys itself might be rate-limiting for Se volatilization. This does not appear to be the case. When we supplied Indian mustard, broccoli, and wetland plants with a crude preparation of SeCys (which contained impurities of selenocystine and dithiothreitol), the plants showed low rates of Se volatilization, similar to plants supplied with inorganic Se (selenate or selenite), in spite of accumulating high root Se concentrations (Table 11-4). When SeMet was supplied, on the other hand, it was volatilized at much higher rates than when Se was supplied as selenate, as selenite, or from the crude preparation of SeCys. This suggests that it is the conversion of SeCys to SeMet that is rate-limiting volatilization rather than the availability of SeCys. We do not yet know whether the overexpression of the enzymes believed to be involved in the conversion of SeCyst to SeMet (i.e., cystathionine γ-synthase, cystathionine β-lyase, and homocysteine methyltransferase) enhance volatilization. We have overexpressed cystathionine β-lyase in Indian mustard and are currently analyzing the physiological responses of these plants.

Selenomethionine Volatilization

Plants actively take up SeMet (Abrams et al., 1990) at rates that are much higher than for selenate or selenite (Table 11-4) (Williams and Mayland, 1992). They also volatilize Se from SeMet at much faster rates than from the inorganic forms of Se (Table 11-4). Therefore, at first glance, conversion of SeMet to volatile forms does not appear to be rate-limiting. However, since a SeMet-like compound accumulates in selenite- and SeMet-supplied plants, but only relatively low levels of selenonium compounds, one of the final steps in the Se volatilization pathway could be rate-limiting, specifically, the methylation of SeMet or the conversion of methylselenomethionine to the selenonium salt. Thus, possible approaches for improving Se volatilization from SeMet would be to overexpress enzymes responsible for (1) the methylation of selenomethionine, (2) selenonium salt biosynthesis, or (3) DMSe production from the selenonium salt or directly from methylselenomethionine. Unfortunately, none of these genes has been isolated. Overexpression of these enzymes would also enhance selenite volatilization because selenite-supplied plants accumulated SeMet.

Conceptual Model

By combining the information obtained from the literature and from our own studies, we have developed a conceptual model for the uptake, metabolism, and volatilization of Se from selenate, selenite, and selenomethionine (Fig. 11-5). Selenate uptake into the root is an active process mediated by sulfate permease. Selenate uptake is faster in the presence of rhizosphere bacteria. Thus, bacteria appear to promote selenate uptake, possibly via a separate pathway. Once selenate has been taken up, it is rapidly translocated to the shoot. Although selenate accumulates in both roots and shoots, it is metabolized only slowly to more reduced Se forms. Selenite is taken up by plants more slowly than selenate, possibly via a passive process. Bacteria do not play a major role in selenite uptake. Selenite taken up by the root is metabolized very rapidly to an

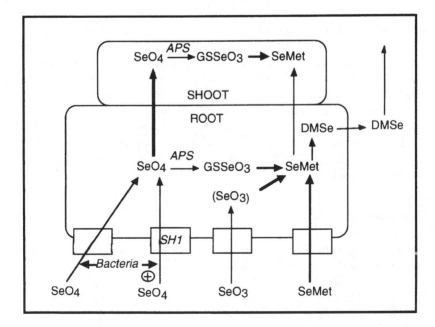

Figure 11-5 Physiological model for Se uptake, metabolism, and volatilization by plants and their associated microbes. Biochemical steps with thick arrows indicate that the reaction proceeds at a faster rate than do steps with thin arrows.

organic form, probably SeMet. Selenite is poorly translocated to the shoot (~10%). Uptake of SeCys and SeMet into the root is very fast and does not appear to involve bacteria.

With regard to volatilization, SeMet is volatilized much faster in the root than is SeCys, selenite, or selenate. Bacteria increase Se volatilization from selenate by increasing selenate uptake. Bacteria enhance volatilization from selenite, possibly by converting selenite to SeMet. Volatilization of SeCys and SeMet does not appear to involve bacteria.

CONCLUSIONS

In the biochemical pathway of Se assimilation and volatilization by plants, there are several possible rate-limiting steps. We have shown that genetic engineering of plants and the manipulation of plant–microbe interactions can be used to overcome some of these rate-limiting steps, thereby increasing the efficiency of phytoremediation. Transgenic plants with enhanced abilities to accumulate and remove Se and with improved Se tolerance can be used for the phytoremediation of soils or waters with Se levels that are too toxic for growth of natural (wild-type) plants. The use of transgenic plants and plant–microbe combinations with improved abilities for Se phytoremediation may considerably reduce the time necessary to clean up a polluted upland site, or, the constructed wetland area necessary to treat a given volume of wastewater.

ACKNOWLEDGMENTS

We thank Dr. Andrew Hanson for reviewing the manuscript and for providing constructive comments. Our work was supported by grants from the Electric Power Research Institute. EAHPS was supported by a fellowship from the Dutch Organization for Scientific Research (NWO).

REFERENCES

Abrams, M. M. et. al., 1990. Selenomethionine uptake by wheat seedlings. *Agron. J.* **82:**1127–1130.

Abu-Erreish, G. M. et. al., 1968. Evolution of volatile selenium from soils. *Soil Sci.* **106:**415–420.

Allen, K. N., 1991. Seasonal variation of selenium in outdoor experimental stream-wetland systems. *J. Environ. Qual.* **20:**865–868.

Anderson, J. W., 1993. Selenium interactions in sulfur metabolism. In *Sulfur Nutrition and Assimilation in Higher Plants—Regulatory, Agricultural and Environmental Aspects*, L. J. De Kok, ed., SPB Academic Publishing.

AVRIL ROBARTS LRC

Anderson, J. W., and A. R. Scarf, 1983. Selenium and plant metabolism. In *Metals and Micronutrients: Uptake and Utilization by Plants*, D. A. Robb and W. S. Pierpoint, eds., Academic Press, London, pp. 241–275.

Ansede, J. H., and D. C. Yoch, 1997. Comparison of selenium and sulfur volatilization by dimethylsulfoniopropionate lyase (DMSP) in two marine bacteria and estuarine sediments. *FEMS Microbiol. Ecol.* **23:**315–324.

Arvy, M. P., 1993. Selenate and selenite uptake and translocation in bean plants (*Phaseolus vulgaris*). *J. Exper. Bot.* **44:**1083–1087.

Asher, C. J. et. al., 1967. Collection and partial characterization of volatile selenium compounds from *Medicago sativa* L. *Austral. J. Biol. Sci.* **20:**737–748.

Atkinson, R. et. al., 1990. Kinetics of the atmospherically important reactions of dimethyl selenide. *Environ. Sci. Technol.* **24:**1326–1332.

Azaizeh, H. et. al., 1997. Microbial selenium volatilization in rhizosphere and bulk soils from a constructed wetland. *J. Environ. Qual.* **26:**666–672.

Bange, G. G. J., 1973. Diffusion and absorption of ions in plant tissue. The role of the root cortex cells in ion absorption. *Acta Bot. Neerl.* **22:**529–542.

Banuelos, G. S., and D. W. Meek, 1990. Accumulation of selenium in plants grown on selenium-treated soil. *J. Environ. Qual.* **19:**772–777.

Banuelos, G. S., and G. Schrale, 1989. Plants that remove selenium from soils. *Calif. Agric.* (May/June):19–20.

Barfield, D. G., and E.-C. Pua, 1991. Gene transfer in plants of *Brassica juncea* using *Agrobacterium tumefaciens*-mediated transformation. *Plant Cell Rep.* **10:**308–314.

Bell, P. F. et. al., 1992. Contrasting selenate-sulfate interactions in selenium-accumulating and nonaccumulating plant species. *Soil Sci. Soc. Am. J.* **56:**1818–1824.

Biggar, J. W., and G. R. Jayaweera, 1993. Measurement of selenium volatilization in the field. *Soil Sci.* **155:**31–35.

Blaylock, M. J., and B. R. James, 1994. Redox transformation and plant uptake of Se resulting from root-soil interactions. *Plant Soil* **158:**1–17.

Boerjan, W. et. al., 1994. Distinct phenotypes generated by overexpression and suppression of S-adenosyl-methionine synthetase reveal developmental patterns of gene silencing in tobacco. *Plant Cell* **6:**1401–1414.

Bottino, N. R. et. al., 1984. Selenium containing amino acids and proteins in marine algae. *Phytochemistry* **23:**2445–2452.

Brady, J. M. et. al., 1996. Volatilization of selenite in aqueous medium by a *Penicillium* species. *Mycol. Res.* **100:**955–961.

Brown, T. A., and A. Shrift, 1982. Selenium: Toxicity and tolerance in higher plants. *Biol. Rev.* **57:**59–84.

Chau, Y. K. et. al., 1976. Methylation of selenium in the aquatic environment. *Science* **192:**1130–1131.

Cooke, T. C., and K. W. Bruland, 1987. Aquatic chemistry of selenium: Evidence of biomethylation. *Environ. Sci. Technol.* **21:**1214–1219.

Cossins, E. A., and L. Chen, 1997. Folates and one-carbon metabolism in plants and fungi. *Phytochemistry* **45:**437–452.

Dacey, J. W. H. et. al., 1987. Factors controlling emission of dimethylsulphide from salt marshes. *Nature* **330:**643–645.

Dawson, J. C., and J. W. Anderson, 1988. Incorporation of cysteine and selenocysteine into cystathionine and selenocystathionine by crude extracts of spinach. *Phytochemistry* **27:**3453–3460.

de Souza, M. P., and N. Terry, 1997. Selenium volatilization by rhizosphere bacteria. *Abstr. Am. Soc. Microbiol.* 499.

de Souza, M. P. et. al., 1998. Rate-limiting steps in selenium assimilation and volatilization by Indian mustard. *Plant Physiol.* **117:**1487–1494.

de Souza, M. P. et. al., 1999. Rhizosphere bacteria enhance selenium accumulation and volatilization by Indian mustard. *Plant Physiol.* **119:**563–573.

Dilworth, G. L., and R. S. Bandurski, 1977. Activation of selenate by adenosine 5′-triphosphate sulphurylase from *Saccharomyces cerevisiae. Biochem. J.* **163:**521–529.

Doran, J. W., and M. Alexander, 1977. Microbial formation of volatile Se compounds in soil. *Soil Sci. Soc. Am. J.* **40:**687–690.

Duckart, E. C. et. al., 1992. Selenium uptake and volatilization from plants growing in soil. *Soil Sci.* **153:**94–99.

Evans, C. S. et. al., 1968. Isolation of dimethyl diselenide and other volatile selenium compounds from *Astragalus racemosus* (Pursh.). *Austral. J. Biol. Sci.* **21:**13–20.

Fan, T. W-M. et. al., 1997. Selenium biotransformations by a euryhaline microalga isolated from a saline evaporation pond. *Environ. Sci. Technol.* **31:**569–576.

Fleming, R. W., and M. Alexander, 1972. Dimethylselenide and dimethyltelluride formation by a strain of Penicillium. *Appl. Microbiol.* **24:**424–429.

Foyer, C. H. et. al., 1995. Overexpression of glutathione reductase but not glutathione synthetase leads to increases in antioxidant capacity and resistance to photoinhibition in poplar trees. *Plant Physiol.* **109:**1047–1057.

Francis, A. J. et. al., 1974. Evolution of dimethylselenide from soils. *Appl. Environ. Microbiol.* **28:**248–250.

Frankenberger Jr., W. T., and U. Karlson, 1988. *Dissipation of Soil Selenium by Microbial Volatilization at Kesterson Reservoir.* Final report submitted to U.S. Department of the Interior. Project 7-F.C-20-05240, Nov. 1988. U.S. Department of the Interior, Bureau of Reclamation, Sacramento, CA.

Ganther, H. E. et. al., 1966. Dietary control of selenium volatilization in the rat. *J. Nutr.* **88:**55–60.

Hansen, D. et. al., 1998. Selenium removal by constructed wetlands: Role of biological volatilization. *Environ. Sci. Technol.* **32:**591–597.

Hanson, A. D. et. al., 1994. Biosynthesis of 3-dimethylsulfoniopropionate in *Wollastonia biflora* (L.) DC. *Plant Physiol.* **105:**103–110.

Hell, R., 1997. Molecular physiology of plant sulfur metabolism. *Planta* **202:**138–148.

Hurd-Karrer, A., 1938. Relation of sulfate to selenium absorption by plants. *Am. J. Bot.* **25:**666–675.

Karlson, U., and W. T. Frankenberger, Jr., 1989. Accelerated rates of selenium volatilization from California soils. *Soil Sci. Soc. Am. J.* **53:**749–753.

Karlson, U., and W. T. Frankenberger, Jr., 1990. Volatilization of selenium from agricultural evaporation pond sediments. *Sci. Total Environ.* **92:**41–54.

Lauchli, A., 1993. Selenium in plants: Uptake, functions, and environmental toxicity. *Bot. Acta* **106:**455–468.

Leggett, J. E., and E. Epstein, 1956. Kinetics of sulphate absorption by barley roots. *Plant Physiol.* **31**:222--226.

Leustek, T. et. al., 1994. Cloning of a cDNA encoding ATP sulfurylase from *Arabidopsis thaliana* by functional expression in *Saccharomyces cerevisiae*. *Plant Physiol.* **105**:897–902.

Lewis, B. G., 1971. Volatile selenium in higher plants. Ph.D. thesis. Univ. California, Berkeley.

Lewis, B. G. et. al., 1966. Release of volatile selenium compounds by plants: Collection procedures and preliminary observations. *J. Agric. Food Chem.* **14**:638–640.

Lewis, B. G. et. al., 1974. Volatile selenium in higher plants. The production of dimethyl selenide in cabbage leaves by enzymatic cleavage of Se-methyl selenomethionine selenonium salt. *Plant Soil* **40**:107–118.

McCluskey, T. J. et. al., 1986. Enzyme catalysed a,b-elimination of selenocystathionine and selenocystine and their sulfur isologues by plant extracts. *Phytochemistry* **25**:2063–2068.

McConnell, K. P., and O. W. Portman, 1952. Toxicity of dimethyl selenide in the rat and mouse. *Proc. Soc. Exp. Biol. Med.* **79**:230–231.

McNeal, J. M., and L. S. Balisteri, 1989. *Geochemistry and Occurrence of Selenium; an Overview: Selenium in Agriculture and the Environment*, SSSA Special Publication No. 23, Madison, WI, p. 1013.

Maier, K. J., and A. W. Knight. 1994. Ecotoxicology of selenium in freshwater systems. *Rev. Environ. Contam. Toxicol.* **134**:31.

Mikkelsen, R. L. et. al., 1988. Effect of salinity and its composition on the accumulation of selenium by alfalfa. *Plant Soil* **107**:63–67.

Mikkelsen, R. L. et. al., 1989. Factors affecting selenium accumulation by agricultural crops. In *Selenium in Agriculture and the Environment*, L. W. Jacobs, ed., SSSA Special Publication 23, Madison, WI, pp. 65–94.

Mudd, S. H., and A. H. Datko, 1990. The *S*-methylmethionine cycle of *Lemna paucicostata*. *Plant Physiol.* **93**:623–630.

Neuhierl, B., and A. Bock, 1996. On the mechanism of selenium tolerance in selenium-accumulating plants. Purification and characterization of a specific selenocysteine methyltransferase from cultured cells of *Astragalus bisulcatus*. *Eur. J. Biochem.* **239**:235–238.

Ng, B. H., and J. W. Anderson, 1979. Light-dependent incorporation of selenite and sulphite into selenocysteine and cysteine by isolated pea chloroplasts. *Phytochemistry* **18**:573–580.

Ohlendorf, H. M. et. al., 1986. Embryonic mortality and abnormalities of aquatic birds: Apparent impacts of selenium from irrigation drain water. *Sci. Total Environ.* **52**:49–63.

Pilon-Smits, E. A. H. et. al., 1999a. Overexpression of ATP sulfurylase in Indian mustard leads to increased selenate uptake, reduction, and tolerance. *Plant Physiol.* **119**:123–132.

Pilon-Smits, E. A. H. et. al., 1999b. Selenium volatilization and accumulation by twenty aquatic plant species. *J. Environ. Qual.* **28**:1011–1017.

Rael, R. M., and W. T. Frankenberger, 1996. Influence of pH, salinity, and selenium on the growth of Aeromonas veronii in evaporation pond agricultural drainage water. *Water Res.* **30**:422–430.

Ravanel, S. et. al., 1995. Cloning of an *Arabidopsis thaliana* cDNA encoding cystathionine b-lyase by functional complementation in *Escherichia coli*. *Plant Mol. Biol.* **29**:875–882.

Reamer, D. C., and W. H. Zoller, 1980. Selenium biomethylation products from soil and sewage sludge. *Science* **208**:500–502.

Rosenfeld, I., and O. A. Beath, 1964. *Selenium, Geobotany, Biochemistry, Toxicity, and Nutrition*, Academic Press, New York.

Saiki, M. K., and T. P. Lowe, 1987. Selenium in aquatic organisms from subsurface agricultural drainage water, San Joaquin Valley, California. *Arch. Environ. Contam. Toxicol.* **19**:496–499.

Saito, K. et. al., 1994. Modulation of cysteine biosynthesis in chloroplasts of transgenic tobacco over-expressing cysteine synthase (*O*-acetylserine(thiol)lyase). *Plant Physiol.* **106**:887–895.

Shaw, W. H., and J. W. Anderson, 1972. Purification, properties and substrate specificity of adenosine triphosphate sulphurylase from spinach leaf tissue. *Biochem. J.* **127**:237–247.

Skorupa, J. P., and H. M. Ohlendorf, 1991. Contaminants in drainage water and avian risk thresholds. In *The Economy and Management of Water and Drainage in Agriculture*, A. Dinar and D. Zilberman, eds., Kluwer Academic Publishers, Norwell, MA, p. 345.

Smith F. W. et. al., 1995. Plant members of a family of sulfate transporters reveal functional subtypes. *Proc. Natl. Acad. Sci.* (USA) **92**:9373–9377.

Smith, G. S., and J. H. Watkinson, 1984. Selenium toxicity in perennial ryegrass and white clover. *New Phytol.* **97**:557–564.

Terry, N. et. al., 1992. Rates of selenium volatilization among crop species. *J. Environ. Qual.* **21**:341–344.

Terry, N., and A. M. Zayed, 1994. Selenium volatilization by plants. In *Selenium in the Environment*, W. T. Frankenberger, Jr. and S. Benson, eds., Marcel Dekker, New York.

Thompson-Eagle, E. T. et. al., 1989. Volatilization of selenium by *Alternaria alternata. Appl. Environ. Microbiol.* **55**:1406–1413.

Thompson-Eagle, E. T., and W. T. Frankenberger, 1990. Protein-mediated selenium biomethylation in evaporation pond water. *Environ. Toxicol. Chem.* **9**:1453–1462.

Thompson-Eagle, E. T., and W. T. Frankenberger, 1991. Selenium biomethylation in a saline environment. *Water Res.* **25**:231–240.

Tsang, M. L.-S. , and J. A. Schiff, 1978. Studies of sulfate utilization by algae 18. Identification of glutathione as a physiological carrier in assimilatory sulfate reduction by *Chlorella. Plant Sci. Lett.* **11**:177–183.

Ulrich, J. M., and A. Shrift, 1968. Selenium absorption by excised *Astragalus* roots. *Plant Physiol.* **43**:14–19.

Velinsky, D., and G. A. Cutter, 1991. Geochemistry of selenium in a coastal salt march. *Geochim. Cosmochim. Acta* **55**:179–191.

White, R. H., 1982. Analysis of dimethyl sulfonium compounds in marine algae. *J. Mar. Res.* **40**:529–536.

Wilber, C. G., 1980. Toxicology of selenium: A review. *Clin. Toxicol.* **17**:171–230.

Wilber, C. G., 1983. *Selenium: A Potential Environmental Poison and a Necessary Food Constituent*, Charles Thomas Publisher, Springfield, IL.

Williams, M. C., and H. F. Mayland, 1992. Selenium absorption by twogrooved milkvetch and western wheatgrass from selenomethionine, selenocystine, and selenite. *J. Range. Manage.* **45**:3774–778.

Wu, L. et. al., 1996. Effects of forage plant and field legume species on soil selenium redistribution, leaching, and bioextraction in soils contaminated by agricultural drain water sediment. *Arch. Environ. Contam. Toxicol.* **31:**329–338.

Zayed, A., and N. Terry, 1992. Selenium volatilization in broccoli as influenced by sulfate supply. *J. Plant Physiol.* **140:**646–652.

Zayed, A., and N. Terry, 1994. Selenium volatilization in roots and shoots: Effects of shoot removal and sulfate level. *J. Plant Physiol.* **143:**8–14.

Zayed, A. M. et. al., 1998. Accumulation and volatilization of different chemical species of selenium by plants. *Planta* **206:**284–292.

Zhang, Y., and J. N. Moore, 1997. Environmental conditions controlling selenium volatilization from a wetland system. *Environ. Sci. Technol.* **31:**511–517.

Zieve, R., and P. J. Peterson, 1981. Factors influencing the volatilization of selenium from soil. *Sci. Total Environ.* **19:**277–284.

Zieve, R., and P. J. Peterson, 1984. The accumulation and assimilation of dimethyl selenide by four plant species. *Planta* **160:**180–184.

SECTION IV

BIOLOGY OF METAL PHYTOREMEDIATION

12

METAL-ACCUMULATING PLANTS

Roger D. Reeves and Alan J. M. Baker

INTRODUCTION

During the past 150 years a very large body of data has been gathered concerning the inorganic composition and nutritional needs of many thousands of plant species. In the nineteenth century the main focus was on major elements (particularly calcium, potassium, magnesium, and phosphorus). Through the twentieth century, as analytical techniques became more specific and more sensitive (and particularly with the development of atomic spectroscopic methods such as flame emission, atomic absorption, and plasma emission), more reliable data on minor and trace elements became available. This body of data can be summarized (Table 12-1), indicating the normal concentration ranges expected in dried plant shoot material. In addition to the obvious constituents of organic matter (C, H, O, N, S), elements that have been shown to be essential for plant growth include B, Ca, Cl, Cu, Fe, K, Mg, Mn, Mo, P, and Zn. These are routinely incorporated into laboratory plant growth media. Some other elements, such as Ni, have been shown to be essential for at least some species, in very small concentrations that would normally be present and available, both in soils and as impurities in the constituents used for nutrient solutions. Other elements, such as Se, Co, and V, apparently have a beneficial effect on some plants, although they may not be universally essential.

Metalliferous soils, with abnormally high concentrations of some of the elements that are normally present only at minor (e.g., 200–2000 mg/kg) or trace levels (e.g., 0.1–200 mg/kg), vary widely in their effects on plants. These effects depend on plant species, the enriched elements, and the physical and chemical characteristics of the

Phytoremediation of Toxic Metals: Using Plants to Clean Up the Environment, edited by Ilya Raskin and Burt D. Ensley.
ISBN: 0-471-19254-6 ©2000 John Wiley & Sons, Inc.

TABLE 12-1   Ranges of Element Concentrations in Dried Plant Leaves (mg/kg)

Element	Low	Normal	High
Ca	800	3000–30,000	60,000
Mg	400	1000–6000	10,000
K	1000	5000–20,000	50,000
Na	400	1000–4000	10,000[a]
P	300	800–3000	5000
Fe	10	60–600	2500[b]
Mn	5	20–400	2000
Zn	5	20–400	2000
Cd	0.03	0.1–3	20
Pb	0.01	0.1–5	100
Ni	0.2	1–10	100
Co	0.01	0.03–2	20
Cr	0.05	0.2–5	50
Cu	1	5–25	100
Se	0.01	0.05–1	10

[a]Sodium levels in some maritime and salt-marsh plants can reach 60,000 mg/kg.
[b]Iron levels in a few species can be up to 35,000 mg/kg.

soil, and may range from complete toxicity to all higher plant life, to the encouragement of the development of a characteristic local flora of metal-tolerant species. The inorganic composition of most metal-tolerant species shows them to be either indifferent to the metal-rich environment or at least only minimally affected, both chemically and physiologically, by the uptake of a slightly higher than normal concentration of the elements with which the soil is enriched. The "high" values for minor and trace elements in Table 12-1 are those sometimes found in plants tolerant of metal-enriched soils.

In some cases, species have evolved that occupy only the ecological niche presented to them by the metalliferous environment. Such species, endemic to areas of well-defined soil composition, have been used as geobotanical indicators in mineral exploration. In recent times, however, it has become apparent that some of the world's metal-tolerant plant species are characterized by the ability to accumulate very high concentrations of certain elements, far in excess of normal physiological requirements (if any), and far in excess of the levels found in the majority of species, even those tolerant of metalliferous soils. These plants represent one of the most important resources available to researchers investigating various strategies of phytoremediation, and are discussed in more detail below.

HISTORICAL RECORDS OF METAL-ACCUMULATING PLANTS

It is not certain when highly abnormal plant uptake of an element generally present only at "trace" levels was first recorded, but an early qualitative observation of this

TABLE 12-2   Zinc Content of Plants from the Calamine Area near Aachen, Germany

Species	Organ	ZnO (dry wt.) %	ZnO (ash) %	Zn (dry wt.) %	Zn (ash) %
Thlaspi alpestre	Root	0.167	2.66[a]	0.134	2.14
var. *calaminare*	Leaves	1.50	13.12	1.205	10.54
Viola calaminaria	Root	0.085	1.52	0.068	1.22
(*V. tricolor*)	Leaves	0.11	1.16	0.088	0.93
Armeria vulgaris	Root	0.17	3.58	0.137	2.88
	Leaves	0.11	1.17	0.088	0.94
Silene inflata	Root	0.02	0.74	0.016	0.59
	Upper parts	0.22	1.92	0.177	1.54

[a]Corrected from original data.

Source: After Sachs (1865).

kind is that of Bellingrodt in Monheim's laboratory in Germany and a subsequent report from the analytical chemist A. Braun, given by Forchhammer (1855, p. 92): "Hieran knüpft sich die von dem Hrn. A. Braun neulich bekannt gemachte, hochst interessante Beobachtung, daß die *Viola calaminaria*, die so charakteristisch für die Zinklager bei Aachen ist, daß man neue Fundorte dieses Metalls nach dem Standorte der Pflanze aufgesuchte hat, Zink enthälte."[*]

Analytical data on this plant and on other species characteristic of the zinc-rich ("calamine") soils in the Aachen area (straddling the modern-day border between Germany and Belgium), including the so-called *Thlaspi alpestre* var. *calaminare*, were provided by F. Risse for inclusion in a volume by J. Sachs (1865) that formed part of Hofmeister's *Handbuch der Physiologischen Botanik*. The data of Risse, given in Table 12-2, include values for ZnO concentrations; equivalent Zn concentrations are also shown.

Although the initial observation of Zn accumulation referred to *V. calaminaria*, the most notable data of Risse concern the leaf tissue of *T. alpestre* var. *calaminare*, exceeding 10,000 mg Zn/kg (1%) in the dry matter, or 10% Zn in the ash. This plant is discussed in more detail below.

One can only speculate about what led Bellingrodt and Braun to notice the high Zn content of *V. calaminaria*. It is possible that they used their knowledge of the zinc-rich nature of the soils in the Altenberg–Aachen area together with the well-known unique behavior of zinc oxide, which undergoes a reversible change to bright yellow on heating. Alternatively, the zinc may have been detected through conventional chemical separations, based on the insolubility of the sulfide in alkaline solution.

Observations of unusual accumulation of other metals have been made only during the twentieth century. There are records of abnormal levels of lead dating from the

---

[*]"Linked with this is the newly announced and most interesting observation of Herr A. Braun, that *Viola calaminaria*, which is so characteristic of the zinc deposits near Aachen that new discoveries of this metal have been sought from the location of this plant, contains zinc."

1920s, selenium from the 1930s, nickel from the 1940s, cobalt and copper from the 1960s, and cadmium and manganese from the 1970s.

Hyperaccumulation

The term *hyperaccumulator*, describing a plant with a highly abnormal level of metal accumulation, appears to have been first applied by Jaffré et al. (1976) in the title of their paper on nickel concentrations in the New Caledonian plant *Sebertia acuminata*. The specific use of the term to denote a defined concentration (>1000 mg Ni/kg) was introduced by Brooks et al. (1977a) in discussing Ni concentrations in species of *Homalium* and *Hybanthus* from various parts of the world. A similar concept had, however, been used previously by Jaffré and Schmid (1974), who referred to certain high-nickel plants from the ultramafic soils of New Caledonia as *hypernickelophores*, or "extreme nickel-bearers." This was itself a modification of terminology proposed by Duvigneaud and Denaeyer-De Smet (1970, 1973) in discussing mineral nutrition and phytochemical classification.

Brooks et al. (1977a) demonstrated the possibility of detecting the occurrence of ultramafic rocks and soils from the nickel content of leaf fragments from herbarium specimens of plants of certain genera of wide distribution. 'Hyperaccumulation' was used to describe accumulation of Ni to >1000 mg/kg in dry leaf tissue because this was a level 100–1000 times higher than that normally found in plants on soils not of ultramafic origin, and 10–100 times higher than that found for most other plants on Ni-rich ultramafic soils. Although there was some arbitrariness in the choice of this criterion, it has been noted that in many ultramafic floras, Ni concentrations of 100–1000 mg/kg are quite rare, and accumulation to >1000 mg/kg seems to represent a distinct form of plant response, implying some characteristic and unusual metabolic behavior.

An attempt to give greater precision to the definition of hyperaccumulation was made by Reeves (1992) for Ni: *A hyperaccumulator of Ni is a plant in which a Ni concentration of at least 1000 mg/kg has been recorded in the dry matter of any aboveground tissue in at least one specimen growing in its natural habitat.* This indicates that the criteria should not be based on analyses of whole plants or subterranean plant parts, largely because of the difficulty of ensuring that the samples are free of soil contamination, but also because plants that immobilize metals in the root system and fail to translocate them further (Baker, 1981), are of less interest for many purposes than those that actively accumulate metals into all their tissues.

This detailed definition also clarifies questions surrounding cases where (1) some specimens of a species may be found with >1000 mg/kg and others with <1000 mg/kg, (2) plant tissue other than leaves (e.g., latex) are found with very high metal concentrations, or (3) a species has been found to take up high levels of a metal under artificial conditions, such as through massive metal-salt amendments to an experimental soil or a nutrient solution. Plants qualify as hyperaccumulators under cases (1) and (2), but not under (3). In the last of these cases, the "forced" metal uptake may lead to plant mortality, and may have no relevance to the continuing life cycle of naturally occurring metallophyte populations, even though it may be of interest for some phytoremediation strategies.

Care should also be taken in the interpretation of some literature reports of high metal concentrations, which prove on close examination to have been expressed on an ash weight basis. Such data, often obtained by analytical methods employing ashed solid samples directly (e.g., arc emission spectroscopy, X-ray fluorescence), give values that are generally at least 10-fold higher than the dry-weight basis, and have led to some exaggerated claims of metal accumulation by plants that are in fact behaving normally. For example, it has been claimed that *Stachys recta* var. *serpentini* is a nickel accumulator; but the figure of 2700 mg/kg (Lisanti, 1952) quoted in several other publications is for NiO in plant ash, and probably corresponds to about 120–150 mg/kg Ni in the dry matter, which is at the upper end of the range of normal plant responses to a serpentine soil.

The concepts and definitions of hyperaccumulation have been extended to elements other than Ni. Malaisse et al. (1978) used the 1000-mg/kg criterion for Cu accumulation, and Brooks et al. (1980) argued for this to be applied also to Co. Reeves and Brooks (1983b) used the same criterion in discussing Pb, but for Mn and Zn, which are normally present at higher and more widely varying concentrations (about 20–400 mg/kg), a 10,000-mg/kg threshold was suggested by Baker and Brooks (1989), following the use of the term *hypermanganèsophore* for plants with this level of Mn accumulation (Jaffré, 1980).

The criteria of Baker and Brooks (1989) may in some cases be seen as unduly restrictive. In particular, for Zn, which is present in most plants on zinc-rich soils at 50–500 mg/kg, it might be more appropriate to regard Zn levels above 3000 mg/kg as remarkable, and deserving of being described as hyperaccumulation. This would draw attention to records such as that of a Zn concentration of over 9000 mg/kg in *Gomphrena canescens* in Queensland, Australia (Cole et al., 1968). It would also help to prevent some very strongly Zn-accumulating species from being described as non-accumulators, as has happened recently (Shen et al., 1997) with *Thlaspi ochroleucum*, a species that can be found with up to 5200 mg/kg Ni from serpentine (Reeves, 1988), and with up to 6300 mg/kg Zn and 5–100 mg/kg Cd from zinc-rich soils (Reeves, 1988, and unpublished). For phytoremediation purposes, there should be interest in any species that consistently shows Zn levels above 3000 mg/kg, particularly if Cd levels are also abnormally high. Normal Cd levels are so low (Table 12-1) that species capable of concentrating this element to >100 mg/kg should be regarded as Cd hyperaccumulators (Baker et al., 1994a), with significant phytoremediation potential.

Soils Hosting Hyperaccumulators

The most important types of metalliferous soil hosting hyperaccumulator plants are (1) so-called serpentine soils, derived from Fe- and Mg-rich ultramafic rocks, which occur in many parts of the world, and which are also enriched in Ni, Cr, and Co; (2) soils derived from various Se-rich rock types, particularly Cretaceous shales of the

TABLE 12-3 Element Concentrations in Some Typical Metalliferous Soils[a]

Soil Type and Locality	K	Na	Ca	Mg	Fe	As	Cd	Co	Cr	Cu	Mn	Ni	P	Pb	Zn
			(%)								(mg/kg)				
Serpentine															
Yamanigüey, Cuba	0.08	0.07	0.29	13.55	15.7	ND	ND	293	5,220	52	2,465	3,602	195	ND	194
Mt. Koghi, New Caledonia	0.05	ND	0.68	5.75	25.8	ND	ND	810	3,236	56	6,107	5,900	329	ND	166
Niquelândia, Brazil	0.08	0.28	0.16	2.59	24.8	ND	ND	555	2,594	3,783	3,769	11,260	836	ND	165
Glen Geddes, Queensland	0.30	1.92	1.13	5.41	11.1	ND	ND	211	8,450	16	1,790	3,240	122	ND	178
Chinese Camp, California	0.09	0.07	0.15	15.57	13.3	ND	ND	188	1,691	38	1,215	4,483	358	ND	109
Lead/zinc															
Balya maden, Turkey	1.63	0.06	8.82	0.40	7.6	1510	317	4	90	1,260	2,160	19	499	31,650	45,440
Le Bleymard, France	0.28	0.13	12.92	7.41	3.8	50	297	21	120	828	834	63	955	40,880	70,480
Charterhouse, Somerset, UK	1.00	0.20	3.65	0.43	8.1	442	11	35	80	84	2,660	82	2,130	49,910	18,490
Kamariza-Lavrion, Greece	0.66	ND	17.88	1.71	4.4	1224	66	21	86	244	2,694	73	398	7,375	13,690
Bleiberg, Austria	1.88	0.09	10.60	4.27	2.5	47	78	11	47	71	312	29	352	3,870	22,910

Copper/cobalt

Mine de l'Etoile, Zaïre	ND	ND	ND	4.2	ND	ND	8,800	ND	15,000	3,500	ND	ND	ND	260
Shinkolobwe, Zaïre	ND	ND	ND	2.6	ND	ND	9,900	ND	5,500	5,600	9,800	ND	ND	200
Shinkolobwe, Zaïre	ND	ND	ND	3.1	ND	ND	1,100	ND	50,900	1,500	200	ND	ND	200

[a]ND = not determined.

Sources: Serpentine and lead/zinc, R. D. Reeves (unpublished); copper/cobalt, Malaisse and Grégoire (1978), Malaisse et al. (1994).

midwestern United States, in which Se concentrations are often >10 mg/kg and can sometimes exceed 50 mg/kg; (3) soils enriched in Pb and Zn—"calamine" soils (usually containing elevated Cd also, and sometimes bearing high concentrations of As and/or Cu); and (4) Cu- and Co-containing soils, particularly those of the Shaban Copper Arc in the Democratic Republic of Congo, derived from argillites and dolomites containing metal sulfides, together with important secondary minerals (oxides, basic carbonates, silicates).

Types 1 and 2 occur over very extensive areas, and the characteristic metal-tolerant floras (including hyperaccumulators of Ni and Se, respectively) may be correspondingly widely distributed. Natural surface exposures of Pb/Zn minerals are generally more limited in area, but where mining, ore transport, and smelting have occurred, more extensive development of mineral-rich soils has taken place, and the distribution of hyperaccumulators of Zn and Pb may reflect this sequence of activities. Many of the areas of waste from these processes have soils containing >1% of either or both of these elements. In the central African copper–cobalt areas, soils can be found with 1000–60,000 mg/kg Cu and 300–15,000 mg/kg Co, although there are wide variations over short distances, and high values for the two elements are not necessarily found together.

Table 12-3 shows some typical soil analyses for relevant elements in serpentine, lead/zinc, and copper/cobalt soils, respectively.

ANALYTICAL PROBLEMS AND THE RELIABILITY OF HYPERACCUMULATION DATA

In connection with reports of elevated metal concentrations in plant material, it must be recognized that there are several problems of an analytical nature, needing critical assessment. In addition to the usual dangers of contamination of samples and containers at all stages of the analytical work, there are other sources of error and confusion.

First, the question of whether the data have been expressed on an ash-weight or a dry-weight basis, noted above, must be examined. Second, it is important to examine the analytical methodology critically. There have been cases in the literature where the apparent identification of an abnormally high concentration of an element has been due, for example, to misidentification of a spectral line (in emission spectrography), or to failure to correct for some form of analytical interference. Several reports of gold-accumulating plants have resulted from this kind of analytical error; the mythology of this subject has been discussed in detail by Brooks et al. (1981a).

Finally, and most importantly, there is the question of contamination of plant material from soil or from aerial fallout. Contamination by metalliferous soil is a major problem when the soil contains secondary weathering products, such as various carbonates, silicates, or sulfates. Such minerals as cerussite and anglesite (Pb), smithsonite and hemimorphite (Zn), and malachite (Cu) may be important forms of these elements in the soils adjacent to mining operations, or may even be the primary target of the mining. These materials are more soluble than are primary minerals such as sulfides, and much of any contamination of plant material from this source is likely to pass into solution under the conditions used for plant analysis.

The problem is particularly acute when pure or nearly pure particles of secondary minerals form on the surface of small rock fragments, as the metal may constitute 50–80% of the mass of the particle. In such cases, it needs only imperceptible levels of contamination of a plant sample to give a large elevation of its metal concentration. For example, 0.2 mg of malachite included as a dust with 100 mg of plant leaf or stem genuinely containing 10 mg/kg Cu is enough to raise the apparent Cu concentration to more than 1150 mg/kg. Some of this type of contamination may be removed by brief agitation of the plant material for a few minutes in water with detergent, but it is impossible to be sure in all cases that the extraneous matter has been removed, because some plants have particularly sticky or hairy leaves. In any case, extended agitation and washing of plant material may also lead to some leaching of metal ions or metal complexes genuinely contained in the plant. A similar problem exists in the sampling of plant material adjacent to smelter operations, where the fallout may contain vapor and very fine dusts, which may deposit on plant material and produce a kind of contamination much more concentrated than that provided by dusts from the soil.

To provide evidence that a measured high plant–metal concentration genuinely exists within the plant, several workers have used data for associated elements that are present in significant concentrations in the soil but would not be expected to be present in high concentrations in the plant. The validity of this approach depends largely on whether these "soil-contamination monitors" are effectively extracted from the soil *under the conditions used for the plant analysis*. Elements used for this purpose have included Fe, Ti, Al, and Cr. The use of Ti or Al, in particular, is open to the objection that these elements may not extract completely from the contaminating soil when plant material or plant ash is dissolved in dilute acids, in which case significant soil contamination may not make itself clearly apparent. In the case of ultramafic soils, high concentrations of both Fe (>1000 mg/kg) and Cr (>25 mg/kg) in the plant analysis should cause the plant data (and any associated high values for elements such as Ni and Co) to be viewed with suspicion. Many other metalliferous soils contain sufficient Fe (>2%) that this element can again be used as a monitor of soil contamination, subject only to the caution that a few plant species genuinely take up Fe to concentrations above 1000 mg/kg.

In the most extreme cases of hyperaccumulation (e.g., most of the instances where Ni is >10,000 mg/kg, and those where Se is >100 mg/kg), the soil has a lower concentration of the accumulated element than is found in the plant material, in which case any soil contamination will actually cause a negative analytical error.

HYPERACCUMULATION OF PARTICULAR ELEMENTS

The following discussion outlines instances of hyperaccumulation of Ni, Zn, Cd, Pb, Co, Cu, Mn, Cr, and Se, which have been studied more extensively than most other trace elements, and for which a substantial body of reliable plant analysis data exists. Other elements now receiving increased attention, for various reasons (toxicological, economic, or simply as a result of improved analytical techniques), include As, Hg, Tl, Au, and the lanthanide elements. It seems likely that, with more comprehensive data for thousands of plant species from different geologic environments, criteria will eventually be defined for hyperaccumulation of these elements, and examples of hyperaccumulators will emerge.

Nickel

Following the discovery of unprecedented Ni concentrations in the Italian serpentine plant *Alyssum bertolonii*, which accumulates up to about 10,000 mg/kg or 1% Ni (Minguzzi and Vergnano, 1948), there appears to have been relatively little further interest for about 25 years. Two additional Ni-accumulating *Alyssum* species—*A. murale* from Armenia (Doksopulo, 1961) and *A. serpyllifolium* subsp. *lusitanicum* from Portugal (Menezes de Sequeira, 1969)—were reported. These were followed by studies in Zimbabwe (Wild, 1970) and two independent discoveries of high Ni concentrations (3000–9800 mg/kg) in *Hybanthus floribundus* from Western Australia (Severne and Brooks, 1972; Cole, 1973).

From 1974, concerted attempts have been made by T. Jaffré and co-workers in New Caledonia, and by R. R. Brooks, R. D. Reeves and co-workers, studying the serpentine vegetation of many parts of the world, to establish the range of Ni hyperaccumulation, both geographically and in terms of distribution in the plant kingdom. Brooks et al. (1977a) identified several species of *Homalium* and *Hybanthus* of New Caledonia as hyperaccumulators, and a comprehensive survey of the 170 species of the genus *Alyssum* (Brooks and Radford, 1978; Brooks et al. 1979) established that there were about 48 Ni hyperaccumulators, all within one section (Odontarrhena) of the genus, distributed from Portugal across Mediterranean Europe to Turkey and Armenia. Most are serpentine-endemic species, and many are of very restricted distribution.

Further work by various groups has focused on other genera of the Brassicaceae found on Ni-rich soils of the Mediterranean region, on the species of serpentine outcrops of the European Alps, southern Africa, Newfoundland, and the Pacific northwest of the United States, on the large serpentine flora of New Caledonia, and on plants of tropical serpentine soils of Brazil, Queensland, and Southeast Asia (especially various islands of Indonesia and the Philippines). Detailed accounts of this work have been given by Brooks (1987) and Reeves (1992).

The most notable recent development has been the discovery that the ultramafic soils of Cuba are host to the largest number of Ni hyperaccumulators found in any one country (Reeves et al., 1996, 1999). A survey of much of the Caribbean serpentine flora has revealed 130 such species in Cuba, mainly from the genera *Buxus, Phyllanthus, Leucocroton, Euphorbia, Pentacalia, Senecio, Psychotria, Ouratea,* and *Tetralix*, as well as a species of *Phyllanthus* from the Dominican Republic.

The most recent information brings the worldwide total of known Ni hyperaccumulators to about 318. The plant families most strongly represented are the Euphorbiaceae (83), Brassicaceae (82), Asteraceae (27), Flacourtiaceae (19),

Buxaceae (17), and Rubiaceae (12). The other 78 species are from 31 different families, distributed widely throughout the plant kingdom, emphasizing the way in which this property has evolved independently many times. Ni hyperaccumulators are shown in Table 12-4. Earlier summaries (Brooks, 1987; Baker and Brooks, 1989) and the most recent work (Reeves et al., 1996, 1999) can be consulted for details of species in genera containing many hyperaccumulators, and for authorities for the species names.

It may be asked what factors are responsible for the large number of hyperaccumulators of Ni, compared with those of other elements. This may be partly the result of the concerted attention to analytical work on ultramafic floras, but there is a more fundamental reason. Of all the various types of metalliferous soils and rock outcrops, the Ni-enriched ultramafics are by far the most widespread on a global scale, and the extent of individual ultramafic tracts is often very large. Continuous ultramafic areas of tens or even hundreds of square kilometers can be found in countries such as New Caledonia, Cuba, and Turkey. Where such areas have been continuously available for plant colonization for many millions of years (not having suffered from glaciation, complete erosion, or burial beneath volcanic or sedimentary material), this appears to have presented a major opportunity for the evolution of a characteristic flora with numerous endemic species, including some that have developed Ni accumulation as a particular form of tolerance and response to the high-Ni soils.

It is the view of the present authors that many further Ni hyperaccumulators remain undiscovered. These include species not yet discovered or described, and known species that have never been analyzed. In particular, there are insufficiently studied ultramafic areas in the Philippines, several islands of Indonesia, Central America, and possibly West Africa, that may support hyperaccumulators. Although it is clear from Table 12-4 that Ni-accumulating ability has evolved independently in a very wide range of plant families and geographic areas, special attention should be paid to widespread families such as the Asteraceae, Euphorbiaceae, Ochnaceae, Rubiaceae, and Violaceae.

Zinc, Lead, and Cadmium

The discovery of zinc accumulation by certain *Viola* and *Thlaspi* species about 1855 is described at the beginning of this chapter. Further work, particularly on the *Thlaspi* from the German and Belgian calamine soils and from British mine wastes, has been carried out at intervals since then, notably by Baumann (1885), Maquinay and Ramaut (1960), Ernst (1968a, 1968b), Denaeyer-De Smet (1970), Shimwell and Laurie (1972), and Garcia-Gonzalez and Clark (1989). The *Thlaspi* species at the center of attention has been variously named since 1753. Nomenclature has been complicated by its variability, arising from its disjunct distribution and inbreeding behavior, the use of different ranks for local variants, and the use of illegitimate names. The more familiar name, *T. alpestre*, widely applied between about 1763 and 1990, has proved to be illegitimate (Gutermann, 1975), and the use of *T. caerulescens*, the first

AVRIL ROBARTS LRC

TABLE 12-4   Hyperaccumulators of Nickel

Family Species	Location	Maximum Ni Concentration (mg/kg)	References
Acanthaceae			
Blepharis acuminata	Zimbabwe	2,000	31
Justicia lanstyakii	Brazil (Goiás)	2,690	39,45
Lophostachys villosa	Brazil (Goiás)	1,890	39,45
Phidiasia lindavii	Cuba	1,850	51
Ruellia geminiflora	Brazil (Goiás)	3,330	45
Adiantaceae			
Adiantum sp.	Brazil (Goiás)	3,540	39
Anacardiaceae			
Rhus wildii	Zimbabwe	1,600	31
Asteraceae			
Berkheya coddii	South Africa	11,600	37
Chromolaena sp. cf. *meyeri*	Brazil (Goiás)	1,100	39,45
Dicoma niccolifera	Zimbabwe	1,500	4,31,46
Gochnatia crassifolia	Cuba	1,120	51
G. recurva	Cuba	1,010	51
Koanophyllon grandiceps	Cuba	6,240	51
K. prinodes	Cuba	6,170	51
Leucanthemopsis alpina	Italy	3,200	24,34
Pentacalia (10 species)	Cuba	16,600	51
Senecio (6 species)	Cuba	11,000	51
Senecio pauperculus	Canada (Newfoundland)	1,900	43
Shafera platyphylla	Cuba	1,890	51
Solidago hispida	Canada (Newfoundland)	1,020	43
Boraginaceae			
Heliotropium sp.	Brazil (Goiás)	2,020	45
Brassicaceae			
Alyssum (48 taxa, all from Sect. Odontarrhena)	Mediterranean Europe, Turkey, Armenia, Iraq, Syria	1,280–29,400	1–3,15,16,19,21, 27,28,33,40
Bornmuellera (6 taxa)	Greece, Albania, Turkey	11,400–31,200	28
Cardamine resedifolia	Italy	3,270	19,24
Cochlearia aucheri	Turkey	17,600	35
C. sempervivum	Turkey	3,140	35
Peltaria emarginata	Greece	34,400	22
Streptanthus polygaloides	USA (CA)	14,800	25,49

continued

TABLE 12-4   (Continued)

Family Species	Location	Maximum Ni Concentration (mg/kg)	References
Thlaspi (23 taxa)	Central Europe, Greece, Turkey, Japan, USA (CA, OR, WA)	2,000–31,000	19,24,27,29,30, 34,35
Buxaceae			
Buxus (17 taxa)	Cuba	1,320–25,420	26,50
Campanulaceae			
Campanula scheuchzeri	Italy	1,090	34
Caryophyllaceae			
Arenaria (3 species)	USA (CA, WA), Canada (Newfoundland)	2,330–2,370	43,47
Minuartia laricifolia	Italy	2,710	24,34
M. verna	Italy	1,390	34
Clusiaceae			
Garcinia bakeriana	Cuba	7,440	51
G. polyneura	Cuba	1,280	51
G. revoluta	Cuba	1,250	51
G. ruscifolia	Cuba	1,640	51
Convolvulaceae			
Merremia xanthophylla	Zimbabwe	1,400	30
Cunoniaceae			
Geissois (7 species)	New Caledonia	1,000–34,000	17,23
Pancheria engleriana	New Caledonia	6,300	23
Dichapetalaceae			
Dichapetalum gelonioides			
ssp. *tuberculatum*	Philippines, Sabah	26,600	42,44
ssp. *andamanicum*	South Andaman Island	3,160	44
Dipterocarpaceae			
Shorea tenuiramulosa	Sabah	1,000	36
Escalloniaceae			
Argophyllum grunowii	New Caledonia	1,380	23
A. laxum	New Caledonia	1,900	23
Euphorbiaceae			
Baloghia sp.	New Caledonia	5,380	23
Bonania (3 taxa)	Cuba	1,620–1,850	50
Cleidion viellardii	New Caledonia	9,900	23
Cnidoscolus sp. cf. *bahianus*	Brazil (Goiás)	1,020[a]	39,45
Euphorbia (3 taxa)	Cuba	4,430–9,340	50

continued

TABLE 12-4 (Continued)

Family Species	Location	Maximum Ni Concentration (mg/kg)	References
Gymnanthes recurva	Cuba	4,070	50
Leucocroton (28 taxa)	Cuba	2,260–27,240	26,50
Phyllanthus (41 taxa)	New Caledonia, Sabah, Philippines (Palawan, Mindanao), Cuba, Dominican Republic	1,090–60,170	18,23,41,44,50
Sapium erythrospermum	Cuba	1,000	50
Savia (3 taxa)	Cuba	2,940–4,890	50
Fabaceae			
Anthyllis sp.	Italy	4,600	24
Pearsonia metallifera	Zimbabwe	15,350	4,31
Trifolium pallescens	Italy	1,990	24,34
Flacourtiaceae			
Casearia silvana	New Caledonia	1,490	20
Homalium (7 species)	New Caledonia	1,160–14,500	7,11,23
Xylosma (11 species)	New Caledonia	1,000–3,750	20,23
Juncaceae			
Luzula lutea	Italy	2,050	24,27,34
Meliaceae			
Walsura monophylla	Philippines (Palawan)	7,090	42,44
Myristicaceae			
Myristica laurifolia	Indonesia (Obi)	1,100	12
Myrtaceae			
Mosiera araneosa	Cuba	5,030	51
M. ekmanii	Cuba	1,460	51
M. x miraflorensis	Cuba	1,210	51
M. ophiticola	Cuba	1,800	51
Psidium araneosum	Cuba	1,360	51
P. havanense	Cuba	1,060	51
Ochnaceae			
Brackenridgea palustris ssp. *foxworthyi*	Philippines (Palawan)	7,600	42,44
ssp. *kjellbergii*	Indonesia (Celebes)	1,050	44
Ouratea nitida	Cuba	1,700	51
O. striata	Cuba	8,190	51
Oleaceae			
Chionanthus domingensis	Cuba	1,570	51

continued

TABLE 12-4   (Continued)

Family Species	Location	Maximum Ni Concentration (mg/kg)	References
Oncothecaceae			
Oncotheca balansae	New Caledonia	2,500	23
Poaceae			
Trisetum distichophyllum	Italy	1,710	34
Ranunculaceae			
Ranunculus glacialis	Italy	1,260	34
Rubiaceae			
Ariadne shaferi ssp. *shaferi*	Cuba	13,070	51
ssp. *moaensis*	Cuba	22,360	51
Mitracarpus sp.	Brazil (Goiás)	1,000	45
Phyllomelia coronata	Cuba	25,540	51
Psychotria clementis	Cuba	19,090	51
P. costivenia	Cuba	38,530	51
P. douarrei	New Caledonia	19,900	7,8,32
P. glomerata	Cuba	10,250	51
P. osseana	Cuba	12,780	51
P. vanhermanii	Cuba	35,720	51
Rondeletia (2 undescr. spp.)	Cuba	8,730	51
Sapotaceae			
Planchonella oxyedra	Indonesia (Obi)	19,600	12
Sebertia acuminata	New Caledonia	11,700[b]	9
Saxifragaceae			
Saxifraga (3 species)	Italy	2,970–3,840	24,34
Scrophulariaceae			
Esterhazya sp.	Brazil (Goiás)	1,060	39
Linaria alpina	Italy	1,990	24,27,34
Stackhousiaceae			
Stackhousia tryonii	Australia (Queensland)	21,500	38
Tiliaceae			
Tetralix brachypetalus	Cuba	13,610	51
T. cristalensis	Cuba	9,010	51
T. jaucoensis	Cuba	3,280	51
T. moaensis	Cuba	3,380	51
T. nipensis	Cuba	2,670	51
Trichospermum kjellbergii	Indonesia (Obi)	3,770	12
Turneraceae			
Turnera subnuda	Brazil (Goiás)	6,130	39,45
Velloziaceae			
Vellozia sp.	Brazil (Goiás)	3,080	39
Violaceae			
Agatea deplanchei	New Caledonia	2,500	23
Hybanthus (5 taxa)	New Caledonia, Western Australia	3,000–17,600	5–8,10,11

continued

TABLE 12-4   (Continued)

Family Species	Location	Maximum Ni Concentration (mg/kg)	References
Rinorea bengalensis	Southeast Asia	17,500	13
R. javanica	Borneo	2,170	14
Rinorea sp.	Indonesia (Karakelong)	1,830	48

[a]Latex contains up to 1.35% Ni (dry-weight basis).
[b]Latex contains up to 26% Ni (dry-weight basis).

References: (1) Minguzzi and Vergnano (1948), (2) Doksopulo (1961), (3) Menezes de Sequeira (1969), (4) Wild (1970), (5) Severne and Brooks (1972), (6) Cole (1973), (7) Jaffré and Schmid (1974), (8) Brooks et al. (1974), (9) Jaffré et al. (1976), (10) Lee et al. (1977b), (11) Brooks et al. (1977a), (12) Wither and Brooks (1977), (13) Brooks and Wither (1977), (14) Brooks et al. (1977b), (15) Brooks and Radford (1978), (16) Brooks et al. (1979), (17) Jaffré et al. (1979a), (18) Kersten et al. (1979), (19) Vergnano Gambi and Gabbrielli (1979), (20) Jaffré et al. (1979b), (21) Vergnano Gambi et al. (1979), (22) Reeves et al. (1980), (23) Jaffré (1980), (24) Vergnano Gambi and Gabbrielli (1981), (25) Reeves et al. (1981), (26) Berazain Iturralde (1981), (27) Vergnano Gambi et al. (1982), (28) Reeves et al. (1983a), (29) Reeves and Brooks (1983a), (30) Reeves et al. (1983b), (31) Brooks and Yang (1984), (32) Baker et al. (1985), (33) Dudley (1986a, 1986b), (34) Gabbrielli et al. (1987), (35) Reeves (1988), (36) Proctor et al. (1989), (37) Morrey et al. (1989), (38) Batianoff et al. (1990), (39) Brooks et al. (1990), (40) Kelepertsis et al. (1990), (41) Schmid (1991), (42) Homer et al. (1991), (43) Roberts (1992), (44) Baker et al. (1992), (45) Brooks et al. (1992), (46) Proctor and Cole (1992), (47) Kruckeberg et al. (1993), (48) Proctor et al. (1994), (49) Kruckeberg and Reeves (1995), (50) Reeves et al. (1996), (51) Reeves et al. (1999).

legitimate name applied to this polymorphic species, is now generally recommended, as the name *T. alpestre* properly applies to a different plant (Akeroyd, 1993).

Several discoveries have widened the interest in *Thlaspi* species as metal accumulators. Rascio (1977) showed that *T. rotundifolium* subsp. *cepaeifolium* from zinc-polluted soils near the border of Italy and Austria was also a hyperaccumulator of zinc. Reeves and Brooks (1983a) observed that a number of European *Thlaspi* species also occurred on serpentine soils, from which hyperaccumulation of Ni occurred, and that Zn levels above 1000 mg/kg were common in many species, not necessarily from areas of known Zn mineralization. A more complete study of the genus (Reeves et al., 1983b; Reeves, 1988) identified high concentrations of Ni and/or Zn in *Thlaspi* species from the Pacific Northwest of the United States, and from Turkey, Cyprus, and Japan.

Reeves and Baker (1984) demonstrated that the ability of the Austrian species *T. goesingense* to accumulate nickel and zinc was an innate property, unrelated to the geochemistry of the area from which the seed originated, and Baker et al. (1994b) showed that *T. caerulescens*, when grown in amended nutrient solutions, had the ability to accumulate to high concentrations a wide variety of elements (Zn, Cd, Co, Mn, and Ni throughout the plant; Al, Cr, Cu, Fe, and Pb largely into the root system). These observations are of particular significance for multielement phytoremediation (see text below).

Records of hyperaccumulation of Zn by plants (as determined by the 10,000 mg/kg criterion noted earlier) are presented in Table 12-5.

Lead is neither essential nor beneficial to plant growth. It is present in most soils and rocks at concentrations below 50 mg/kg, and the associated vegetation generally contains <10 mg/kg. Even where concentrations of 1–10 mg/kg are measured in aboveground plant parts, it is likely that a substantial component of this comes from various forms of environmental and laboratory contamination. The very low solubility at neutral pH of many naturally occurring lead compounds in rocks, minerals, and soils (sulfide, sulfate, carbonate), together with the insolubility of lead salts of many organic compounds that may be exuded by plant root systems, combine to restrict severely the uptake of this element under normal conditions. Even when it enters the roots, significant translocation to the upper parts of the plant is uncommon. The few reports of very high lead concentrations in plants in their natural habitats (invariably taken from areas of Pb/Zn mineralization, or mine or smelter wastes) are shown in Table 12-5.

Cadmium occurs in most rock types at concentrations between 0.1% and 1% of those of zinc, and is obtained as a byproduct of the extraction of the latter metal from its ores. Elevated Cd levels (10–200 mg/kg, sometimes even higher) are generally found in soils containing waste materials from the mining of Zn sulfides and other Zn ores, but may also occur in soils treated with industrial wastes or with certain Cd-rich phosphatic fertilizers. Plant Cd is generally <3 mg/kg (Table 12-1), but may reach 20 mg/kg or more in the flora of Cd-enriched soils. A plant concentration of >100 mg/kg may be regarded as exceptional, even on a Cd-contaminated site, if it can be demonstrated not to be the result of aerial deposition. On Pb/Zn mine waste sites, *Thlaspi caerulescens* has been found by the present authors to contain >100 mg/kg Cd frequently, and >1000 mg/kg occasionally, with very large variations between sites and populations, and considerable intrasite variability. In view of the interest and practicality of reducing Cd concentrations in contaminated soils by the use of *T. caerulescens*, there is a need for more detailed field and laboratory studies of the factors governing Cd uptake by this species.

Cobalt and Copper

These two elements are considered together because the majority of cases of hyperaccumulation reported have been from the metalliferous soils of the Democratic Republic of Congo (formerly Zaïre), where the two metals occur together at elevated levels in the soils, although in widely varying proportions. Elsewhere, there have been occasional records of plants with >1000 mg/kg Cu from copper-mineralized areas (Blissett, 1966; Dykeman and De Sousa, 1966; Ernst, 1966). These reports, and the plant species involved, are worthy of more detailed investigation.

Normal concentrations of cobalt and copper in plants are in the ranges 0.03–2 and 5–25 mg/kg, respectively (Table 12-1). The behavior of the black gum of the southeastern United States (*Nyssa sylvatica* var. *biflora* and var. *sylvatica*) in accumulating as much as 845 mg/kg Co from normal soils may be noted as

TABLE 12-5 Some Hyperaccumulators of Zinc, Cadmium, and Lead

Family / Species	Locality	Maximum Concentration (mg/kg)[a]			References
		Zn	Cd	Pb	
Aceraceae					
Acer pseudoplatanus	UK			1,955	12
Brassicaceae					
Cardaminopsis halleri	Germany	13,620			6
Thlaspi avalanum	Thasos, Greece			1,250	18
T. brachypetalum	France	15,300			14
T. caerulescens	Western and Central Europe	43,710	2,130	2,740	1–4,6–9,14
T. ochroleucum	Thasos, Greece	4,130		1,210	16
T. rotundifolium subsp. cepaeifolium	Southern Austria, Northern Italy	18,500		8,200	10,15
T. praecox	Bulgaria	15,500			14
T. stenopterum	Spain	11,500			14
T. tatrense	Slovakia	20,100			14
Caryophyllaceae					
Minuartia verna	Yugoslavia; UK	11,400		20,000	9,12,13
Polycarpaea synandra	Queensland, Australia	6,960		1,044	5

	Location			Ref
Cistaceae				
Cistus incanus ssp. *creticus*	Thasos, Greece		2,300	18
Dichapetalaceae				
Dichapetalum gelonioides[b]	Sumatra; Mindanao; Sabah	30,000		17
Plumbaginaceae				
Armeria maritima var. *halleri*	Germany		1,600	3
Poaceae				
Agrostis stolonifera	UK		1,965	12
A. tenuis	UK		13,490	11,13
Arrhenatherum elatius	France		1,500	19
Festuca ovina	UK		11,750	13
Polygonaceae				
Rumex acetosa	UK	11,000	5,450	12
Violaceae				
Viola calaminaria	Belgium, Germany	10,000		1,3

[a]Maximum values for various elements, not all found in the same specimen.
[b]Subspecies *tuberculatum*, *sumatranum*, and *pilosum*.

References: (1) Sachs (1865), (2) Baumann (1885), (3) Linstow (1924), (4) Ernst (1966), (5) Cole et al. (1968), (6) Ernst (1968a), (7) Shimwell and Laurie (1972), (8) Denaeyer-De Smet and Duvigneaud (1974), (9) Ernst (1974), (10) Rascio (1977), (11) Williams et al. (1977), (12) Johnston and Proctor (1977), (13) Barry and Clark (1978), (14) Reeves and Brooks (1983a), (15) Reeves and Brooks (1983b), (16) Kelepertsis and Bibou (1991), (17) Baker et al. (1992), (18) Konstantinou and Babalonas (1992), (19) Deram and Petit (1997).

exceptional (Beeson et al., 1955; Kubota et al., 1960; Brooks et al., 1977c). Even on cobalt-enriched soils, such as those derived from ultramafic rocks, Co in plants rarely exceeds 20 mg/kg. Similarly, plant Cu concentrations are controlled within a remarkably narrow range, and plant Cu concentrations above 100 mg/kg are rare, even in the presence of high soil Cu.

Duvigneaud (1959) drew attention to the anomalous behavior of *Crotalaria cobalticola* in accumulating 354 mg/kg Co from cobalt-rich soils in Zaïre, and later work (Brooks et al., 1980) showed even higher concentrations in this species. Extensive studies of the vegetation of many sites of mining and smelting activity throughout Zaïre by F. Malaisse, R. R. Brooks, and their co-workers have identified 30 hyperaccumulators of cobalt and 32 of copper, with 12 species being common to the two lists, as shown in Table 12-6. In using this table, it should be noted that it is rare to find both Co and Cu >1000 mg/kg in the same plant specimen.

One of the most notable features of the cobalt and copper data obtained for the Zaïrean species is their variability. This behavior contrasts strongly with that of most Ni accumulators from serpentine soils, where the Ni concentrations in both the soils and the hyperaccumulator plants fall within fairly restricted ranges. (Many Ni accumulators, for example, have only been found on soils with 1000–5000 mg/kg Ni, and the plants have always been found with 1000–10,000 mg/kg Ni.) The wide variation in Co and Cu accumulation may be due in part to large variations of the soil concentrations of the two elements, occurring over short distances, but may also relate to other factors such as pH and soil moisture, the season of the year, individual genotypic variability, and varying degrees of soil contamination. In spite of the suspicion that the last of these possibilities is sometimes involved, there is little doubt that many of the records of Co and Cu hyperaccumulation represent genuine uptake by the plant from the soil: Malaisse et al. (1994), for example, have presented Fe data that indicate little likelihood of soil contamination (e.g., *Anisopappus davyi* with 3504 mg/kg Cu, 3 mg/kg Co, and 67 mg/kg Fe).

Typical of the variation in uptake behavior is that recorded for *Pandiaka metallorum* (Amaranthaceae). This was found with 740 mg/kg Cu by Duvigneaud and Denaeyer-De Smet (1963); four field and herbarium specimens from Fungurume analyzed by Morrison (1980) had 101–570 mg/kg Co and 629–6270 mg/kg Cu. The latter values were tabulated by Malaisse et al. (1979), although the Cu values were erroneously discussed in the text of that paper as Co. Six further analyses of plants from four other locations (Brooks et al. 1987) gave 31–2131 mg/kg Co and 72–2839 mg/kg Cu. Similar wide variations can be found in species such as *Bulbostylis pseudoperennis* (7–2127 mg/kg Co, 169–7783 mg/kg Cu), *Gutenbergia pubescens* (published as *G. cupricola*) (156–2309 mg/kg Co, 82–5095 mg/kg Cu), *Haumaniastrum robertii* (98–10230 mg/kg Co, 46–2070 mg/kg Cu), *Crepidorhopalon* (formerly *Lindernia*) *perennis* (25–9322 mg/kg Cu), and *Sopubia neptunii* (14–2476 mg/kg Co) (Brooks et al., 1987; F. Malaisse, personal communication). These wide variations indicate that great care needs to be taken in selecting seed of any of these species for studies connected with their potential for phytoremediation.

Few of the Co- and Cu-accumulating species appear to be absolutely restricted to metalliferous soils, although some have had local or regional uses as indicator plants. It should also be noted that some of the species that have been used as ore indicator plants, such as *Becium homblei*, do not take up significant amounts of either metal.

Plants that hyperaccumulate Ni from serpentine soils show a wide variation in their concomitant Co-accumulating ability in their natural habitats, ranging from <1 mg/kg in some cases to occasional values in the 10–100-mg/kg range and >1000 mg/kg in one instance known to the present authors. One specimen of the Cuban *Phyllanthus williamioides* (Table 12-6) has been found with 430 mg/kg Ni, 4200 mg/kg Mn, and 1140 mg/kg Co, although elsewhere this species occurs with >1% Ni and much lower Co. Co concentrations above 1000 mg/kg have been induced in leaves of Ni hyperaccumulators in pot trials with Co-amended soils or in Co-enriched nutrient solutions (Reeves and Baker, 1984; Homer et al., 1994; Baker et al., 1994b).

Manganese

Normal levels of manganese in plant dry matter fall within the rather wide range of 20–500 mg/kg, and occasionally exceed 1000 mg/kg in plants on normal soils, attaining 5270 mg/kg in *Vaccinium myrtillus* (Denaeyer-De Smet, 1966). Symptoms of deficiency can often be detected when the leaf concentration is <10 mg/kg, whereas toxic levels fall in the range 1000–12,000 mg/kg, depending on species. Manganese is an activator of a number of enzymes involved in the tricarboxylic acid cycle, and a central role for manganese cluster complexes in the oxidation of water to oxygen has been demonstrated.

Relatively little work has been done on the manganese concentrations of plants in areas of manganiferous soils, although some information is available on plants of ultramafic soils. Some plant species have been found with 1000–5000 mg/kg Mn on soils with manganese mineralization (bearing >1% Mn) and on soils of lower Mn concentration. Ultramafic soils may have 1000–5000 mg/kg Mn, which is not regarded as strongly abnormal, but it is from these soils that the main reports of high plant Mn have come. Jaffré (1980) found that 98 out of 445 species (22%) on ultramafic soils of New Caledonia had mean Mn concentrations above 1000 mg/kg; 6 species had means exceeding 10,000 mg/kg, and 9 species had at least one specimen above this level. Baker and Brooks (1989) chose this level to define Mn hyperaccumulators, and listed 8 such species. Table 12-7 lists those in which the maximum value found has exceeded 10,000 mg/kg (1%). Comparable values have not been found in studies of the ultramafic vegetation of Queensland, Cuba, and Brazil.

Chromium

Serpentine soils commonly contain 500–5000 mg/kg Cr, and there are occasional reports in the literature of plants that have accumulated this element. On normal soils, plant Cr is often below 1 mg/kg and rarely exceeds 5 mg/kg, and even on serpentine soils, levels above 50 mg/kg are so uncommon that this may be used as a criterion of

LIVERPOOL JOHN MOORES UNIVERSITY
LEARNING & INFORMATION SERVICES

TABLE 12-6   Hyperaccumulators of Copper and Cobalt [All Species from Democratic Republic of Congo (Zaïre) Unless Otherwise Noted]

Family / Species	Maximum Concentration(mg/kg) Co	Cu	References
Amaranthaceae			
Pandiaka metallorum	2,131	6,270	1,9,11,15
Celosia trigyna		2,051	15
Asteraceae			
Anisopappus chinensis[a]		1,657	12,14,15
A. davyi	2,889	3,504	9,11,15,16
Gutenbergia pubescens[a]	2,309	5,095	15
Millotia myosotidifolia[b]		2,400	2
Vernonia petersii		1,555	12,13,15
Caryophyllaceae			
Minuartia verna ssp. *hercynica*[c]		1,033	4
Silene cobalticola		1,660	1
Commelinaceae			
Commelina zigzag		1,214	9,11
Cyanotis longifolia	4,197		9,11
Convolvulaceae			
Ipomoea alpina		12,300	9
Crassulaceae			
Crassula alba	1,625		9,11
C. vaginata	1,712		9,11,15
Cyperaceae			
Ascolepis metallorum	1,118	1,211	1,15,16
Bulbostylis cupricola[a]		1,523	11
B. pseudoperennis[a]	2,127	7,783	10,11,15
Euphorbiaceae			
Monadenium cupricola		1,234	15
Phyllanthus williamioides[d]		1,140	18
Fabacaeae			
Crotalaria cobalticola	3,010		10,15
Vigna dolomitica		3,000	6
Iridaceae			
Gladiolus gregarius		1,412	16
Lamiaceae			
Aeollanthus subacaulis var. *linearis*[a,e]	4,300	13,700	6–8,10
A. homblei[a]		1,113	11
A. saxatilis	1,343		8,10,15
A. subacaulis var. *ericoides*	1,429		15
var. *linearis*	5,176	2,574	15
Becium grandiflorum var. *vanderystii*[a]		1,135	15
Haumaniastrum homblei	2,633		9,15
H. katangense	2,241	9,222	10–12,15,17
H. robertii	10,232	2,070	1,5,9,15,17
H. rosulatum		1,089	15
Malvaceae			
Hibiscus rhodanthus	1,527		15

continued

TABLE 12-6   (Continued)

| Family Species | Maximum Concentration(mg/kg) | | References |
	Co	Cu	
Pinaceae			
Abies balsamea[f]		1,120	3
Poaceae			
Eragrostis racemosa[a]		2,800	6
Rendlia altera[a]		1,560	14
Sporobolus congoensis		1,671	16
Pteridaceae			
Actiniopteris sp.		3,535	15,16
Scrophulariaceae			
Alectra sessiliflora var. *senegalensis*	2,782		11,15
Buchnera henriquesii	2,435	3,520	10,15
Crepidorhopalon tenuis[a,g]	1,113		6,15
C. perennis[a,g]	2,300	9,322	6,15
Sopubia mannii[a]	1,767		6,9,15
S. metallorum	1,742		15
S. neptunii	2,476		15
Striga hermontheca		1,105	15
Tiliaceae			
Triumfetta dekindtiana		1,283	15
T. digitata		1,057	10,11,15
T. welwitschii var. *descampii*	2,380		15
Velloziaceae			
Xerophyta retinervis var. *equisetoides*	1,520		15

[a]Nomenclature and/or identification changed since original publication (F. Malaisse, personal communication).
[b]From Williamstown, South Australia.
[c]From Könnern, Germany.
[d]From El Toldo, Moa, Cuba.
[e]Spelling of generic name revised as shown; Cu up to 13,700 mg/kg in whole plant or corms, 2150–3500 mg/kg in leaves and stems.
[f]From New Brunswick, Canada.
[g]Formerly in genus *Lindernia*.

References: (1) Duvigneaud and Denaeyer-De Smet (1963), (2) Blissett (1966), (3) Dykeman and De Sousa (1966), (4) Ernst (1966), (5) Brooks (1977), (6) Malaisse and Grégoire (1978), (7) Malaisse et al. (1978), (8) Brooks et al. (1978), (9) Malaisse et al. (1979), (10) Brooks et al. (1980), (11) Morrison (1980), (12) Brooks et al. (1982), (13) De Plaen et al. (1982), (14) Brooks and Malaisse (1985), (15) Brooks et al. (1987), (16) Malaisse et al. (1994), (17) Paton and Brooks (1995), (18) Reeves et al. (1999).

soil contamination. Nevertheless, attention should be drawn to the report of Wild (1974) concerning much higher concentrations of Cr in leaves of *Dicoma niccolifera* and *Sutera fodina* from near a chrome mine in Zimbabwe. The reported ash-weight concentrations are probably equivalent to about 1500 and 2400 mg/kg, respectively, on a dry-weight basis. The work of Brooks and Yang (1984) on the same species has

TABLE 12-7 Hyperaccumulators of Manganese (All Species from New Caledonia)

Family Species	Maximum (Mean) Concentration(mg/kg)	References
Apocynaceae		
Alyxia rubricaulis	14,000 (10,270)	1,3,4
Celastraceae		
Maytenus bureaviana	33,750 (19,230)	1,3
M. pancheriana	16,370 (16,370)	3
M. sebertiana	22,500 (22,500)	3
Clusiaceae		
Garcinia amplexicaulis	10,500 (2,230)	3
Myrtaceae		
Eugenia clusioides	10,880 (6,810)	3
Proteaceae		
Beaupreopsis paniculata	12,000 (6,900)	3
Macadamia angustifolia	11,590 (11,110)	3
M. neurophylla	55,200 (40,730)	2,3

References: (1) Jaffré (1977), (2) Jaffré (1979), (3) Jaffré (1980), (4) Brooks et al. (1981b).

yielded Cr values at least 20 times lower, and the earlier data may have been affected by windblown ore dust. Other reports of Cr >1000 mg/kg have proved to refer to an ash-weight basis, or are believed to have resulted from incorrect identification of Cr spectrographic emission lines. There is doubt, for example, about the figure of 20,000 mg/kg quoted for Cr in the ash of *Leptospermum scoparium* from serpentine soil at Dun Mountain, New Zealand (Lyon et al., 1968). More than 20 specimens of the same species collected by the present authors from serpentine at North Cape, NZ, and analyzed by inductively coupled plasma (ICP) emission spectroscopy, contained a normal range (for serpentine plants) of 1–30 mg/kg Cr on a dry-weight basis.

Unusually high Cr concentrations (1800–7500 mg/kg ash weight, or 180–720 mg/kg dry weight) have also been reported in the epiphytic moss *Aerobryopsis longissima* from ultramafic areas of New Caledonia (Lee et al., 1977a). The hosts of this moss include the Ni hyperaccumulating shrub *Homalium guillainii*, which contains much lower Cr concentrations. Soil contamination was minimized by careful washing, and the high rainfall, dense understory, and absence of strong winds and of direct ground contact all combine to make dust contamination unlikely. Further work on this moss does not appear to have been carried out.

Selenium

Selenium is an essential element for animal and human health, with a remarkably narrow range between the levels required to prevent deficiency diseases and those which produce symptoms of toxicity. The selenium content of soils is usually 0.01–2 mg/kg, but can greatly exceed this range, reaching concentrations of several hundred milligrams per kilogram in soils derived from certain Cretaceous shales, or where the soil has become Se-enriched through the disposal of materials such as fly-ash. Selenium-rich soils occur naturally in areas of the western part of the United States, and in Ireland, Queensland (Australia), Colombia, and Venezuela. In plant dry matter, selenium concentrations are generally below 1 mg/kg, and may even be below 0.01 mg/kg in areas of low-selenium soils.

From the early 1900s it was suspected that certain disease symptoms in horses and cattle in localized areas of the Great Plains and Rocky Mountain states of the United States were the result of selenium toxicosis. Extensive analytical work in the 1930s and 1940s (Beath et al., 1935, 1941; Byers, 1935, 1936; Miller and Byers, 1937) showed that several species of the very large leguminous genus *Astragalus* were capable of accumulating selenium to concentrations in excess of 1000 mg/kg from soils that are rich in this element. Such plants were described as "primary Se indicators" or Se "converter plants." Specimens of *A. bisulcatus* ("two-grooved loco weed" or "two-grooved milk vetch"), *A. pectinatus* and *A. racemosus*, for example, were found with selenium concentrations in the range 1000–8000 mg/kg, making these plants extremely toxic to stock. Further investigation of plants from these areas (extending into Canada and Mexico) has shown that more than 20 other species of *Astragalus* accumulate Se to some extent (30–3000 mg/kg), as do a few species from other families, such as the "alkali Prince's plume," *Stanleya pinnata* (Brassicaceae), and composites from the genera *Haplopappus* (Sect. *Oonopsis*) and *Machaeranthera* (Sect. *Xylorhiza*). Subsequent work has added other examples from the flora of seleniferous areas of Queensland, and the chemically remarkable nut of the Venezuelan "monkey nut" tree, *Lecythis ollaria*, has been found to contain as much as 18,200 mg/kg selenium, explaining its toxicity to humans and animals. Examples of highly seleniferous plants are given in Table 12-8.

These discoveries stimulated a considerable amount of chemical work in the 1960s and 1970s demonstrating the presence in these plants of very high concentrations of the selenium analogs of several well-known sulfur-containing amino acids, such as selenocystathionine and Se-methylselenocysteine. As in the case of Ni, the occurrence of Se accumulation in a variety of unrelated genera and families makes it likely that this property has evolved independently many times, rather than from a single Se-accumulating ancestor (Peterson, 1983).

An unusual feature of the Se levels in *Astragalus* species is the wide range of accumulation exhibited by the accumulator species, even at a single seleniferous location. Specimens of the most strongly accumulating species (Table 12-8) show an approximately lognormal distribution of Se levels, with medians of about 150–500 mg/kg, but with occasional values below 40 mg/kg and others in the 1000–5000-mg/kg range (Rosenfeld and Beath, 1964, p. 92). No clear explanation for this wide range seems to have been given, apart from genotypic variation and the rather wide variations in soil Se concentrations that occur on even a local scale. The high values are certainly genuine, and not the result of soil and dust contamination,

TABLE 12-8   Some Hyperaccumulators of Selenium

Family Species	Locality	Maximum Concentration (mg/kg)	References
Asteraceae			
Haplopappus[a] condensata	Midwest USA	9,120	1
H. fremontii	Midwest USA	4,800	5,13
Machaeranthera[b]			
glabriuscula	Midwest USA	1,800	13
M. parryi	Midwest USA	5,390	1
M. ramosa	Midwest USA	1,345	13
M. venusta	Utah, Colorado (USA)	3,486	13
Brassicaceae			
Stanleya pinnata	Midwest USA	1,190	5,13
S. bipinnata	Midwest USA	2,380	5,10,13
Chenopodiaceae			
Atriplex confertifolia	Midwest USA	1,734	13
Lecythidaceae			
Lecythis ollaria	Venezuela	18,200[c]	14
Leguminosae			
Acacia cana	Queensland (Australia)	1,121	12
Astragalus bisulcatus	Midwest USA, Canada	8,840	1,2,9,13
A. grayi	Wyoming	4,450	1
A. osterhoutii	Midwest USA	3,000[c]	13
A. pattersonii	Midwest USA	>2,560	13
A. pectinatus	Midwest USA, Canada	5,170	1,3,6,8,13
A. racemosus	Midwest USA, Canada	14,920	2,4,7,10,13
Neptunia amplexicaulis	Queensland (Australia)	4,334	12
Rubiaceae			
Morinda reticulata	Queensland (Australia)	1,141	11
Scrophulariaceae			
Castilleja chromosa	Midwest USA	1,812	13

[a]Sect. Oonopsis.
[b]Sect. Xylorhiza.
[c]Concentration in seed.

References: (1) Byers (1935), (2) Byers (1936), (3) Miller and Byers (1937), (4) Knight and Beath (1937), (5) Byers et al. (1938), (6) Williams (1938), (7) Moxon et al. (1939), (8) Williams et al. (1940), (9) Lakin and Byers (1948), (10) Moxon et al. (1950), (11) Knott et al. (1958), (12) McCray and Hurwood (1963) , (13) Rosenfeld and Beath (1964), (14) Aronow and Kerdel-Vegas (1965).

because the plant Se is often higher than the soil Se, and because it is possible to extract large concentrations of well-defined organoselenium compounds from the plants.

All the recorded instances of elevated Se levels in plants have apparently not been completely compiled to date. Since the normal levels are below 2 mg/kg, it might be reasonable to consider any plant with more than 100 mg/kg as a hyperaccumulator of this element, especially since levels above 10 mg/kg are considered capable of being toxic to animals. This would add many more examples to those listed in Table 12-8, including additional *Astragalus* species; several American species of *Aster, Atriplex, Grindelia*, and *Gutierrezia* (Rosenfeld and Beath, 1964); and species from other countries, such as the legume *Prosopis faracata* from Israel, which has been found with 135–311 mg/kg Se (Ravikovitch and Margolin, 1957).

HYPERACCUMULATORS AND PHYTOREMEDIATION

A major attraction of the use of hyperaccumulators in phytoremediation is the possibility of employing species that remove large amounts of a particular element from the soil without significant chemical intervention, other than the application of conventional fertilizers.

The key to the use of hyperaccumulators (or any other plant species) in phytoremediation lies in the rate of biomass production, coupled with the concentration of the element transferred to the plant matter. If the rate of dry biomass production is B kg/ha per year, and an average concentration of C mg/kg of the remediated element can be achieved in the plant dry matter, then its rate of removal from the soil is $10^{-6}BC$ kg/ha per year, or $10^{-7}BC$ g/m^2 per year. For a plant just meeting the hyperaccumulator criterion of 1000 mg/kg (e.g., for Cu, Co, Ni, Pb), the rate of removal is $10^{-4}B$ g/m^2 per year, and removal of 1 g/m^2 requires a dry matter production of 10,000 kg/ha (about 4 tons per acre). A crop of hyperaccumulator produced at 25,000 kg/ha, averaging 10,000 mg/kg of the accumulated element, could bring about an annual reduction of 125 mg/kg in the element concentration in the uppermost 0.2-m depth of the soil (assuming a bulk soil density of the order of 1000 kg/m^3). This estimate forms the basis of calculations designed to find how many successive crops would be required to bring about the desired reduction in soil concentration, assuming that the high concentration in the plant can be maintained as the soil concentration is progressively reduced.

The hyperaccumulators discussed in the previous section cover the full range of plant life forms. Many are herbs or small woody-based shrubs, although some (especially in the tropical floras) are large shrubs 1–3 m in height, and a few of the tropical accumulators are trees of 4–10 m (e.g., *Sebertia, Shorea, Myristica* species). Few hyperaccumulator species have been studied agronomically, and it is not known what yields of plant dry matter might be achieved under optimum conditions of climate, nutrition and plant density, particularly since in their natural habitats they are clearly growing under adverse nutritional conditions. Experimental work on remediation of soil with fairly low levels of contamination from industrial sewage sludge, using low-growing hyperaccumulators of Zn and Cd, such as *Thlaspi caerulescens* (Baker et al., 1994a), have shown that even at a soil Zn concentration of 444 mg/kg, the mature plants can contain Zn at 5000–7000 mg/kg, and that removal of 30 kg/ha is possible with a single crop. Similarly, Cd could be removed at 0.143 kg/ha from soil with 13.6 mg/kg Cd. Nickel-accumulating *Alyssum* species, producing about 23 tonne/ha dry matter, could remove 1.34 kg/ha of Ni from a low soil concentration of 35 mg/kg. Considerably higher rates of removal can be expected from more heavily contaminated soils, such as those encountered in mine and smelter wastes.

One important question to be addressed is the specificity of metal accumulation. The specificity of uptake shown by many hyperaccumulators may be both an advantage and a disadvantage: an advantage where removal of a specific contaminant is required, and a disadvantage if multielement removal is the goal. Most hyperaccumulators exhibit remarkable single-element specificity in their natural habitats. For example, in serpentine soils, the ratio (total Ni:total Co) is often about 10:1, but most Ni hyperaccumulators show great selectivity for Ni relative to Co, such that the plant ratio is usually in the range 100:1 to 5000:1. This is true of the *Alyssum* Ni hyperaccumulators, for example, in which pronounced Co uptake occurs only when this element is provided in solution form to a soil without high Ni.

Similarly, the Ni accumulators rarely show strong tendency to accumulate Zn, even from soils with elevated Zn concentrations. Ni hyperaccumulators from *Alyssum* were greatly inferior to *Thlaspi caerulescens* in their Zn uptake when grown on the same soil (Baker et al., 1994a). Apart from the tendency of some Zn accumulators to show elevated Cd levels, and of the African Co and Cu accumulators to take up either element (but seldom together to high concentrations), single-element hyperaccumulation seems to be the rule. Exceptions to this may be found in the genus *Thlaspi*, where several species (not the common annuals) can accumulate high levels of Ni when the soil is Ni-rich, and Zn when the soil is Zn-rich (Reeves and Brooks, 1983a). There is growing evidence from this work, and subsequent studies by the present authors, that several *Thlaspi* species also accumulate high Zn concentrations even when the soil Zn is within the normal range. Furthermore, the soil culture experiments of Reeves and Baker (1984) and the solution culture experiments of Baker et al. (1994b) show that species such as *Thlaspi goesingense* and *T. caerulescens* do have a multielement hyperaccumulation capability, the behavior depending on the precise composition of the soils or the nutrient solutions. It is not known whether this behavior extends to many other genera and species, although it has been observed (Baker et al., 1992) that various subspecies of the Ni hyperaccumulator *Dichapetalum gelonioides* from the Philippines, Sabah, Sumatra, and Malaysia, can take up Ni to >1% from serpentine soils, but on other (unspecified) substrata can be found with >1% Zn. This is an aspect of the behavior of this species that needs further detailed investigation.

One further possible use of hyperaccumulators, not widely considered under the heading of phytoremediation, may be mentioned here. There are some soils that are deficient in essential elements, such as Zn, Se, or Co, with deleterious effects on crops or livestock. This situation might be remedied in various ways by using plants with elevated metal-accumulating ability, rather than by direct application of the deficient element to the soil. Banuelos (1996) has discussed the transport of Se-enriched fodder crops from areas of high-Se soils to areas where the soils and crop plants are Se-deficient. Since some *Thlaspi* species accumulate Zn from normal soils, they are probably also able to extract more Zn than most plants from Zn-deficient soils, and crops of such a species, returned to decompose in the soil, may be useful in increasing Zn availability to other species such as pasture grasses and horticultural crops.

ENDEMISM AND CONSERVATION OF METAL-ACCUMULATING PLANTS

It has been noted (Reeves, 1992) that many plant species found on serpentine soils, whether hyperaccumulators or not, are rare species. They are often endemic to a few small areas of metalliferous soil; some are known from only a single site, and specimens of some have been collected on only a few occasions. The Ni hyperaccumulator *Alyssum pinifolium*, for example, occurring at a few sites in western Turkey, has been collected only twice in the nineteenth century and by two other collectors in the twentieth century, and several others in this genus are of comparable rarity. The Ni hyperaccumulator *Bornmuellera baldaccii* ssp. *markgrafii* (Reeves et al., 1983a) now appears to be accessible only in the form of a single herbarium specimen in Vienna. Several Cuban hyperaccumulators of Ni are known from only one or two localities or collected specimens. Vigorous efforts must be made to locate these species in their natural habitats again, and to provide them with proper protection. There are also serpentine endemic plants, such as the Cuban *Phyllacanthus grisebachianus*, collected only once or twice, and not subsequently rediscovered at the same site or anywhere else, which must now be presumed extinct (A. Borhidi, personal communication). Even widespread metal hyperaccumulators may be locally under threat; populations of the Zn-accumulating *Thlaspi caerulescens* are dense in some localities, but in danger of being exterminated in others. There can be sufficient variation in metal-accumulating properties among isolated populations of such a sporadically distributed species that preservation of all its populations should be regarded as vital.

Metal-tolerant species in general, and the hyperaccumulators in particular, represent a remarkable biological resource of great potential utility, and vigorous research effort is now needed to improve our field knowledge and distribution records of these plants in many parts of the world. This includes tracts of serpentine soils, especially in the tropics, and areas of wastes (e.g., from mining of Cu, Zn, and Pb) in many countries. This information will then be valuable in the formulation of strategies designed to conserve rare species and threatened populations.

Major threats include the following: (1) mining, (2) agricultural expansion, (3) fire, (4) urban development, (5) recreational development, and (6) the activities of overzealous plant collectors. Specific examples of the effects of these threats on serpentine plants and Ni hyperaccumulators have been given elsewhere (Kruckeberg, 1992; Reeves, 1992), but the same comments apply to many other species listed in the tables in this chapter, and are relevant to any metalliferous site where a characteristic flora has developed.

In some cases, especially on ultramafic soils, new mining activities (for nickel ores, chromite, asbestos, talc, magnesite, etc.) may threaten an existing metallophyte flora, for which a proper inventory may or may not exist. Full plant surveys, including chemical analysis, should be part of the preliminary investigations of such sites, and appropriate conservation steps taken before land is cleared. Where a metallophyte flora has developed over existing mine or smelter wastes, as often occurs in Zn/Pb or Cu mining areas, similar steps need to be taken if the area is to be reworked by physical or chemical processes, or even if phytoremediation is being considered.

Agricultural expansion into metalliferous areas often occurs, either incidentally or deliberately. Chemical modification of metalliferous soils can render them suitable for limited types of agriculture or horticulture, but this can also be at the expense of a characteristic adapted flora. Preliminary surveys are again necessary to ensure that nothing unique is being lost in the process.

Some of the rarest metallophytes, especially those known from only a single site covering areas from a few hectares up to a few square kilometers, are very vulnerable to the effects of brush fires or forest fires, which may occur naturally or may be deliberately set. A single incident could thus eliminate a species completely. The establishment of a seedbank for species open to this threat should be seen as a high conservation priority. Improved information about the total distribution of hyperaccumulator species would also help to assess the risk to the plant communities posed by continuing urban and associated development, including the use of mine waste areas and serpentine 'barrens' for various types of recreation.

Finally, some restrictions on access to sites of rare plants, even by members of the botanical community, may be necessary. Removal of the rarest species from the field for herbarium collections, or of their seed for commercial exploitation as horticultural items or for phytoremediation trials, all constitute threats to the continued existence of the natural resource unless these activities are strictly and responsibly controlled.

REFERENCES

Akeroyd, J. R., 1993. *Thlaspi*. In *Flora Europaea*, Vol. 1, 2nd ed., Cambridge Univ. Press, Cambridge, UK, pp. 384–388.

Aronow, L., and F. Kerdel-Vegas, 1965. Seleno-cystathionine, a pharmacologically active factor in the seeds of *Lecythis ollaria*. *Nature* **205**:1185–1186.

Baker, A. J. M., 1981. Accumulators and excluders—strategies in the response of plants to heavy metals. *J. Plant Nutr.* **3**:643–654.

Baker, A. J. M., and R. R. Brooks, 1989. Terrestrial higher plants which hyperaccumulate metallic elements—a review of their distribution, ecology and phytochemistry. *Biorecovery* **1**:81–126.

Baker, A. J. M. et al., 1985. Accumulation of nickel by *Psychotria* species from the Pacific Basin. *Taxon* **34**:89–95.

Baker, A. J. M. et al., 1992. Hyperaccumulation of nickel by the ultramafic flora of Palawan, Republic of the Philippines. In *The Vegetation of Ultramafic (Serpentine) Soils*, J. Proctor et al., eds., Intercept Ltd., Andover, UK, pp. 291–304.

Baker, A. J. M. et al., 1994a. The possibility of *in situ* heavy metal decontamination of soils using crops of metal-accumulating plants. *Resources, Conserv. Recycl.* **11**:41–49.

Baker, A. J. M. et al., 1994b. Heavy metal accumulation and tolerance in British populations of the metallophyte *Thlaspi caerulescens* J. & C. Presl (Brassicaceae). *New Phyt.* **127**:61–68.

Banuelos, G., 1996. The use of different plants to lower selenium concentrations in California soils. *Phytoremediation Conference Abstracts*, Washington, DC, May 8–10.

Barry, S. A. S., and S. C. Clark, 1978. Problems of interpreting the relationship between the amounts of lead and zinc in plants and soil on metalliferous wastes. *New Phyt.* **81**:773–783.

Batianoff, G. N. et al., 1990. *Stackhousia tryonii* Bailey: A nickel-accumulating serpentinite-endemic species of central Queensland. *Austral. J. Bot.* **38**:121–130.

Baumann, A., 1885. Das verhalten von zinksalzen gegen pflanzen und im boden. *Die Landwirtsch. Versuchs-Stat.* **31**:1–53.

Beath, O. A. et al., 1935. Selenium and other toxic minerals in soils and vegetation. *Wyoming Agric. Exper. Stn. Bull.* **206**:1–55.

Beath, O. A. et al., 1941. The use of indicator plants in locating seleniferous areas in western United States. IV. Progress Report. *Am. J. Bot.* **28**:887–900.

Beeson, K. C. et al., 1955. Some plant accumulators of the micronutrient elements. *Ecology* **36**:155–156.

Berazain Iturralde, R., 1981. Sobre el endemismo de la florula serpentinicola de Lomas de Galindo, Canasi, Habana. *Revista. Jardin Bot. Nacional* (Cuba) **2**:29–59.

Blissett, A. H., 1966. Copper tolerant plants from the Ukaparinga copper mine, Williamstown. *Quart. Geol. Notes. Geol. Survey S. Austral.* **18**:1–3.

Brooks, R. R., 1977. Copper and cobalt uptake by *Haumaniastrum* species. *Plant Soil* **48**:541–544.

Brooks, R. R., 1987. *Serpentine and Its Vegetation: A Multidisciplinary Approach.* Dioscorides Press, Portland, OR.

Brooks, R. R., and F. Malaisse, 1985. *The Heavy Metal-Tolerant Flora of Southcentral Africa.* Balkema, Rotterdam.

Brooks, R. R., and C. C. Radford, 1978. Nickel accumulation by European species of the genus *Alyssum. Proc. Roy. Soc. Lond.* **B200**:217–224.

Brooks, R. R., and E. D. Wither, 1977. Nickel accumulation by *Rinorea bengalensis* (Wall.) O.K. *J. Geochem. Explor.* **7**:295–300.

Brooks, R. R., and X. H. Yang, 1984. Elemental levels and relationships in the endemic serpentine flora of the Great Dyke, Zimbabwe, and their significance as controlling factors for this flora. *Taxon* **33**:392–399.

Brooks, R. R. et al., 1974. Some New Zealand and New Caledonian plant accumulators of nickel. *J. Ecol.* **62**:493–499.

Brooks R. R. et al., 1977a. Detection of nickeliferous rocks by analysis of herbarium specimens of indicator plants. *J. Geochem. Explor.* **7**:49–57.

Brooks, R. R. et al., 1977b. Cobalt and nickel in *Rinorea* species. *Plant Soil* **47**:707–712.

Brooks, R. R. et al., 1977c. Cobalt and nickel uptake by the Nyssaceae. *Taxon* **26**:197–201.

Brooks, R. R. et al., 1978. Copper and cobalt in African species of *Aeolanthus* Mart. (Plectranthinae, Labiatae). *Plant Soil* **50**:503–507.

Brooks, R. R. et al., 1979. Hyperaccumulation of nickel by *Alyssum* Linnaeus (Cruciferae). *Proc. Roy. Soc. Lond.* **B203**:387–403.

Brooks, R. R. et al., 1980. Hyperaccumulation of copper and cobalt—a review. *Bull. Soc. Roy. Bot. Belg.* **113**:166–172.

Brooks, R. R. et al., 1981a. Horsetails (*Equisetum*) as indicators of gold mineralization. *J. Geochem. Explor.* **16**:21–26.

Brooks, R. R. et al., 1981b. Studies on manganese-accumulating *Alyxia* from New Caledonia. *Taxon* **30**:420–423.

Brooks, R. R. et al., 1982. Phytogéochimie de l'anticlinal de Kasonta (Shaba, Zaïre). *Geo-Eco-Trop* **6**:219–228.

Brooks, R. R. et al., 1987. The elemental content of metallophytes from the copper/cobalt deposits of Central Africa. *Bull. Soc. Roy. Bot. Belg.* **119**:179–191.

Brooks, R. R. et al., 1990. The Brazilian serpentine plant expedition (BRASPEX), 1988. *Nat. Geog. Res.* **6**:205–219.

Brooks, R. R. et al., 1992. The serpentine vegetation of Goiás State, Brazil. In *The Vegetation of Ultramafic (Serpentine) Soils*, J. Proctor et al., eds., Intercept Ltd., Andover, UK, pp. 67–81.

Byers, H. G., 1935. Selenium occurrence in certain soils in the United States, with a discussion of related topics. *USDA Tech. Bull.* **482**:1–47.

Byers, H. G., 1936. Selenium occurrence in certain soils in the United States, with a discussion of related topics. II. *USDA Tech. Bull.* **530**:1–78.

Byers, H. G. et al., 1938. Selenium occurrence in certain soils in the United States, with a discussion of related topics. Third Report. *USDA Tech. Bull.* **601**:1–74.

Cole, M. M., 1973. Geobotanical and biogeochemical investigations in the sclerophyllous woodland and scrub associations of the eastern goldfields area of Western Australia, with particular reference to the role of *Hybanthus floribundus* (Lindl.) F. Muell. as nickel indicator and accumulator plant. *J. Appl. Ecol.* **10**:269–320.

Cole, M. M. et al., 1968. Geobotany, biogeochemistry and geochemistry in the Bulman–Waimuna Springs area, Northern Territory, Australia. *Trans. Inst. Mining Metallurgy Sect. B* **77**:81–104.

Denaeyer-De Smet, S., 1966. Note sur un accumulateur de manganèse, *Vaccinium myrtillus*. *Bull. Soc. Roy. Bot. Belg.* **99**:331–343.

Denaeyer-De Smet, S., 1970. Considérations sur l'accumulation du zinc par les plantes poussant sur sols calaminaires. *Bull. Inst. Roy. Sci. Nat. Belg.* **46**:1–13.

Denaeyer-De Smet, S., and P. Duvigneaud, 1974. Accumulation de métaux lourds dans divers écosystèmes terrestres pollués par les retombées d'origine industrielle. *Bull. Soc. Roy. Bot. Belg.* **107**:147–156.

De Plaen, G. et al., 1982. The copper flowers of Central Africa and their significance for archaeology and mineral prospecting. *Endeavour* **6**:72–77.

Deram, A., and D. Petit, 1997. Ecology of bioaccumulation in *Arrhenatherum elatius* L. (Poaceae) populations—applications of phytoremediation of zinc, lead and cadmium contaminated soils. *J. Exper. Bot.* **48**(spec. suppl.):98.

Doksopulo, E. P., 1961. *Nickel in Rocks, Soils, Water and Plants Adjacent to the Talc Deposits of the Chorchanskaya Group*. Izdatel vo Tbiliskovo Universitet, Tbilisi.

Dudley, T. R., 1986a. A new nickelophilous species of *Alyssum* (Cruciferae) from Portugal, *Alyssum pintodasilvae* T. R. Dudley, sp. nov.. *Feddes Repertor.* **97**:135–138.

Dudley, T. R., 1986b. A nickel hyperaccumulating species of *Alyssum* from Spain: *Alyssum malacitanum* (Rivas Goday) T. R. Dudley, comb. et stat. nov. *Feddes Repertor.* **97**:139–142.

Duvigneaud, P., 1959. Plantes cobaltophytes dans le Haut Katanga. *Bull. Soc. Roy. Bot. Belg.* **91**:111–134.

Duvigneaud, P., and S. Denaeyer-De Smet, 1963. Cuivre et végétation au Katanga. *Bull. Soc. Roy. Bot. Belg.* **96**:93–231.

Duvigneaud, P., and S. Denaeyer-De Smet, 1970. Phytogéochimie des groupes écosociologiques forestiers de Haute Belgique. 1. Essai de classification phytochimique des espèces herbacées. *Oecol. Plant.* **5**:1–32.

Duvigneaud, P., and S. Denaeyer-De Smet, 1973. Considérations sur l'écologie de la nutrition minérale des tapis végétaux naturels. *Oecol. Plant.* **8**:219–246.

Dykeman, W. R., and A. S. De Sousa, 1966. Natural mechanisms of copper tolerance in a copper swamp. *Can. J. Bot.* **44**:871–878.

Ernst, W. H. O., 1966. Ökologisch-soziologische Untersuchungen an Schwermetallpflanzengesellschaften Südfrankreichs und des östlichen Harzvorlandes. *Flora, Jena* **B156**:301–318.

Ernst, W. H. O., 1968a. Das Violetum calaminariae westfalicum, eine Schwermetallpflanzengesellschaft bei Blankenrode in Westfalen. *Mitteil. Floristisch. Arbeit.* **13**:263–268.

Ernst, W. H. O., 1968b. Zur Kenntnis der Soziologie und Ökologie der Schwermetallvegetation Grossbritanniens. *Ber. Deutsch. Bot. Gesell.* **81**:116–124.

Ernst, W. H. O., 1974. *Schwermetallvegetation der Erde*. Fischer, Stuttgart.

Forchhammer, J. G., 1855. Über den Einfluss des Kochsalzes auf die Bildung der Mineralien. *Poggendorf's Annal. Phys. Chem.* **91**:60–96.

Gabbrielli, R. et al., 1987. Ulteriori dati sulla composizione minerale della vegetazione degli affioramente ofiolitici dell'alta Valle d'Ayas. *Revue Valdôtaine Hist. Nat.* **41**:99–110.

Garcia-Gonzalez, A., and S. C. Clark, 1989. The distribution of *Minuartia verna* and *Thlaspi alpestre* in the British Isles in relation to 13 soil metals. *Vegetatio* **84**:87–98.

Gutermann, W., 1975. Notulae nomenclaturales 1-18. *Phyton* (Austria) **17**:31–50.

Homer, F. A. et al., 1991. Characterization of the nickel-rich extract from the nickel hyperaccumulator *Dichapetalum gelonioides*. *Phytochemistry* **30**:2141–2145.

Homer, F. A. et al., 1994. Comparative studies of nickel, cobalt and copper uptake by some nickel hyperaccumulators of the genus *Alyssum*. *Plant Soil* **138**:195–205.

Jaffré, T., 1977. Accumulation du manganèse par les espèces associées aux terrains ultrabasiques de Nouvelle Calédonie. *Comptes Rend. Acad. Sci.* (Paris), *Séries D* **284**:1573–1575.

Jaffré, T., 1979. Accumulation du manganèse par les Proteacées de Nouvelle Calédonie. *Comptes Rend. Acad. Sci.* (Paris), *Séries D* **289**:425–428.

Jaffré, T., 1980. *Etude Écologique du Peuplement Végétal des Sols Dérivés de Roches Ultrabasiques en Nouvelle Calédonie*. Vol. 124, Travaux et Documents de l'ORSTOM, (Paris).

Jaffré, T., and M. Schmid, 1974. Accumulation du nickel par une Rubiacée de Nouvelle Calédonie, *Psychotria douarrei* (G. Beauvisage) Däniker. *Comptes Rend. Acad. Sci. (Paris), Séries D* **278**:1727–1730.

Jaffré, T. et al., 1976. *Sebertia acuminata*: A hyperaccumulator of nickel from New Caledonia. *Science* **193**:579–580.

Jaffré, T. et al., 1979a. Hyperaccumulation of nickel by *Geissois* species. *Plant Soil* **51**:157–162.

Jaffré, T. et al., 1979b. Nickel uptake by the Flacourtiaceae of New Caledonia. *Proc. Roy. Soc. Lond.* **B205**:385–394.

Johnston, W. R. and J. Proctor, 1977. A comparative study of metal levels in plants from two contrasting lead-mine sites. *Plant Soil* **46**:251–257.

Kelepertsis, A. E., and A. Bibou, 1991. Heavy metal contamination of soils at old mining sites on Thasos Island, Greece. *Environ. Geochem. Health* **13**:23–28.

Kelepertsis, A. E. et al., 1990. The use of the genus *Alyssum* as a reliable geobotanical-biogeochemical indicator in geological mapping of ultrabasic rocks in Greece. *Praktika tis Akademias Athinon* **65**:170–176.

Kersten, W. J. et al., 1979. Nickel uptake by New Caledonian species of *Phyllanthus*. *Taxon* **28**:529–534.

Knight S. H., and O. A. Beath, 1937. The occurrence of selenium and seleniferous vegetation in Wyoming. *Wyoming Agric. Exper. Stn. Bull.* **221**.

Knott, S. G. et al., 1958. Selenium poisoning in horses in North Queensland. *Queensland Dept. Agric., Div. Animal Ind., Bull.* **41**:1–16.

Konstantinou, M., and D. Babalonas, 1992. Geobotanische Untersuchung auf Galmei-Erzhalden auf der Insel Thassos (Griechenland). *J. Agron. Crop Sci.* **168**:1–9.

Kruckeberg, A. R., 1992. Serpentine biota of western North America. In *The Vegetation of Ultramafic (Serpentine) Soils*, J. Proctor et al., eds., Intercept Ltd., Andover, UK, pp. 19–33.

Kruckeberg, A. R., and R.D. Reeves, 1995. Nickel accumulation by serpentine species of *Streptanthus* (Brassicaceae): Field and greenhouse studies. *Madroño* **42**:458–469.

Kruckeberg, A. R. et al., 1993. Hyperaccumulation of nickel by *Arenaria rubella* (Wahlenb.) J.E. Smith (Caryophyllaceae) from Washington, USA. *Madroño* **40**:25–30.

Kubota, J. et al., 1960. The study of cobalt status of soils in Arkansas and Louisiana using the black gum as the indicator plant. *Soil Sci. Proc.* **24**:527–528.

Lakin, H. W., and H. G. Byers, 1948. Selenium occurrence in certain soils in the United States, with a discussion of related topics. 7th Report. *USDA Tech. Bull.* **950**:1–36.

Lee, J. et al., 1977a. Chromium-accumulating bryophyte from New Caledonia. *Bryologist* **80**: 203–205.

Lee, J. et al., 1977b. Plant-soil relationships in a New Caledonian serpentine flora. *Plant Soil* **46**:675–680.

Linstow, O. von, 1924. Die natürliche Anreicherung der Metallsalzen und anderen anorganischen Verbindungen in den Pflanzen. *Feddes Repertor.* **31**:1–151.

Lisanti, E. L., 1952. Contributo allo studio delle morfosi che si riscontrano sui serpentini. *Nuovo Giorn. Bot. Ital.* **14**:349–360.

Lyon, G. L. et al., 1968. Trace elements in a New Zealand serpentine flora. *Plant Soil* **29**:225–240.

Malaisse, F., and J. Grégoire, 1978. Contribution á la phytogéochimie de la Mine de l'Étoile (Shaba, Zaïre). *Bull. Soc. Roy. Bot. Belg.* **111**:252–260.

Malaisse, F. et al., 1978. *Aeolanthus biformifolius*: A hyperaccumulator of copper from Zaïre. *Science* **199**:887–888.

Malaisse, F. et al., 1979. Copper and cobalt in vegetation of Fungurume, Shaba Province, Zaïre. *Oikos* **33**:472–478.

Malaisse, F. et al., 1994. Diversity of vegetation communities in relation to soil heavy metal content at the Shinkolobwe copper/cobalt/uranium mineralization, Upper Shaba, Zaïre. *Belg. J. Bot.* **127**:3–16.

Maquinay, A., and J. L. Ramaut, 1960. La teneur en zinc des plantes du *Violetum calaminariae*. *Nat. Belg.* **41**:265–273.

Menezes de Sequeira, E., 1969. Toxicity and movement of heavy metals in serpentinitic rocks (North-Eastern Portugal). *Agron. Lusit.* **30**:115–154.

McCray, C. W. R., and I. S. Hurwood, 1963. Selenosis in northwestern Queensland associated with a marine Cretaceous formation. *Queensland J. Agric. Sci.* **20**:475–498.

Miller, J. T., and H. G. Byers, 1937. Selenium in plants in relation to its occurrence in soils. *J. Agric. Res.* **55**:59–68.

Minguzzi, C., and O. Vergnano, 1948. Il contenuto di nichel nelle ceneri di *Alyssum bertolonii* Desv.. *Atti Soc. Toscana Sci. Nat., Mem. Serie A* **55**:49–77.

Morrey, D. R. et al., 1989. Studies on serpentine flora: Preliminary analyses of soils and vegetation associated with serpentine rock formations in the south-eastern Transvaal. *S. Afr. J. Bot.* **55**:171–177.

Morrison, R. S., 1980. *Aspects of the Accumulation of Cobalt, Copper and Nickel by Plants.* Ph.D. thesis, Massey Univ., New Zealand.

Moxon, A. L. et al., 1939. Selenium in rocks, soils and plants. *S. Dakota Agric. Exper. Stn., Tech. Bull.* **2**.

Moxon, A. L. et al., 1950. Selenium in rocks, soils and plants. *S. Dakota Agric. Exper. Stn., Rev. Tech. Bull.* **2**:1–94.

Paton, A., and R. R. Brooks, 1995. A re-evaluation of *Haumaniastrum* species as geobotanical indicators of copper and cobalt. *J. Geochem. Explor.* **56**:37–45.

Peterson, P. J., 1983. Adaptation to toxic metals. In *Metals and Micronutrients: Uptake and Utilization by Plants,* D. A. Robb and W. S. Pierpoint, eds., Academic Press, London.

Proctor J., and M. Cole, 1992. The ecology of ultramafic areas in Zimbabwe. In *The Ecology of Areas with Serpentinized Rocks—a World View,* B. A. Roberts and J. Proctor, eds., Kluwer Academic Publishers, Dordrecht, pp. 313–331.

Proctor, J. et al., 1989. Ecological studies on Gunung Silam, a small ultrabasic mountain in Sabah, Malaysia. II. Some forest processes. *J. Ecol.* **77**:317–331.

Proctor, J. et al., 1994. A preliminary re-investigation of a plant geographical "El Dorado." *Trop. Biodivers.* **2**:303–316.

Rascio, N., 1977. Metal accumulation by some plants growing on zinc-mine deposits. *Oikos* **29**:250–253.

Ravikovitch, S., and M. Margolin, 1957. Selenium in soils and plants. *Agric. Res. Stn., Rehovot* **7**:41–52.

Reeves, R. D., 1988. Nickel and zinc accumulation by species of *Thlaspi* L., *Cochlearia* L., and other genera of the Brassicaceae. *Taxon* **37**:309–318.

Reeves, R. D., 1992. Hyperaccumulation of nickel by serpentine plants. In *The Vegetation of Ultramafic (Serpentine) Soils*, J. Proctor et al., eds., Intercept Ltd., Andover, UK, pp. 253–277.

Reeves, R. D., and A. J. M. Baker, 1984. Studies on metal uptake by plants from serpentine and non-serpentine populations of *Thlaspi goesingense* Hálacsy (Cruciferae). *New Phytol.* **98**:191–204.

Reeves, R. D., and R. R. Brooks, 1983a. European species of *Thlaspi* L. (Cruciferae) as indicators of nickel and zinc. *J. Geochem. Explor.* **18**:275–283.

Reeves, R. D., and R. R. Brooks, 1983b. Hyperaccumulation of lead and zinc by two metallophytes from a mining area in Central Europe. *Environ. Pollut.* **31**:277–287.

Reeves, R. D. et al., 1980. Nickel accumulation by species of *Peltaria* Jacq. (Cruciferae). *Taxon* **29**:629–633.

Reeves, R. D. et al., 1981. Nickel uptake by Californian *Streptanthus* and *Caulanthus* with particular reference to the hyperaccumulator *S. polygaloides* Gray (Brassicaceae). *Am. J. Bot.* **68**:708–712.

Reeves, R. D. et al., 1983a. Uptake of nickel by species of *Alyssum*, *Bornmuellera* and other genera of Old World Tribus Alysseae. *Taxon* **32**:184–192.

Reeves, R. D. et al., 1983b. Accumulation of nickel by western North American genera containing serpentine-tolerant species. *Am. J. Bot.* **70**:1297–1303.

Reeves, R. D. et al., 1996. Nickel-accumulating plants from the ancient serpentine soils of Cuba. *New Phytol.* **133**:217–224.

Reeves, R. D. et al., 1999. Nickel hyperaccumulation in the serpentine flora of Cuba. *Annals of Bot.* **83**:29–38.

Roberts, B. A., 1992. The ecology of serpentine areas, Newfoundland, Canada. In *The Ecology of Areas with Serpentinized Rocks—a World View*, B. A. Roberts and J. Proctor, eds., Kluwer Academic Publishers, Dordrecht, pp. 75–113.

Rosenfeld, I., and O. A. Beath, 1964. *Selenium—Geobotany, Biochemistry, Toxicity and Nutrition*, Academic Press, New York.

Sachs, J., 1865. *Handbuch der Experimental-Physiologie der Pflanzen. Handbuch der Physiologischen Botanik*, W. Hofmeister ed., Vol. IV, Engelmann, Leipzig. pp. 153–154.

Schmid, M., 1991. *Phyllanthus*. In *Flore de la Nouvelle Calédonie et Dépendances*. Vol. 17, *Euphorbiacées*. Museum National d'Histoire Naturelle, Paris, pp. 39–40.

Severne, B. C., and R. R. Brooks, 1972. A nickel accumulating plant from Western Australia. *Planta* **103**:91–94.

Shen, Z. G. et al., 1997. Uptake and transport of zinc in the hyperaccumulator *Thlaspi caerulescens* and the non-hyperaccumulator *Thlaspi ochroleucum*. *Plant, Cell Environ.* **20**:898–906.

Shimwell, D. W., and A. E. Laurie, 1972. Lead and zinc contamination of vegetation in the southern Pennines. *Environ. Pollut.* **3**:291–301.

Vergnano Gambi, O., and R. Gabbrielli, 1979. Ecophysiological and geochemical aspects of nickel, chromium and cobalt accumulation in the vegetation of some Italian ophiolitic outcrops. *Ofioliti* **4**:199–208.

Vergnano Gambi, O., and R. Gabbrielli, 1981. La composizione minerale della vegetazione degli affioramenti ofiolitici dell'alta Valle d'Ayas. *Revue Valdôtaine Hist. Nat.* **35**:51–61.

Vergnano Gambi, O. et al., 1979. L'accumulo di nichel nelle specie italiane del genere *Alyssum. Webbia* **33**:269–277.

Vergnano Gambi, O. et al., 1982. Nickel, chromium and cobalt in plants from Italian serpentine areas. *Acta Oecol., Oecol. Plant.* **3**:291–306.

Wild, H., 1970. The vegetation of nickel-bearing soils. *Kirkia* **7**(suppl.):1–62.

Wild, H., 1974. Indigenous plants and chromium in Rhodesia. *Kirkia* **9**:233–241.

Williams, K. T., 1938. Selenium and soils. In *Soils and Man. USDA Yearbook 1938.* pp. 831–834.

Williams, K. T. et al., 1940. Selenium occurrence in certain soils in the United States, with a discussion of related topics. Fourth Report. *USDA Tech. Bull.* **702**:1–59.

Williams, S. T. et al., 1977. The decomposition of vegetation growing on metal mine waste. *Soil Biol. Biochem.* **9**:271–275.

Wither, E. D., and R. R. Brooks, 1977. Hyperaccumulation of nickel by some plants of South-East Asia. *J. Geochem. Explor.* **8**:579–583.

13

MECHANISMS OF METAL HYPERACCUMULATION IN PLANTS

DAVID E. SALT AND UTE KRÄMER

INTRODUCTION TO METAL HYPERACCUMULATION

Discovery

The recognition that plants may contain high concentrations of certain metals, including zinc and nickel, dates back to the origins of biogeochemical prospecting. As early as 1885, A. Baumann, a German botanist working near the border of Germany and Belgium, observed that leaves of certain plant species growing on soils naturally enriched in zinc contained extraordinarily high levels of this element (Baumann, 1885). Of particular note were the violet *Viola calaminaria* and the mustard *Thlaspi calaminare*, more recently classified as *Thlaspi caerulescens* (Ingrouille and Smirnoff, 1986), which contained about 1% and 1.7% zinc in dry leaf biomass, respectively. This is about two orders of magnitude higher than zinc concentrations in dried leaves of plants growing on nonmetalliferous soils.

Fifty years later, studies in the United States implicated selenium as the plant component responsible for "alkali disease" in range animals in South Dakota. This observation led to the discovery of plants, notably of the genus *Astragalus*, capable of accumulating up to 0.6% selenium in dry shoot biomass (Byers, 1935, 1936). Shortly thereafter, two Italian botanists (Minguzzi and Vergnano, 1948) discovered plants which accumulate nickel. They observed that dried leaves of *Alyssum bertolonii*

Phytoremediation of Toxic Metals: Using Plants to Clean Up the Environment, edited by Ilya Raskin and Burt D. Ensley.
ISBN: 0-471-19254-6 ©2000 John Wiley & Sons, Inc.

growing on nickel enriched ultramafic soils near Florence, Italy, contained about 1% nickel, over 100–1000 times more than other plants growing nearby.

Since these early observations, plants that accumulate elevated levels of cobalt, and possibly copper, manganese, chromium, iron, and lead have also been described (Baker and Brooks, 1989). At the time of writing, at least 45 plant families are known to contain metal accumulating species, and 397 metal accumulating taxa have been identified to date (Baker et al., 1999). However, the existence of plants that accumulate elevated levels of metals other than nickel, zinc, cobalt, and selenium has been continuously questioned and requires further substantiation.

Definition

In a landmark paper in the study of metal accumulation in plants, Brooks et al. (1977) determined the nickel concentrations in over 2000 herbarium specimens. On the basis of this information, Brooks et al. (1977) first used the term *hyperaccumulator* to describe plants that contain >1000 $\mu g/g$ (0.1%) nickel in their dried leaves, a concentration at least an order of magnitude higher than nickel levels in nonaccumulator species. Threshold values for zinc and manganese hyperaccumulation have now been set at 10,000 $\mu g/g$ (>1%), and for copper and cobalt at 1000 $\mu g/g$ (0.1%) (Baker and Brooks, 1989). These values are also one order of magnitude greater than those occurring in nonaccumulator species.

On the basis of the bimodal distribution of shoot nickel concentrations (ranging from 1 to 300 $\mu g/g$ and 3000 to 30,000 $\mu g/g$) observed in 167 different *Alyssum* species (Brooks et al., 1979), it was suggested that a threshold of 1000 $\mu g/g$ represents a true division between nickel hyperaccumulator and nonaccumulator species. However, this distinction has been called into question as more data on nickel accumulating plants have been collected. It is now clear that shoot nickel concentrations in plants from a number of different ultramafic sites show a more continuous distribution (Reeves, 1992). However, the threshold values defining hyperaccumulation have proved useful by allowing the rapid classification of plants as hyperaccumulators, based on the analysis of metal concentrations in dried shoot material from herbarium and field collected samples.

Nickel, zinc, and cobalt hyperaccumulators are known to accumulate these metals preferentially in the shoot, with lower concentrations in the roots (Baker, 1981; Gabbrielli et al., 1990; Homer et al., 1991a; Baker et al., 1994; Brown et al., 1995; Lloyd-Thomas, 1995; Krämer et al., 1996, 1997b; Lasat et al., 1996; Köhl et al., 1997). This observation was made in field-collected as well as in hydroponically grown plants, irrespective of whether extracellularly bound metal was removed by desorption (Lasat et al., 1996). The shoot : root ratios of metal concentrations in all confirmed hyperaccumulators are therefore generally above one, whereas they are invariably below unity in nonaccumulators. On this basis it has been proposed that metal hyperaccumulator plants can be categorized based on a leaf : root metal concentration ratio of > 1 (Baker, 1981). This definition has a physiological basis and may be more useful in studies designed to investigate the biological mechanisms underlying metal hyperaccumulation.

Ecological Role

The ecological role of metal hyperaccumulation is still not entirely clear. It has been suggested that metal accumulation provides protection against fungal and insect attack (Boyd and Martens, 1992; Reeves et al., 1981). Recent evidence has confirmed the protective function of nickel hyperaccumulation against fungal and bacterial pathogens in *Streptanthus polygaloides* (Boyd et al., 1994), and insect herbivory in *S. polygaloides* and *T. montanum* (Boyd and Martens, 1994; Martens and Boyd, 1994). The antiherbivory effect of zinc has been also demonstrated in the zinc hyperaccumulator *T. caerulescens* (Pollard and Baker, 1997).

BIOLOGICAL MECHANISMS OF METAL HYPERACCUMULATION

A prerequisite for metal hyperaccumulation must be the ability to efficiently tolerate high concentrations of metals within plant tissues and cells. However, it is also possible that metal-hyperaccumulating plants have the ability to solubilize metals from the soil matrix, efficiently absorb metals into the root and translocate them to the shoot. Here we will review the underlying biological mechanisms involved in these processes and discuss their involvement in metal hyperaccumulation.

Rhizospheric Interactions

Physiological data on metal accumulation by hyperaccumulators suggests that hyperaccumulators may enhance metal solubility in the rhizosphere by releasing metal-chelating compounds from the roots. However, there is no direct experimental evidence as yet to confirm this.

Hyperaccumulator species are able to accumulate higher metal concentrations in their shoots than surrounding nonaccumulator plants even from soils containing nonphytotoxic background levels of metals (Baker et al., 1991; Brown et al., 1994, McGrath et al., 1997). One possible mechanism to explain this enhanced metal accumulation could be an enhanced ability to solubilize metals within the rhizosphere of the hyperaccumulator. This is supported by evidence suggesting that the zinc hyperaccumulator *T. caerulescens* has an enhanced ability to extract zinc from the immobile fraction of the soil, although further studies are needed to confirm this (McGrath et al., 1997). In soil, cocultivation of the zinc hyperaccumulator *T. caerulescens* with the nonaccumulator *T. arvense* increased shoot zinc accumulation in the nonaccumulator (Whiting et al., 1997), again suggesting that the hyperaccumulator is able to modify the rhizosphere to enhance metal solubility. However, this effect was not observed for nickel when the nickel hyperaccumulator

T. goesingense was cocultured hydroponically with the nonaccumulator *T. arvense* (Salt et al., 1999a).

The release of specific metal-chelating compounds into the rhizosphere by plant roots is a well-established phenomenon involved in the solubilization and uptake of iron (Ma and Nomoto, 1996) and possibly zinc (Cakmak et al., 1996a, 1996b). However, unequivocal evidence for the involvement of such a mechanism in metal hyperaccumulation remains to be established (Salt et al., 1999a).

Modification of the rhizosphere pH or redox potential by plant roots have also been reported to contribute to the mobilization of plant nutrients in some species (Marschner, 1995). However, comparison of both hyperaccumulator and nonaccumulator species has shown that changes in rhizosphere pH and redox potential are not associated with either nickel (Bernal et al., 1994; Bernal and McGrath, 1994) or zinc (McGrath et al., 1997) hyperaccumulation.

Root Uptake

Hyperaccumulation does not appear to be driven by the enhanced affinity of root uptake systems for the hyperaccumulated metal. However, increased rates of root uptake may be involved.

The dramatic metal accumulation observed in hyperaccumulator species may be driven by an enhanced root metal uptake system. Roots of the zinc hyperaccumulator *T. caerulescens* appear to contain more zinc transporters per gram fresh weight than the nonaccumulator *T. arvense* (Lasat et al., 1996), and this could reflect either a fundamental difference in the expression level of the root zinc transporter in the hyperaccumulator or simply be a consequence of a finer root system. However, the root zinc uptake systems in both the zinc hyperaccumulator and nonaccumulator have equal affinities for zinc (Lasat et al., 1996), suggesting that the root zinc transporter(s) in the hyperaccumulator are not functionally different from the nonaccumulator.

Not only are hyperaccumulating plants able to accumulate high concentrations of metals in shoot tissues; they also show a remarkable degree of selectivity, accumulating only specific metals from the substrate. For example, the nickel-hyperaccumulator *Alyssum bertolonii* preferentially accumulates nickel in shoots over cobalt and zinc (Gabbrielli et al., 1991). In a similar manner, the zinc hyperaccumulator *Thlaspi caerulescens* can hyperaccumulate zinc, manganese, cobalt, nickel, cadmium, and molybdenum from hydroponic culture but is unable to accumulate silver, chromium, copper, aluminum, iron, or lead (Baker et al., 1994). It has been suggested that the mechanism of metal selectivity could be associated with metal transport across the root plasma membrane during either metal uptake into the symplast or metal export into the xylem (Still and Williams, 1980).

The equal accumulation of nickel, cobalt, and zinc, and their competition for uptake, in excised roots of the nickel hyperaccumulator *A. bertolonii*, suggests that metal uptake into roots of hyperaccumulators is not very selective (Gabbrielli et al., 1991). Metal selectivity in hyperaccumulator species may therefore take place at the level of xylem loading.

As more information becomes available on the molecular and biochemical mechanisms of metal ion transport across the root plasma membrane, the role of this transport in hyperaccumulation should start to become clearer.

Root-to-Shoot Metal Translocation

The limited evidence available to date suggests that rates of metal translocation from root to shoot are no higher in hyperaccumulators than in related nonaccumulator species. Instead, hyperaccumulators may lack the ability to restrict metal movement into the shoot that is present in most other plants found on metaliferous soils. However, further work is needed to carefully determine metal translocation rates in additional hyperaccumulator species.

The characteristic observation that shoot : root ratios of metal concentrations are above unity in hyperaccumulators of nickel, zinc, or cobalt has been interpreted to suggest an efficient root-to-shoot translocation system for the hyperaccumulated metals (Baker et al., 1994; Lloyd-Thomas, 1995; Krämer et al., 1996; Shen et al., 1997). The translocation of metal cations from roots to shoots is governed by two main processes: movement into the xylem and volume flux through the xylem. The latter is mediated by root pressure and transpiration. Most likely, some degree of cycling occurs of metal cations from the shoots back to the roots via the phloem. To enter the xylem, solutes have to be taken up into root cells for passage through the root endodermis. There is some evidence for a second, wholly apoplastic pathway for the entry of water and possibly cations into the xylem in certain regions of the root (Marschner, 1995, p. 67). In general, however, plants are likely to have tight control over the solutes entering the shoot via the xylem through solute release and absorption by xylem parenchyma cells.

The process of xylem loading is not well understood, but researchers agree that it is a process separate from ion uptake into the root cells. Evidence has been found that xylem loading is energized by a negative membrane potential generated in xylem parenchyma cells through the operation of proton pumping ATPases (Pitman, 1972; De Boer et al., 1983; Clarkson and Hanson, 1986). It is therefore possible that xylem loading of cations could operate through cation–proton antiport, cation–ATPases, or ion channels (Wegner and Raschke, 1994; Roberts and Tester, 1995, 1997).

In the nickel hyperaccumulator *A. lesbiacum*, xylem transport of nickel occurs at a high rate, since millimolar nickel concentrations were detected in xylem sap of plants grown hydroponically in micromolar concentrations of nickel (Krämer, 1996). Elevated rates of zinc loading into the xylem have been suggested for the hyperaccumulator *T. caerulescens*, from the observation that shoot accumulation of zinc occurs at a higher rate in the zinc hyperaccumulator *T. caerulescens* than in the nonaccumulator *T. arvense* within several days of exposure to zinc (Lasat et al., 1996). Also, xylem sap concentrations of zinc were found to be between 7–10 fold higher in *T. caerulescens* compared to *T. arvense* (Lasat et al., 1998). It has also been proposed that metal selectivity in hyperaccumulators operates at the level of xylem loading (Gabbrielli et al., 1991). Excised roots of the nickel hyperaccumulator *Alyssum bertolonii* accumulate equal concentrations of either nickel, cobalt, or zinc, whereas shoot accumulation in whole seedlings was highest for nickel, lower for cobalt, and lowest for zinc.

Xylem concentrations of the free amino acid histidine and nickel show a linear correlation in several nickel hyperaccumulators in the genus *Alyssum* (Krämer et al., 1996), suggesting that histidine may facilitate xylem loading of nickel, possibly by forming a nickel–histidine complex. Since histidine favors the binding of nickel over other divalent metal cations such as zinc and cobalt, nickel chelation by histidine could also explain the observed selectivity for nickel translocation to the shoot in *Alyssum* hyperaccumulators. A role for histidine in root-to-shoot translocation of nickel is supported by the observation that the addition of L-histidine to excised roots of *A. montanum*, a plant species that does not hyperaccumulate nickel, increases nickel flux into the xylem substantially to rates as high as in the hyperaccumulator *A. lesbiacum* (Krämer et al., 1996). However, since a major effect of histidine is the reduction of nickel toxicity, probably through chelation of the toxic metal, the data may simply reflect the protective effect of histidine on normal xylem loading processes in a nontolerant nonaccumulator.

The mechanistic involvement of an amino acid in the xylem loading of a metal is still highly speculative. Complexation of nickel by histidine in the xylem sap may influence the chemical equilibrium to favor nickel transport into the xylem. Histidine could also have a direct role in membrane transport. There is evidence for the transport of a metal–histidine complex through the plasma membrane during zinc influx into human and rat erythrocytes (Horn et al., 1995).

Recent hydroponic experiments have shown that rates of root-to-shoot nickel translocation (normalized to root biomass) are the same in the hyperaccumulator *T. goesingense* and in the nonaccumulator *T. arvense*, as long as *T. arvense* is unaffected by nickel toxicity (Krämer et al., 1997b). Although directly comparable data is not available for *Alyssum*, in a generalized view, it appears that metal tolerance is the key component in metal hyperaccumulation, and that histidine is therefore primarily involved in affording metal tolerance to hyperaccumulating *Alyssum* species. On metaliferous soils, however, growing side by side with metal hyperaccumulator species, many species can be found which are metal-tolerant and yet do not hyperaccumulate (Brooks, 1987; Reeves, 1992). Consequently, a different metal tolerance strategy, the restriction of root-to-shoot transport of metals, must be proposed for metal-tolerant nonaccumulators. Indeed, this has been done by Rascio (1977), who concluded from field studies on a calamine (zinc-rich) soil, that, in contrast to the zinc hyperaccumulator *Thlaspi cepeaefolium*, various nonaccumulator species immobilized zinc in their roots during the main growing–flowering season.

In future experiments investigating metal translocation, total biomass of all the considered plant tissues should be documented so that root : shoot biomass ratios specific to certain species or stages in plant development can be taken into account and a mass balance of the movement of metals can be established. Furthermore, it is absolutely vital that in a valid experiment aimed at comparing rates of metal translocation from roots to shoots, there should be no signs of metal toxicity in either hyperaccumulator or nonaccumulator species under the experimental conditions. In many publications the data clearly show that the nonaccumulator species was suffering from metal toxicity, which is known to interfere with metal translocation.

Metal Sequestration and Complexation

Hyperaccumulators tend to accumulate highest metal concentrations in epidermal and subepidermal tissues, including leaf trichomes. Cellular locations for metal sequestration are presumably the apoplast and vacuoles. Recent evidence suggests that high-affinity chelation of metals is vital for the tolerance of high tissue metal concentrations in hyperaccumulators.

Tolerance of plant enzymes to metals alone cannot be sufficient to explain tolerance to tissue metal concentrations as high as those withstood by hyperaccumulators. Indeed, evidence has been found for sequestration of metals in the leaves of hyperaccumulators, both at the cellular and tissue level.

Multiple extraction procedures, so far most commonly used to investigate subcellular localization of nickel in hyperaccumulators, have indicated that most of the metal is readily water-soluble. It was concluded that nickel is predominantly localized in the vacuoles (Pelosi et al., 1974; Kelly et al., 1975; Lee et al., 1978; Morrison et al., 1981; Brooks et al., 1981; Homer, 1991; Homer et al., 1991b). Water extractability of nickel is lowest in the leaves of *Hybanthus floribundus* (Farago et al., 1975; Kelly et al., 1975; Farago and Mahmoud, 1983), for which 43% of the nickel was found to be associated with cell wall pectates. During the tissue extraction process, metals, potential chelators, and metal binding sites were mixed from different leaf tissues, cell, and cellular compartments. The extractability of the metal may represent an artifact of the homogenization procedure. Therefore extractability per se allows only tentative conclusions about subcellular localization.

Less invasive techniques have been applied to localize metals in hyperaccumulator leaves. On the basis of histochemical staining and microscopy as well as analysis by electron microprobe, Severne (1974) concluded that nickel was localized predominantly in the epidermal and subepidermal leaf tissues of *Hybanthus floribundus*. Using energy dispersive X-ray microanalysis (EDAX), Vázquez et al. (1992, 1994) reported high zinc concentrations in the vacuoles of epidermal and subepidermal leaf cells of *T. caerulescens* and some zinc in the leaf apoplast. This observation was recently confirmed using single-cell sap extraction from *T. caerulescens* leaf cells (Küpper et al., 1999). Also, using EDAX, Heath et al. (1997) observed that Ni in leaves of *T. montanum* var. *siskiyouense* was mainly localized in the subsidiary cells that surround leaf guard cells. It remains to be demonstrated that freeze substitution, which is necessary to prepare specimen for subcellular metal localization in EDAX, does not cause the movement of metals within the specimens, which could result in localization artifacts.

By directly isolating intact protoplasts and vacuoles, Krämer and Salt (unpublished) have recently found that 75% of the intracellular Ni in leaves of the Ni hyperaccumulator *T. goesingense* is localized within the vacuole. In the same studies it was observed that *T. goesingense* was able to compartmentalize Ni into the vacuole 2-fold more efficiently than the nonaccumulators *T. arvense*. The efficient compartmentalization of metals in the shoots of hyperaccumulators appears to be contrasted by reduced vacuolar compartmentalization in the roots. Based on efflux analysis it has recently been suggested that roots of the nonaccumulator *T. arvense* more efficiently accumulate Zn in the vacuole than the hyperaccumulator *T. caerulescens* (Lasat et al., 1998). So far, localization of metals at the tissue level of gently freeze-dried specimen, using a scanning proton microprobe (SPM), may have produced the most reliable results (Mesjasz-Przybylowicz et al., 1994, Krämer et al., 1997a). Scanning proton microscopy has revealed that leaf nickel is localized in epidermal and subepidermal tissues in the nickel hyperaccumulators *Senecio coronatus* (Mesjasz-Przybylowicz et al., 1994) and *Alyssum lesbiacum* (Krämer et al., 1997a). The highest nickel concentrations were found in the unicellular stellate trichomes covering the leaf surface in *A. Iesbiacum*. Interestingly, several nonaccumulator plant species have also been observed to accumulate cadmium (Salt et al., 1995), copper, nickel, and zinc (Neumann et al., 1995), lead (Martell, 1974), and manganese (Blamey et al., 1986) in leaf trichomes.

The complexation of metal cations by specific high-affinity ligands, or precipitation as an insoluble compound, can reduce the solution concentration of free metal ions (chemical activity), preventing metal toxicity at a cellular and subcellular level. The ability of a plant to control and modify the chemical activity of metal cations in selected cellular compartments or the apoplast would allow thermodynamic control over membrane transport processes involved in the sequestration of the metal. The mechanism of Cd resistance in plants, which is based on the inactivation of cadmium as a cadmium–phytochelatin complex localized in the vacuole, is an example of such a tolerance mechanism (Grill et al., 1985; Zenk, 1996). A large research effort has been dedicated to the identification of high-affinity low-molecular-weight metal chelators, which may provide metal tolerance (Mathys, 1977), and which may even account for the metal specificity of hyperaccumulation (Still and Williams, 1980; Reeves, 1992).

As metal chelators in hyperaccumulators organic acids have received most research attention so far, and the results have been reviewed comprehensively by Reeves (1992). The extraction of leaves with various solvents and subsequent gel filtration has been the method most commonly applied for the identification of metal ligands in plant tissues. Using this technique, Pelosi et al. (1974) and Kelly et al. (1975) obtained unidentified complexes with a molecular weight below 200 from *Alyssum bertolonii* and from several New Caledonian nickel hyperaccumulators. Later, malate and malonate were reported to be the predominant nickel ligands in complexes isolated from *Alyssum* nickel hyperaccumulators (Pancaro et al., 1978; Brooks et al., 1981). High malate concentrations were also reported in the leaves of the zinc hyperaccumulator *T. caerulescens*. Citrate, malate, and mixtures of the two organic acids were also the predominant ligands in complexes isolated from New Caledonian nickel hyperaccumulators (Lee et al., 1977; Kersten et al., 1980; Homer et al., 1991b). Among the other metal ligands considered were amino acids and cell wall pectates (Farago et al., 1975). No evidence was found for a role for chlorophyll in metal binding in hyperaccumulators (Reeves, 1992). An involvement of mustard oil glucosides was proposed, but clear evidence is lacking to date (Mathys, 1977).

The multiple-extraction approach suffers from the problem that during tissue homogenization and extraction, metals can redistribute to become chelated by ligands that were originally present in distant tissues of the leaf or distant compartments in a cell. Consequently, results obtained in these experiments have to be considered with caution.

Many nickel and zinc hyperaccumulators constitutively accumulate high concentrations of organic acids in their leaves (Mathys, 1977; Krämer, 1996; Shen et al., 1997; Tolrà et al., 1996). For example, the organic acid levels found in leaves of hyperaccumulator are comparable to or slightly lower than those found in CAM plants. If variable, leaf organic acid concentrations in hyperaccumulators are correlated with the total leaf cation concentrations in most cases, and not with the tissue concentrations of the hyperaccumulated metal (Tolrà et al., 1996). This indicates that, as in plants in general, organic acids may be accumulated to provide a charge balance for nonspecific cation accumulation in the leaf vacuoles (Osmond, 1976). It is therefore not surprising that Vergnano Gambi and Gabbrielli (1987) measured elevated malate concentrations in the leaves of *A. bertolonii* in response to elevated concentrations of nickel as well as elevated potassium in the rooting medium. Lee et al. (1978) detected a linear correlation between nickel and citrate concentrations in field-collected leaves of a variety of New Caledonian nickel hyperaccumulators, and it is possible that the maximum capacity for organic acid accumulation in leaf tissues may determine the maximum ion accumulation in the vacuole and thereby limit the extent of hyperaccumulation. Organic acids that form precipitates with metal cations, such as oxalate, may also provide an effective way of metal sequestration (Morrison et al., 1981). However, an involvement of organic acids in metal hyperaccumulation would explain "neither the metal-specificity, nor species-specificity of metal hyperaccumulation" (Reeves, 1992). The role of organic acids in metal tolerance is also debatable. At pH values typically found in the cytosol, amino acids, peptide, and other metabolites have a higher affinity for metals than do organic acids, and therefore organic acids would not offer any protection from metal toxicity in the cytoplasm. In vacuoles and apoplast, at pH values typically below 6, coordination of metal cations by amino acids becomes less stable, and coordination with organic acids, especially citrate, could decrease the chemical activity of the metal cation.

In nickel hyperaccumulators in the genus *Alyssum*, free histidine has been identified as a metal chelator able to increase metal tolerance and translocation to the shoot (Krämer et al., 1996). At pH values above 6, histidine forms nickel complexes of greater stability than any other organic or amino acid (Dawson et al., 1986; Smith et al., 1993). The mechanism by which Ni tolerance is achieved through the action of histidine, however, has yet to be established. In the search for specific high-affinity metal chelators in hyperaccumulators, it may be possible to make predictions according to a comparison of published stability constants of metal complexes with different ligands commonly present in plants.

Chemical reduction and/or incorporation into organic compounds is another possible detoxification mechanism, as observed for metalloids such as selenium. Excess selenium is toxic to most plants because it is metabolized to selenocysteine and selenomethionine which replace cysteine and methionine residues in proteins (Shift, 1969). By funneling selenium into the non-protein amino acids methylselenocysteine and selenocystathionine, selenium accumulator species of *Astragalus* are able to reduce the amount of selenium incorporated into proteins, thereby tolerating elevated concentrations of selenium in shoots (Läuchi, 1993). Recently the enzyme responsible for the methylation of selenocysteine in the selenium accumulator *Astragalus bisculatus* has been isolated and characterized, a first step in determining the molecular basis of selenium resistance in plants (Neuhierl and Böck, 1996). It also appears that several selenium-accumulating species are able to selectively exclude selenium from the methionine biosynthetic pathway, thereby avoiding the synthesis of selenomethionine, a toxic selenium derivative of methionine (Burnell, 1981).

Volatilization of selenium from plant tissues may also provide a mechanism of selenium detoxification. Lewis et al. (1966) were the first to show that both selenium nonaccumulator and accumulator species volatilize selenium. The volatile selenium compound released from the selenium accumulator *Astragalus racemosus* was identified as dimethyl diselenide (Evans et al., 1968). Selenium released from alfalfa, a selenium nonaccumulator, was different from the accumulator species and was identified as dimethyl selenide (Lewis et al. 1974).

In an attempt to verify some of the in vitro observations on metal speciation in plants, X-ray absorption spectroscopy has recently been applied to tissues from metal-exposed plants (Salt et al., 1995, 1997, 1999b; Krämer et al., 1996). Prior to analysis, tissues are simply frozen in liquid nitrogen, avoiding many of the problems associated with the extraction and purification of metal complexes. Using this technique, it has been possible to show that a significant amount of nickel is coordinated with histidine in intact tissues and xylem exudates of the nickel hyperaccumulator *Alyssum lesbiacum* (Krämer et al., 1996). It was also demonstrated that phytochelatins are not involved in nickel-binding in *A. lesbiacum*. In a similar, quantitative approach metals were found to be mainly associated with organic acids in shoot tissue and xylem exudates of both the zinc hyperaccumulator *Thlaspi caerulescens* (Salt et al. 1999b) and the nickel hyperaccumulator *Thlaspi goesingense* (Salt, Prince, and Pickering, unpublished). In the roots of both species, however, the coordination spheres of zinc and nickel were significantly different, indicating a histidine or histidine-like ligand (Salt et al., 1999b; Salt, Prince, and Pickering, unpublished). Given the present developments, we can expect more highly attractive in vivo techniques to be available for metal localization and the analysis of metal speciation in the future.

CONCLUSIONS

Although there is considerable research interest in metal hyperaccumulation, as shown by the large number of publications available, only a few researchers have directly addressed the biochemical processes underlying metal hyperaccumulation among the diversity of different hyperaccumulator species. Research has been hindered by the poor availability of hyperaccumulator seed material and by the fact that most of the hyperaccumulator species are very rare and have not yet been studied experimentally.

However, the characterization of metal hyperaccumulators has provided some important leads for research into the molecular mechanisms of metal hyperaccumulation. One of these mechanisms is the induction by nickel of increased concentrations of free histidine in the xylem sap and in bulk tissue of *Alyssum* spp. and the role of histidine in metal tolerance. Once the molecular mechanism of the histidine response is identified, transformation of nonaccumulator plants will show whether the histidine response is merely necessary or is also sufficient for nickel hyperaccumulation. Given the available information on metal sequestration at the tissue and cellular level, membrane transport of metals is a prime candidate for further investigation.

At present, the diversity within metal hyperaccumulators suggests that several mechanisms of hyperaccumulation have evolved even for one metal. Evidence has suggested that metal hyperaccumulation may require several processes: increased root uptake as well as reduced root accumulation, sequestration at the cellular level as well as the tissue level, and, most importantly, metal tolerance. It is therefore possible that numerous genetic alterations may be involved in generating the hyperaccumulator phenotype in each group of related hyperaccumulators.

To reveal molecular mechanisms involved in the remarkable phenomenon of metal hyperaccumulation in plants, research will have to be developed by (1) further biochemical characterization of metal hyperaccumulators, (2) detailed analysis of metal-regulated gene expression in hyperaccumulators, (3) characterization of mutations in the genes for certain enzymes that have been identified to operate differently in hyperaccumulators through comparison with closely related nonaccumulators, (4) molecular genetic approach to hyperaccumulation should be attempted, and (5) further insight into the metabolism of trace elements and its regulation in crop or genetic model plants could help to direct and focus research on metal hyperaccumulators.

REFERENCES

Baker, A. J. M., 1981. Accumulators and excluders—strategies in the response of plants to heavy metals. *J. Plant Nutr.* **3**:643–654.

Baker, A. J. M., and R. R. Brooks, 1989. Terrestrial higher plants which hyperaccumulate metallic elements—a review of their distribution, ecology and phytochemistry. *Biorecovery* **1**:81–126.

Baker, A. J. M. et al., 1991. *In Situ* decontamination of heavy metal polluted soils using crops of metal-accumulating plants —a feasibility study. In *In Situ Bioreclamation*, R. E. Hinchee and R. F. Olfenbuttel, (eds.) Butterworth-Heinemann, Stoneham, MA, pp. 539–544.

Baker, A. J. M. et al., 1994. Heavy metal accumulation and tolerance in British populations of the metallophyte *Thlaspi caerulescens* J. & C. Presl (Brassicaceae). *New Phytol.* **127**:61–68.

Baker, A. J. M. et al., 1999. Metal hyperaccumulator plants: A review of the biological resource for possible exploitation in the phytoremediation of metal-polluted soils. In *Phytoremediation of Contaminated Soil and Water*, N. Terry and G. S. Bañuelos (eds.), Chapter 5, CRC Press LLC, Boca Raton, FL.

Baumann A., 1885. Das Verhalten von Zinksalzen gegen Pflanzen und im Boden. *Landwirtsch. Versuch.* **31:**1–53.

Bernal, M. P., and S. P. McGrath, 1994. Effect of pH and heavy metal concentrations in solution culture on the proton release, growth and elemental composition of *Alyssum murale* and *Raphanus sativus* L. *Plant Soil* **166:**83–92.

Bernal, M. P. et al., 1994. Comparison of the chemical changes in the rhizosphere of the nickel hyperaccumulator *Alyssum murale* with the non-hyperaccumulator *Raphanus sativus* L. *Plant Soil* **164:**251–259.

Blamey, F. P. C. et al., 1986. Role of trichomes in sunflower tolerance to manganese toxicity. *Plant Soil* **91:**171–180.

Boyd, R. S., and S. N. Martens, 1992. The *raison d'être* for metal hyperaccumulation by plants. In *The Vegetation of Ultramafic (Serpentine) Soils*, A. J. M. Baker et al., eds., Intercept, Andover, UK, pp. 279–289.

Boyd, R. S., and S. N. Martens, 1994. Nickel hyperaccumulated by *Thlaspi montanum* var. *montanum* is acutely toxic to an insect herbivore. *OIKOS* **70:**21–25.

Boyd, R. S. et al., 1994. Nickel hyperaccumulation defends *Streptanthus polygaloides* (Brassicaceae) against pathogens. *Am. J. Bot.* **81:**294–300.

Brooks, R. R., 1987. *Serpentine and Its Vegetation*, T. R. Dudley (ed.), Discoroides Press, Portland, OR.

Brooks, R. R. et al., 1977. Detection of nickeliferous rocks by analysis of herbarium specimens of indicator plants. *J. Geochem. Explor.* **7:**49–77.

Brooks, R. R. et al., 1979. Hyperaccumulation of nickel by *Alyssum* Linnaeus (Cruciferae) *Proc. Roy. Soc. Lond. Sec. B,* **203:**387–403.

Brooks, R. R. et al., 1981. The chemical form and physiological function of nickel in some Iberian *Alyssum* species. *Physiol. Plant.* **516:**167–170.

Brown, S. L. et al., 1994. Phytoremediation potential of *Thlaspi caerulescens* and Bladder Campion for zinc- and cadmium-contaminated soil. *J. Environ. Qual.* **23:**1151–1157.

Brown, S. L. et al., 1995. Zinc and cadmium uptake by hyperaccumulator *Thlaspi caerulescens* grown in nutrient solution. *Soil Sci. Soc. Am. J.* **59:**125–133.

Burnell J. N., 1981. Selenium metabolism in *Neptunia amplexicaulis*. *Plant Physiol.* **67:**316–324.

Byers, H. G., 1935. Selenium occurrence in certain soils in the United States with a discussion of related topics. *U.S. Dept. Agr. Tech. Bull.* **482:**1–47.

Byers, H. G., 1936. Selenium occurrence in certain soils in the United States, with a discussion of related topics. Second report. *U.S. Dept. Agr. Tech. Bull.* **530:**1–78.

Cakmak, I. et al., 1996a. Zinc-efficient wild grasses enhance release of phytosiderophores under zinc deficiency. *J. Plant Nutr.* **19:**551–563.

Cakmak, I. et al., 1996b. Phytosiderophore release in bread and duram wheat genotypes differing in zinc efficiency. *Plant Soil* **180:**183–189.

Clarkson, D. T., and J. B. Hanson, 1986. Proton fluxes and the activity of a stelar proton pump in onion roots. *J. Exper. Bot.* **37:**1136–1150.

Dawson, R. M. C. et al., 1986. *Data for Biochemical Research*, 3rd ed., Clarendon Press, Oxford, UK.

De Boer, A. H. et al., 1983. Bi-phasic composition of trans-root electrical potential in roots of *Plantago* species: Involvement of spatially separated electrogenic pumps. *Planta* **157**:259–266.

Evans, C. S. et al., 1968. Isolation of dimethyl diselenide and other volatile selenium compounds from *Astragalus racemosus* (Pursh.). *Austral. J. Biol. Sci.* **21**:13–20.

Farago, M. E., and I. E. D. A. W. Mahmoud, 1983. Metal compounds of pectin. *Inorg. Chim. Acta* **86**:273–278.

Farago, M. E. et al., 1975. The chemistry of plants which accumulate metals. *Coord. Chem. Rev.* **16**:1–8.

Gabbrielli, R. et al., 1990. Comparison of two serpentine species with different nickel tolerance strategies. *Plant Soil* **122**:271–277.

Gabbrielli, R. et al., 1991. Accumulation mechanisms and heavy metal tolerance of a nickel hyperaccumulator. *J. Plant Nutr.* **14**:1067–1080.

Grill, E. et al., 1985. Phytochelatins: The principal heavy-metal complexing peptides of higher plants. *Science* **230**:674–676.

Heath, S. M. et al., 1997. Localization of nickel in epidermal subsidiary cells of leaves of *Thlaspi montanum* var. *siskiyouense* (Brassicaceae) using energy-dispersive X-ray microanalysis. *Int. J. Plant Sci.* **158**:184–188.

Homer, F. A., 1991. *Chemical Studies on some Plants that Hyperaccumulate Nickel.* D.Phil. thesis, Massey University, New Zealand.

Homer, F. A. et al., 1991a. Comparative studies of nickel, cobalt, and copper uptake by some nickel hyperaccumulators of the genus *Alyssum. Plant Soil* **138**:195–205.

Homer, F. A. et al., 1991b. Characterization of the nickel-rich extract from the nickel hyperaccumulator *Dichapetalum gelonioides. Phytochemistry* **30**:2141–2145.

Horn, N. M. et al., 1995. The effect of histidine and cysteine on zinc influx into rat and human erythrocytes. *J. Physiol.* **489**:73–80.

Ingrouille, M. J., and N. Smirnoff, 1986. *Thlaspi caerulescenes* J. & C. Presl. (*T. alpestre* L.) in Britain. *New Phytol.* **102**:219–233.

Kelly, P. C. et al., 1975. Preliminary observations on the ecology and plant chemistry of some nickel accumulating plants from New Caledonia. *Proc. Roy. Soc. Lond. Ser. B* **189**:69–80.

Kersten, W. J. et al., 1980. Nature of nickel complexes in *Psychotria douarrei* and other nickel-accumulating plants. *Phytochemistry* **19**:1963–1965.

Köhl, K. I. et al., 1997. Defining a metal-hyperaccumulator plant: the relationship between metal uptake, allocation and tolerance. Abstract. *Plant Physiol.* **114**:124.

Krämer, U., 1996. *Nickel Hyperaccumulation in the Genus Alyssum.* D.Phil. thesis, Univ. Oxford, UK.

Krämer, U. et al., 1996. Free histidine as a metal chelator in plants that accumulate nickel. *Nature* **379**:635–638.

Krämer, U. et al., 1997a. Micro-PIXE as a technique for studying nickel localization in leaves of the hyperaccumulator plant *Alyssum lesbiacum. Nucl. Instr. Meth. Phys. Res. B* **130**:346–350.

Krämer, U. et al., 1997b. The role of metal transport and tolerance in nickel hyperaccumulation by *Thlaspi goesingense* Halacsy. *Plant Physiol.* **115**:1641–1650.

Küpper, H. et al., 1999. Cellular compartmentation of zinc in leaves of the hyperaccumulator *Thlaspi caerulescens*. *Plant Physiol.* **119:**305–311.

Lasat, M. M. et al., 1996. Physiological characterization of root Zn^{2+} absorption and translocation to shoots in Zn hyperaccumulator and nonaccumulator species of *Thlaspi*. *Plant Physiol.* **112:**1715–1722.

Lasat, M. M., et al., 1998. Altered Zn compartmentation in the root symplast and stimulated Zn absorption into the leaf as mechanisms involved in Zn hyperaccumulation in *Thlaspi caerulescens*. *Plant Physiol.* **118:**875–883.

Läuchli, A., 1993; Selenium in plants: Uptake, functions, and environmental toxicity. *Bot. Acta* **106:**455–468.

Lee, J. et al., 1977. Isolation and identification of a citrato-complex of nickel from nickel-accumulating plants. *Phytochemistry* **16:**1503–1505.

Lee, J. et al., 1978. The relation between nickel and citric acid in some nickel-accumulating plants. *Phytochemistry* **17:**1033–1035.

Lewis, B. G. et al., 1966. Release of volatile selenium compounds by plants. Collection procedures and preliminary observations. *J. Agric. Food Chem.* **14:**638–640.

Lewis, B. G. et al., 1974. Volatile selenium in higher plants. The production of dimethyl selenide in cabbage leaves by enzymic cleavage of Se-methyl selenomethionine selenonium salt. *Plant Soil* **40:**107–118.

Lloyd-Thomas, D. H., 1995. *Heavy Metal Hyperaccumulation by Thlaspi caerulescens J. & C. Presl.* Ph.D. thesis, Univ. Sheffield, UK.

Ma, J. F., and K. Nomoto, 1996. Effective regulation of iron acquisition in graminaceous plants—the role of mugineic acids as phytosiderophores. *Physiol. Plant.* **97:**609–617.

Marschner, H., 1995. *Mineral Nutrition of Higher Plants*, 2nd ed., Academic Press, London.

Martell, E. A. 1974. Radioactivity of tobacco trichomes and insoluble cigarette smoke particles. *Nature* **249:**215–217.

Martens, S. N., and R. S. Boyd, 1994. The ecological significance of nickel hyperaccumulation: a plant chemical defense. *Oecologia* **98:**379–384.

Mathys, W., 1977. The role of malate, oxalate and mustard oil glucosides in the evolution of zinc-resistance in herbage plants. *Physiol. Plant* **40:**130–136.

McGrath, S. P. et al., 1997. Heavy metal uptake and chemical changes in the rhizosphere of *Thlaspi caerulescens* and *Thlaspi ochroleucum* grown in contaminated soils. *Plant Soil* **188:**153–159.

Mesjasz-Przybylowicz, J. et al., 1994. Proton microprobe and X-ray fluorescence investigations of nickel distribution in serpentine flora from South Africa. *Nucl. Instr. Meth. B* **89:**208–212.

Minguzzi C., and O. Vergnano, 1948. Il contenuto di nichel nelle ceneri di *Alyssum bertolonii* Desv. *Mem. Soc. Tosc. Sci. Nat. Ser. A,* **55:**49–77.

Morrison, R. S. et al., 1981. The diverse chemical forms of heavy metals in tissue extracts of some metallophytes from Shaba province, Zaïre. *Phytochemistry* **20:**455–458.

Nenhierl B., and A. Bröck, 1996. On the mechanism of selenium tolerance in selenium-accumulating plants. Purification and characterization of a specific selenocysteine methyltransferase from cultured cells of *Astragalus bisculatus*. *Eur. J. Biochem.* **239:**235–238.

Neumann, D. et al, 1995. How does *Armeria maritima* tolerate high heavy metal concentrations? *J. Plant Physiol.* **146:**704–717.

Osmond, C. B., 1976. Ion absorption and carbon metabolism in cells of higher plants. In *Encyclopedia of Plant Physiology* (New Series), Vol. 2B, U. Lüttge, and M. G. Pitman (eds.), Springer-Verlag, Berlin, pp. 346–372.

Pancaro, L. et al., 1978. Ulteriori indagini sul rapporto tra nichel e acidi malico e malonico in *Alyssum. Giorn. Bot. Ital.* **112:**141–146.

Pelosi, P. et al., 1974. Sulla nature dei composti del nichel presenti in *Alyssum bertolonii* Desv.—nota I. *Agric. Ital.* **29:**1–5.

Pitman, M. G., 1972. Uptake and transport of ions in barley seedlings. II. Evidence for two active stages in transport to the shoot. *Austral. J. Biol. Sci.* **25:**243–257.

Pollard, J. A., and A. J. M. Baker, 1997. Deterrence of herbivory by zinc hyperaccumulation in *Thlaspi caerulescens* (Brassicaceae). *New Phytol.* **135:**655–658.

Rascio, N., 1977. Metal accumulation by some plants growing on zinc-mine deposits. *Oikos* **29:**250–253.

Reeves, R. D., 1992. The hyperaccumulation of nickel by serpentine plants. In *The Vegetation of Ultramafic (Serpentine) Soils*, A. J. M. Baker et al. (eds.), Intercept, Andover, Hampshire, UK, pp. 253–277.

Reeves, R. D. et al., 1981. Nickel uptake by Californian Streptanthus and Caulanthus with particular reference to the hyperaccumulator *S. polygaloides* Gray (Brassicaceae). *Am. J. Bot.* **68:**708–712.

Roberts, S. K., and M. Tester, 1995. Inward and outward K^+-selective currents in the plasma membrane of protoplasts from maize root cortex and stele. *Plant J.* **8:**811–825.

Roberts, S. K., and M. Tester, 1997. Permeation of Ca^{2+} and monovalent cations through an outwardly rectifying channel in maize root stelar cells. *J. Exper. Bot.* **48:**839–846.

Salt, D. E. et al., 1995. Mechanisms of cadmium mobility and accumulation in Indian Mustard. *Plant Physiol.* **109:**1427–1433.

Salt, D. E. et al., 1997. Metal accumulation by aquacultured seedlings of Indian mustard. *Environ. Sci. Technol.* **31:**1636–1644.

Salt, D. E., et al., 1999a. The role of root exudates in nickel hyperaccumulation and tolerance in accumulator and nonaccumulator species of *Thlaspi*. In *Phytoremediation of Contaminated Soil and Water*, N. Terry, G. S. Bañuelos (eds.), Chapter 10, CRC Press LLC, Boca Raton, FL.

Salt, D. E. et al., 1999b. Zinc ligands in the metal hyperaccumulator *Thlaspi caerulescens* as determined using X-ray absorption spectroscopy. *Environ. Sci. Technol.* **33:**713–717.

Severne, B. C., 1974. Nickel accumulation by *Hybanthus floribundus*. *Nature* **248:**807–808.

Shen, Z. G. et al., 1997. Uptake and transport of zinc in the hyperaccumulator *Thlaspi caerulescens* and the non-hyperaccumulator *Thlaspi ochroleucum*. *Plant Cell Environ.* **20:**898–906.

Shift A., 1969. Aspects of selenium metabolism in higher plants. *Annu. Rev. Plant Physiol.* **20:**475–495.

Smith, R. M. et al., 1993. *NIST Critical Stability Constants of Metal Complexes Database*, version 1.0. U.S. Department of Commerce, National Institute of Standards and Technology, Gaithersburg, MD.

Still, E. R., and R. J. P. Williams, 1980. Potential methods for selective accumulation of nickel(II)ions by plants. *J. Inorg. Biochem.* **13:**35–40.

Tolrà, R. P. et al., 1996. Zinc hyperaccumulation in *Thlaspi caerulescens.* II. Influence on organic acids. *J. Plant Nutr.* **19:**1541–1550.

Vázquez, M. D. et al., 1992. Localization of zinc and cadmium in *Thlaspi caerulescens* (Brassicaceae), a metallophyte that can hyperaccumulate both metals. *J. Plant Physiol.* **140:**350–355.

Vázquez, M. D. et al., 1994. Compartmentation of zinc in roots and leaves of the zinc hyperaccumulator *Thlaspi caerulescens* J. & C. Presl. *Bot. Acta* **107:**243–250.

Vergnano Gambi, O., and R. Gabbrielli, 1987. The response of plants to heavy metals: organic acid production. *Giorn. Bot. Ital.* **121:**269–277.

Wegner, L. H., and K. Raschke, 1994. Ion channels in the xylem parenchyma of barley roots. *Plant Physiol.* **105:**799–813.

Whiting, S. N. et al., 1997. Changes in phytoavailability of zinc to plants sharing a rhizosphere with the zinc hyperaccumulator *Thlaspi caerulescens* J. & C. Presl. In *Proceedings of Extended Abstracts from the Fourth International Conference on the Biogeochemistry of Trace Elements,* I. K. Iskandar et al. (eds.), Berkeley, CA pp. 469–470.

Zenk, M. H., 1996. Heavy metal detoxification in higher plants—a review. *Gene* **179:**21–30.

14

MECHANISMS OF METAL RESISTANCE: PHYTOCHELATINS AND METALLOTHIONEINS

Christopher S. Cobbett and Peter B. Goldsbrough

INTRODUCTION

The development of phytoremediation as a useful process will require a thorough understanding of the biochemical and physiological processes involved in the uptake and metabolism and/or detoxification of heavy metals by plants. Plants, like other organisms, have adaptive mechanisms whereby they are able to respond to both nutrient deficiencies and toxicities. In particular, heavy metals pose the biological problem of, in some cases being both essential and toxic, and in other cases being only toxic to living cells. The two major heavy metal-binding compounds in plant cells are the phytochelatin peptides (PCs) and metallothioneins (MTs). A clear understanding of the roles of these two classes of compounds in the binding and sequestration of heavy metals will be an important contribution to the ultimate success of phytoremediation.

Metallothionein was first identified as a cadmium-binding protein isolated from horse kidney. Since this discovery, similar proteins have been identified in a large number of animals and, more recently, in plants and bacteria (Kägi, 1991). MTs have the following general characteristics: (1) low molecular weight (<10 kD), (2) large fraction of cysteine residues, and (3) high metal content with coordination of metal ions in metal–thiolate clusters. A formal classification system for MTs (and PCs) has been adopted where Class I includes MTs from mammals and other organisms with a

Phytoremediation of Toxic Metals: Using Plants to Clean Up the Environment, edited by Ilya Raskin and Burt D. Ensley.
ISBN: 0-471-19254-6 ©2000 John Wiley & Sons, Inc.

highly conserved arrangement of cysteine residues, and Class II includes all other MT proteins. Class III comprises cysteine-rich, metal-binding peptides that are not produced by translation of a mRNA on ribosomes (Kägi, 1991) and therefore includes PCs. In this chapter, however, we continue to refer to Classes I and II as MTs and Class III as PCs.

Prior to the discovery of PCs in plants, there were a number of reports of MTs in plants. However, these proteins were not well characterized, and primary amino acid sequences were not obtained. In many cases these metal-binding complexes may indeed have been comprised of PCs as they were identified in studies of plant responses to Cd. After the structure of PCs had been elucidated, and these peptides were shown to be widely distributed in the plant kingdom and induced by a large number of essential and nonessential metals, it was proposed that PCs were the functional equivalent of MTs and that plants did not contain gene-encoded MT proteins (Grill et al., 1987). Shortly after this proposal was made, however, proteins with all of the characteristics of Class II MTs were purified from wheat embryos (Lane et al., 1987). This MT was identified as a wheat embryo protein that incorporated a large amount of cysteine during early stages of germination, giving rise to the name Ec. The amino acid sequence of the wheat Ec protein is indicative of a MT and the protein was subsequently isolated from embryos as a Zn-binding MT. In the past 10 years (at the time of writing) a large number of MT genes have been isolated from a variety of plant species. It now appears that plants are equipped with at least two cysteine-based metal ligand systems, namely, PCs and MTs. Our current level of understanding of the roles of these compounds in plants is reviewed here. PCs and MTs are considered separately since we currently have no clear understanding of any interactions or redundancy of their functions in metal metabolism.

PHYTOCHELATINS

Structures

Determination of the amino acid composition and subsequent detailed analysis of Cd-binding peptides from plant tissues (Wagner, 1984) and cell cultures (Grill et al., 1985; Steffens et al., 1986; Jackson et al., 1987) and from the yeast *Schizosaccharomyces pombe* (Kondo et al., 1983), using a variety of techniques, demonstrated the structure of the peptides. They were composed of only three amino acids; glutamate, cysteine, and glycine, and comprised a family of related structures described as $(\gamma\text{-GluCys})_n\text{-Gly}$ (where $n > 1$) that are clearly related to the tripeptide glutathione (GSH) for which $n = 1$. These peptides have been referred to variously as cadystins (from *S. pombe*), poly-$\gamma(EC)_nG$ peptides, Cd-binding peptides, and PCs and are broadly classified as Class III MTs. The demonstration that the Glu and Cys residues in PCs are linked through a γ-carboxylamide bond indicated that, unlike the Classes I and II MTs, these peptides were not encoded directly by genes but were likely to be the products of a biosynthetic pathway, presumably using GSH as a substrate.

Native PC–Cd complexes are generally identified by using gel-filtration chromatography. In some cases both low- and high-MW complexes can be resolved (see text below). PCs can be quantitated by using the thiol reagent 5,5'-dithiobis-(2-nitrobenzoic acid) (DTNB) (Grill et al., 1991; Rauser, 1991). HPLC followed by derivatization using DTNB allows the different PC isoforms to be resolved and quantitated with a detection limit of 0.2 nmol thiol. Up to 100-fold greater sensitivity can be obtained using the fluorescent compound monobromobimane coupled with an appropriate detector. Using these detection methods, PCs have been identified in a wide variety of plant species including monocots, dicots, gymnosperms, and algae (Gekeler et al., 1989), in addition to various fungal species (Mehra et al., 1988; Kneer et al., 1992) and marine diatoms (Morelli and Pratesi, 1997).

A number of structural variants of PCs have been identified in different species. The des-Gly variant, $[(\gamma\text{-GluCys})_n]$ has been found in Cd-binding complexes from maize (Meuwly et al., 1995) and some yeasts. For a number of legumes, which contain not GSH but a related compound called homoGSH (γ-GluCys-β-Ala), the analogous family of PCs or homoPCs $[(\gamma\text{-GluCys})_n\text{-}\beta\text{-Ala}]$ are synthesized in response to exposure to Cd (Grill et al., 1986). Also in maize the structure $(\gamma\text{-GluCys})_n\text{-Glu}$ has been detected (Meuwly et al., 1995), whereas in other monocots, *Agrostis* and rice, for example, $(\gamma\text{-GluCys})_n\text{-Ser}$ peptides were identified (Klapheck et al., 1994). These variants are all involved in heavy-metal binding and are likely be functionally analogous, although this has not been unequivocally demonstrated.

Biosynthesis of PCs

In plant tissues or cultures of yeasts or plant cells grown in media containing trace levels of only essential heavy metals, low levels of PCs can be detected (Grill et al., 1988; Kneer and Zenk, 1992; Howden et al., 1995a). Exposure to Cd, generally in the 1–100-µM range, depending on the species or conditions of cultivation, leads to the rapid accumulation of PCs with the progressively larger oligomers accumulating in succession (Grill et al., 1987; Klapheck et al., 1994; Meuwly et al., 1995). In plant cell cultures PCs can be detected within minutes (Grill et al., 1987; Scheller et al., 1987), whereas in intact plants induction can be detected within hours of exposure to heavy metals (Meuwly et al., 1995). Cell cultures of *R. serpentina* have been used to assess the capacity of a wide range of metal ions to induce PC biosynthesis: Cd, Ni, Cu, Zn, Ag, Sn, Sb, Te, W, Au, Hg, Pb, and Bi cations and arsenate and selenite anions were effective to varying degrees (Grill et al., 1987). No induction was observed in the presence of Na, Mg, Al, Ca, V, Cr, Mn, Fe, Co, Cs, and Mo ions. Cd was the most effective inducer under the conditions used. However, because the applied concentrations of the metal ions tested varied and the different metals may have been accumulated in the cells at different rates, the level of "effectiveness" cannot be clearly measured. A similar wide range of metals ions induced PC biosynthesis in roots of *Rubia tinctorum* (Maitani et al., 1996).

Various studies have confirmed that PCs are derived from GSH (Fig. 14-1). The induction of PCs by the addition of metal ions to plant cell cultures coincided with a decrease in the levels of GSH (Grill et al., 1987; Scheller et al., 1987) and the exposure of whole plants or plant cell cultures to an inhibitor of GSH biosynthesis, buthionine sulfoximine (BSO), also inhibited the induction of PCs and/or conferred hypersensitivity to metal ions (Scheller et al., 1987; Steffens et al., 1986; Reese and Wagner, 1987, Howden and Cobbett, 1992). Furthermore, the addition of GSH to cell cultures treated with BSO restored the biosynthesis of PCs (Mendum et al., 1990). These physiological studies have been confirmed by the demonstration that mutants of both *S. pombe* and *A. thaliana*, which have defects in genes encoding GSH biosynthetic enzymes, are deficient in PCs and are hypersensitive to metal ions. By screening for Cd-sensitive mutants of *S. pombe*, GSH-deficient mutants have been identified (Mutoh and Hayashi, 1988). Conversely, mutants isolated on the basis of GSH deficiency were Cd-sensitive (Glaeser et al., 1991). These mutants are unable to produce PCs. In *Arabidopsis*, the *cad2-1* mutant contains only 30% of wild-type levels of GSH. This mutant also has reduced accumulation of Cd-binding complexes, is

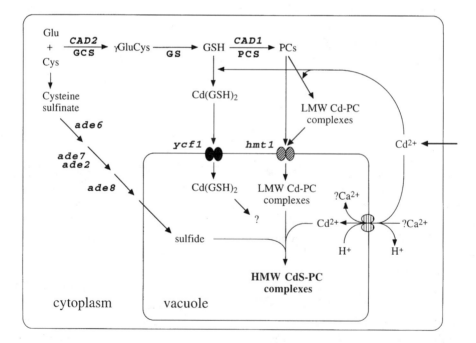

Figure 14-1 Genes and functions contributing to Cd detoxification in plants and fungi. The figure is a composite of different functions described in different organisms. (GCS, γ-glutamylcysteine synthetase; GS, glutathione synthetase; PCS, phytochelatin synthase). Refer to the text for a more detailed description of the various functions. Gene loci are shown in italics.

deficient in PC biosynthesis, and has a Cd-sensitive phenotype, emphasizing the link between GSH and PC biosynthesis (Howden et al., 1995b).

In plants the biosynthesis of GSH from its constituent amino acids has been well characterized and is similar to that of animals and microrganisms (Alscher, 1989; May et al., 1998; Noctor and Foyer, 1998). This is a two-step pathway catalyzed successively by the ATP-dependent enzymes, γ-glutamylcysteine synthetase (GCS) and glutathione synthetase (GS) (Fig. 14-1). These enzymes have been purified from plant tissues and some studies have suggested the existence of different isozymes of each in chloroplastic and cytosolic fractions (Hell and Bergmann, 1988; 1990; Klapheck et al., 1988; Law and Halliwell, 1986). These observations have been interpreted as indicating the existence of separate GSH biosynthetic pathways in these two cellular compartments. Genes encoding these enzymatic activities have been cloned from *Arabidopsis* (May and Leaver, 1994; Rawlins et al., 1995; Wang and Oliver, 1996; Ullman et al., 1996) and tomato (P. Goldsbrough; unpublished). For GCS and GS in *Arabidopsis*, only a single gene has been identified. This may indicate there is in fact only a single GSH biosynthetic pathway. The *cad2-1* mutant of *Arabidopsis*, which contains only about 30% of wild-type levels of GSH, is deficient in GCS activity. The *cad2-1* phenotype is complemented by a genomic clone of the GSHA gene encoding GCS, and the allele of this gene from the mutant contains a 6-basepair (6-bp) deletion within an exon (Cobbett et al., 1998). It is likely that this mutant gene product has partial activity that would account for the residual level of GSH.

Grill et al. (1989) identified an enzymatic activity from cultured cells of *Silene cucubalis* that synthesized PCs from GSH by transferring a γGluCys moiety from a donor to an acceptor molecule. Both donor and acceptor could be either GSH or an extant PC molecule. In reactions catalyzed by this enzyme, the successive appearance of PC_2, PC_3, and PC_4 was observed. The reaction was presumed to involve the transpeptidation of the γGluCys moiety of GSH onto initially a second GSH molecule to form PC_2 or, in later stages of the incubation, onto a PC molecule to produce an *n* + 1 oligomer. This γGluCys dipeptididyl transpeptidase (EC 2.3.2.15) has been termed *PC synthase*. The enzyme was a tetramer of M_r 95,000 with a K_m for GSH of 6.7 mM. In vitro, the activity of the partially purified enzyme was active only in the presence of metal ions. The best activator tested was Cd followed by Ag, Bi, Pb, Zn, Cu, Hg, and Au cations. It is noteworthy that these all induce PC biosynthesis in vivo in *R. serpentina*. In these reactions PC biosynthesis continued until the activating metal ions were chelated either by the PCs formed or by the addition of a metal chelator such as EDTA (Loeffler et al., 1989). This provides a mechanism to autoregulate the biosynthesis of PCs in which the product of the reaction chelates the activating metal, thereby terminating the reaction. A dipeptidyl transferase activity was also detected in *S. pombe* (Hayashi et al., 1991). However, the activity did not require Cd. How PC biosynthesis is induced by Cd in this organism is not apparent. Hayashi et al. (1991) also proposed an alternative pathway for PC biosynthesis in which γGluCys was polymerized and a terminal Gly added by GS.

Similar PC synthase activities have been detected in pea, tomato, and *Arabidopsis*. In *Arabidopsis* the activity was detected in crude extracts and was activated by Cd

(Howden et al., 1995a). Mutants at the *cad1* locus in *Arabidopsis*, similar to the *cad2-1* mutant, are Cd-sensitive and deficient in the formation of Cd-binding complexes and in PC biosynthesis, particularly the *cad1-3* mutant in which no PCs were detected after exposure to Cd. In contrast to the *cad2-1* mutant, the *cad1* mutants have wild-type levels of GSH, suggesting a defect in PC synthase. Consistent with this, PC synthase activity in crude extracts from *cad1* mutants was less than 1% of the level in both wild-type and the *cad2-1* mutant (Howden et al., 1995a). It is likely that *CAD1* is the structural gene for PC synthase (Fig. 14-1). All five independent Cd-sensitive, PC synthase-deficient mutants isolated are *cad1* alleles, and although it is formally possible that *CAD1* encodes a regulatory protein for PC synthase expression, it is unlikely that some mutants at the PC synthase structural gene locus would not have been isolated if this were the case. The absence of PCs in the *cad1-3* mutant argues against the presence of a second pathway of PC biosynthesis in plants, similar to that observed for *S. pombe*. If it exists, such a pathway must be either relatively ineffective or also dependent on the *CAD1* gene product.

Crude enzyme preparations from the roots of pea, which normally contains both GSH and homoGSH, could use GSH efficiently, but hGSH or γGluCysSer less efficiently, as substrates for PC synthesis. In the presence of both GSH and hGSH, for example, the synthesis of hPCs was enhanced (Klapheck et al., 1995). These observations were interpreted to indicate the enzyme had a γGluCys donor site that was relatively specific for GSH but a less specific acceptor site able to use each of the three substrates. In these extracts the enzyme was activated by Cd and Cu but only poorly by Zn ions.

Kinetic studies using plant cell cultures exposed to Cd demonstrated that PC biosynthesis occurs within minutes and is independent of de novo protein synthesis, consistent with the observation of enzyme activation in vitro. The enzyme activity appears to be constitutively expressed. It has been detected in *S. cucubalis* cells grown in culture medium (Grill et al., 1989), and in tomato (Chen et al., 1997) and *Arabidopsis* (Howden et al., 1995a) plants grown in soil or agar medium in the presence of only trace levels of essential heavy metals. Together these observations indicate that PC synthase activity is exclusively regulated at the level of enzymatic activity in response to heavy metals.

Little is known about the tissue specificity of PC synthase expression and/or PC biosynthesis. Callus tissue derived from a *cad1* mutant of *Arabidopsis* was, similar to the intact plant, more sensitive to Cd than callus derived from the wild type (Howden and Cobbett, 1992). This indicates that PCs confer a cellular function that is not dependent on the differentiation of plant tissues to be expressed. In the only study of tissue-specific PC synthase expression to date, the activity was detected in the roots and stems of tomato plants but not in leaves or fruits (Chen et al., 1997).

Structure of PC–Metal Complexes

The nature of PC–metal complexes formed in vivo has concentrated on Cd-binding complexes largely because Cd is an effective inducer of PC biosynthesis and had been used in many of the earlier physiological studies of heavy-metal-binding compounds. In extracts of *S. pombe* two complexes [referred to as high and low molecular weight (HMW and LMW)] can be clearly resolved using gel-filtration chromatography. In contrast, in extracts of plant cell cultures or tissues exposed to Cd, LMW and HMW complexes have seldom been resolved, although there are some examples in which clear or partial resolution has been observed (Kneer and Zenk, 1992; Howden et al., 1995a). It is likely that there is a transition from the LMW to the HMW complexes that may be dependent on the period and level of exposure to Cd and that may differ between organisms. In *Arabidopsis*, the HMW and LMW complexes could be clearly resolved by gel-filtration chromatography. However, exposure to a higher concentration of Cd resulted in a higher proportion of the bound Cd appearing in the HMW complexes, whereas in the "leaky" *cad1-1*, PC-deficient mutant, small amounts of only LMW complexes were detected, suggesting that the amount of HMW complexes formed may be dependent on the total amount of PCs synthesized (Howden et al., 1995a). In addition, the resolution of complexes may be influenced by the particular experimental conditions used for extraction of plant tissues and for gel-filtration chromatography.

In *S. pombe* and *C. glabrata*, the HMW complexes contain mostly the PC_2 oligomers and both Cd and acid-labile sulfide. Those complexes with a comparatively high ratio of S^{2-}:Cd consist of aggregates of 20-Å-diameter particles that themselves consist of a CdS crystallite core coated with PCs (Dameron et al., 1989; Mehra et al., 1994). The incorporation of sulfide into the HMW complexes increases both the amount of Cd per molecule and the stability of the complex (Reese and Winge, 1988). Genetic evidence for the importance of sulfide in the function of PCs has been obtained from the analysis of Cd-sensitive mutants of *S. pombe* that are deficient in PC–Cd complexes. In one case the gene mutated in the Cd-sensitive derivative, when cloned, was identified as a gene involved in adenine biosynthesis. Subsequent genetic analysis demonstrated that different single or double mutants deficient in steps in the adenine biosynthetic pathway lacked HMW complexes (Speiser et al., 1992b) (Fig. 14-1). Biochemical characterization of the enzymes encoded by these genes indicated that, in addition to catalyzing the conversion of aspartate to intermediates in adenine biosynthesis, this pathway could also utilize cysteine sulfinate, a sulfur-containing analog of aspartate, to form other sulfur-containing compounds that are believed to be intermediates or carriers in the pathway of sulfide incorporation into HMW complexes (Juang et al., 1993). Together these observations confirm the importance of sulfide in the mechanism of PC detoxification of Cd. Whether sulfide is involved in the detoxification of other metal ions by PCs is unknown.

In plants, labile sulfide can also be detected in PC–Cd complexes. Where HMW and LMW complexes have been resolved, in *Brassica juncea* (Speiser et al., 1992a) and tomato (Reese et al., 1992), for example, the S^{2-}:Cd ratio was higher in the HMW complex than in the LMW complex, as had been observed in yeast species. CdS crystallites have been observed in HMW–Cd complexes from tomato (Reese et al., 1992). Interestingly, the *cad* mutants of *Arabidopsis* when exposed to Cd form a brown pigment in their roots and culture medium that on preliminary analysis appears to be a mixture of metal sulfides (C. Dameron, unpublished). In these mutants the overproduction and excretion of sulfide may be a compensatory mechanism resulting in the precipitation of Cd and other metal ions in the absence of PCs.

For metals other than Cd there are few studies demonstrating the formation of PC–metal complexes either in vitro or in vivo. Mehra et al. (1995, 1996a, 1996b) have characterized complexes formed in vitro between PCs and Pb, Hg, Ag, and copper ions. Maitani et al. (1996) have used inductively coupled plasma atomic emission spectroscopy in combination with HPLC separation of native PC–metal complexes in the roots of *Rubia tinctorum*. PCs were induced to varying levels by a wide range of metal ions tested. The most effective appeared to be Ag, As^{3+}, Cd, Cu, Hg, and Pb ions. However, the only PC complexes identified in vivo were with Cd, Ag, and Cu ions. PC complexes formed in response to Pb and As^{3+}, but these complexes contained copper ions and not the metal ion used for induction of synthesis.

Cellular Localization

Vogeli-Lange and Wagner (1990) showed that almost all of both the Cd and PCs accumulated in mesophyll protoplasts derived from tobacco plants exposed to Cd was confined to the vacuole. The importance of sequestration in the vacuole is also supported by studies of the Cd-sensitive mutant of *S. pombe*, *hmt1*, which is unable to form HMW complexes on exposure to Cd. The *hmt1* gene encodes a member of the family of ATP-binding cassette membrane transport proteins that is located in the vacuolar membrane (Ortiz et al., 1992) (Fig. 14-1). Both the HMT1 protein and ATP were required for the transport of both PCs in the absence of Cd and LMW PC–Cd complexes into vacuolar membrane vesicles. The HMT1 protein did not transport Cd alone, and the transport of PCs and PC–Cd complexes was not dependent on the proton gradient established across the vacuolar membrane by the vacuolar proton-ATPase (Ortiz et al., 1995). In plants a similar ATP-dependent, proton gradient-independent activity capable of transporting both PCs and PC–Cd complexes into tonoplast vesicles derived from oat roots has been identified (Salt and Rauser, 1995) (Fig. 14-1). In addition, in both *S. pombe* vacuolar membrane and oat tonoplast vesicles there is a Cd/H^+ antiporter activity that is dependent on the proton gradient but which is distinct from the PC transporter activity (Salt and Wagner, 1993; Ortiz et al., 1995). Furthermore, a vacuolar membrane transporter in yeast, YCF1, has been identified which transports both GSH conjugates and $(GSH)_2Cd$ complexes and is able to confer resistance to Cd when overexpressed (Szczypka et al., 1994; Li et al., 1997) (Fig. 14-1). Homologues of this gene that encode GSH-conjugate transporters have been identified in *Arabidopsis* (Lu et al., 1997, 1998). However, their role in transporting GSH–Cd complexes has not been reported.

Functions of PCs

Whereas both induction of PCs in vivo and activation of PC synthase in vitro is conferred by a range of metal ions, there is only limited evidence supporting the role of PCs in the detoxification of such a wide range of metal ions. GSH-deficient mutants of *S. pombe* are more sensitive than the wild type to Ag, Pb and Hg, in addition to Cd, but not more sensitive to copper ions suggesting that PCs may be required for the detoxification of some but not other heavy metals (Glaeser et al., 1991). Similarly, plant cell cultures exposed to BSO, thereby inhibiting GSH and consequently PC biosynthesis, were rendered sensitive to Cd but not to Cu or Zn (Mendum et al., 1990; Reese and Wagner, 1987). The clearest evidence for the role of PCs in heavy-metal detoxification comes from characterization of the PC-deficient, *cad1-3* mutant of *Arabidopsis*. The sensitivity of *cad1-3* compared to wild-type plants to various heavy metals has been tested by using, in each case, a range of metal ion concentrations increasing in 2- or 2.5-fold steps (C. Cobbett, unpublished data). By comparing a wild type with a mutant one can semiquantitatively estimate the difference in concentration required to give the same degree of growth inhibition. For Cd and the AsO_4 anion, that difference was about 40-fold and 20-fold, respectively, indicating that PCs have a significant role in detoxification of those metal ions. In addition, the mutant was clearly more sensitive to Ag and Pb; however this has not been quantitated. In contrast, the difference for copper ions and Hg was about two-fold, whereas no difference in growth inhibition was observed between mutant and wild-type plants in the presence of Zn, SeO_3, and Ni, indicating that, for these metal ions, PCs have comparatively little or no role in metal detoxification.

From the preceding discussion it is apparent that the mechanism of metal detoxification is more complex than simply the chelation of the metal ion by PCs. The metal ion must activate PC synthase, be chelated by the PCs synthesized, and then, presumably, be transported to the vacuole and possibly form a more complex aggregation in the vacuole with, for example, sulfide or organic acids. In the case of Cd, the inability to carry out any of these steps decreases the capacity for detoxification and thereby confers a sensitive phenotype on the organism. Whether the same sequence of events occurs and is required for other metal ions is not clear. For metal ions that do not appear to be detoxified by PCs in *Arabidopsis*, such as Zn, Ni, and SeO_3, it is not known at which step in the sequence the mechanism is ineffective. For example, in *Arabidopsis* PCs are induced by Zn, presumably because PC synthase is activated by Zn. However, it is unknown whether PCs effectively chelate Zn in vivo or, if formed, whether PC–Zn complexes cannot be sequestered in the vacuole. Alternatively, other metals that are known to induce PC biosynthesis in other plants may be ineffective activators of the *Arabidopsis* PC synthase, and consequently PCs could have no role in their detoxification. It may be that for some metal ions there is an alternative, more effective detoxification mechanism, such as MTs or histidine in the case of Zn and Ni, respectively. Thus even if PCs could detoxify these metal ions, in the presence of an alternative, more effective mechanism, a deficiency in PCs may have no observable effect on detoxification. Whatever the answer to these questions, the metal sensitivity profile of the *cad1-3* mutant has provided the first definitive indication of the in vivo role of PCs in the detoxification of different metals.

One approach to assess the role of PCs in heavy-metal tolerance mechanisms has been to examine both naturally evolved heavy-metal-tolerant plant varieties found growing in soils of high concentrations of metals and laboratory-selected tolerant cell lines able to grow in normally inhibitory concentrations of metal ions. In contrast to the observations of the different *Arabidopsis* or *S. pombe* mutants in which Cd sensitivity results from decreased PC biosynthesis or function, these studies have not demonstrated a clear correlation between increased resistance and increased PC production. Among a number of Cd-tolerant plant cell cultures PCs accumulated at higher levels than in the nontolerant parent culture (Jackson et al., 1987; Steffens et al., 1986; Gupta and Goldsbrough, 1991). Two independently selected Cd-tolerant tomato cell lines have increased GCS activity (Chen and Goldsbrough, 1994; Steffens and Williams, 1989). Since this activity is believed to be the rate-limiting step in GSH biosynthesis, it is possible that this is able to confer tolerance by increasing the synthesis of GSH, and consequently PC biosynthesis. Nevertheless, it is difficult, in the study of cell culture variants, to demonstrate a direct role in a tolerance mechanism for any such differences in physiology or metabolism. Both Cd-tolerant and non-Cd-tolerant cell cultures of *Datura innoxia* accumulated PCs at similar rates but the rate of formation of PC–Cd metal complexes was increased in the tolerant line (Delhaize et al., 1989). In whole plants, Cu tolerance in a naturally selected line of *Mimulus guttatus* appeared to be attributable to increased PC formation (Salt et al., 1989). In contrast, tolerance to Cd, Zn, and Cu ions in different lines of *Silene vulgaris* is not due to increased PC production (Harmens et al., 1993; de Knecht et al., 1994). Indeed, in hydroponic culture conditions, a metal-tolerant ecotype produced lower levels of PCs than did sensitive plants in the presence of the same external concentrations of metal ions.

The observations that PCs can be detected in plant tissues and cell cultures exposed only to trace levels of essential metals and that the level of PCs observed in cell cultures correlates with the depletion of metal ions from the medium have been interpreted to indicate a role for PCs in the homeostasis of essential metal ion metabolism (Grill et al., 1988; Zenk, 1996). In addition, in vitro experiments have shown that PC-Cu and PC-Zn complexes could reactivate the apo forms of the copper-dependent enzyme diamino oxidase and the Zn-dependent enzyme carbonic anhydrase, respectively (Thumann et al., 1991). While these experiments demonstrate that PC–metal complexes are capable of donating metal ions to metal-requiring enzymes, in each case the Cu or Zn complexes were no more effective than the free metal sulfate salt. There is, as yet, no direct evidence that PCs play a role in essential metal homeostasis. Genes encoding many of the components of the mechanism of copper homeostasis in both the yeast *S. cerevisiae* and in mammals have been identified. Genes analogous to these have now been identified in plants, particularly *Arabidopsis*. It is likely that copper homeostasis at the cellular level in plants is not greatly different from that in yeast. While PCs may play some role in this process it is likely that it is a redundant one.

METALLOTHIONEINS

Gene Sequence and Structure

The vast majority of data on the structure of plant MTs comes from DNA sequences of cloned genes. The only MT proteins to have been purified from plants are the wheat

Ec protein (Lane et al., 1987) and a number of MTs from *Arabidopsis* (Murphy et al., 1997). All plant MTs fall into Class II of the MT nomenclature system. Robinson et al. (1993) proposed a further subclassification of plant MTs based on conserved features of amino acid sequences and this has proved useful. Type 1 plant MTs contain a total of 12 cysteine residues present in Cys–Xaa–Cys motifs in two clusters at the amino- and carboxy-termini of the protein. In the majority of Type 1 MTs these two clusters are separated by a spacer of approximately 40 amino acids. However, Type 1 MTs in *Arabidopsis* and *Brassica napus* lack this large spacer and contain an additional cysteine residue. Type 2 MTs also contain amino- and carboxy-terminal clusters of cysteines but are distinguished from Type 1 MTs by the presence of cysteines in Cys–Cys and Cys–Xaa–Xaa–Cys motifs. The majority of plant MT genes described to date can be categorized as either Type 1 or Type 2. However, as more MT genes have been sequenced it has become apparent that not all plant MT genes can be accommodated easily within these groups. On the basis of the predicted amino acid sequences, there appear to be at least two additional categories of MTs present in many plant species. For consistency these will be designated Types 3 and 4. Type 3 MTs have only four cysteine residues in the amino-terminal domain, and at least six in the carboxy terminus. Examples of this type have been identified in Arabidopsis (Murphy et al., 1997), kiwifruit (Ledger and Gardner, 1994) and banana (Clendennen and May 1997). Type 4 MTs contain three distinct cysteine-rich domains and include the wheat Ec protein as well as MTs predicted from genes that have been identified in *Arabidopsis* (Bundithya and Goldsbrough, unpublished observations) and maize (White and Rivin, 1995). A number of other genes have been sequenced that encode proteins with characteristics typical of MTs but that do not fit readily into any of these four groups.

The amino acid sequences of the seven MT genes that are known to be expressed in *Arabidopsis* are shown in Figure 14-2. They are grouped into the four types described above. In the *Arabidopsis* genome, MT Types 1, 2, and 4 are represented by two expressed genes of each type, whereas there is only a single gene encoding a Type 3 MT (Zhou and Goldsbrough, 1995). There is at least one pseudogene related to the Type 1 MTs. The complement of MT genes in *Arabidopsis* is the most extensive described to date for a plant species, in part a result of the various efforts under way to describe the *Arabidopsis* genome in its entirety. However, there is growing evidence that the size and complexity of the *Arabidopsis* MT gene family is not unusual. Several plant species have been shown to contain either multiple MT genes of a single type or more than one type of MT gene. Most animal genomes also contain large MT gene families.

The structure of *Arabidopsis* MT genes suggests that at least some share a common ancestral gene. Types 1 and 2 MT genes contain a single intron located just upstream of the last cysteine codon in the amino-terminal domain. This intron position is also conserved in MT genes from pea, tomato, and cotton. While the intron positions in MT genes of other types are not conserved, there are still striking similarities in the cysteine-rich domains both within individual MTs and between different MTs, suggesting that this domain may have been duplicated initially within a single MT gene (Fig. 14-3). Subsequent gene amplification has produced at least some members

```
MT1a    MADSNCGCGS  SCKCGDSCSC  EKNYNKECDN  CSCGSNCSCG  SNCNC
MT1c    MAGSNCGCGS  SCKCGDSCSC  EKNYNKECDN  CSCGSNCSCG  SSCNC

MT2a    MSCCGGNCGC  GSGCKCGNGC  GGCKMYPDLG  FSGETTTET   FVLGVAPAMK  NQYEASGESN  NAESDACKCG  SDCKCDPCTC  K
MT2b    MSCCGGSCGC  GSACKCGNGC  GGCKRYPDL.  ...ENTATET  LVLGVAPAMN  SQYEASGETF  VAENDACKCG  SDCKCNPCTC  K

MT3     MSSNCGSCDC  ADKTQCVKKG  TSYTFDIVET  QESYKEAMIM  DVGAEENNAN  CKCKCGSSCS  CVNCTCCPN

MT4a    AGCNDSCGCP  SPCPGGNSCR  CRMREASAGE  .QGHMVCPCG  EHCGCNPCNC  PKTQTQTSDK  GCTCGEGCTC  ASCDT
MT4b    ASCNDRCGCP  SPCPGGESCR  CKMMSEASGG  DQEHNTCPCG  EHCGCNPCNC  PKTQTQTSAK  GCTCGEGCTC  ATCAA
```

Figure 14-2 Amino acid sequence alignment of *Arabidopsis* MTs. The predicted amino acid sequences of the seven MT genes that are expressed in *Arabidopsis* are shown. Cysteine residues are shown in bold. The four groups correspond to the types of MT gene found in plants as described in the text. The MT4 group is only expressed in seeds; the others are expressed in vegetative tissues. The initiating methionine of *Arabidopsis* Type 4 MTs has not been unambiguously identified.

of the current MT gene family found in Arabidopsis and other plants. A number of MT genes have been cloned as tightly linked gene clusters. A genomic clone isolated from cotton contained three MT genes within 20 kbp (kilobasepairs) (Hudspeth et al., 1996), and two *Arabidopsis* Type 1 MT genes are separated by less than 3 kbp (Zhou and Goldsbrough, 1995). These observations point to gene duplication as an important process in the evolution of the MT gene family in plants.

Plant Metallothionein Proteins

The wheat embryo Ec protein was the first MT to be purified from a plant tissue and was characterized as a Zn-binding protein (Lane et al., 1987). This was the first demonstration that plants could also produce these metal-binding proteins in addition to phytochelatins. Discovery of the wheat MT protein was followed by the cloning of MT genes from many plant species. A number of these proteins have been expressed in a variety of microbial hosts, as either native proteins or translational fusions. However, until recently there was a conspicuous lack of information on the identification and characterization of MT proteins from vegetative tissues of plants. Antibodies raised against *Arabidopsis* MTs expressed in *E. coli* failed to detect the predicted proteins under standard immunoblotting conditions (Zhou and Goldsbrough, unpublished observations). This led to discussion as to whether the relatively abundant MT RNAs were in fact translated into proteins. The controversy was resolved only when Murphy et al. (1997) purified a number of copper-binding proteins from *Arabidopsis* and showed that these included proteins with the amino acid sequences predicted for MT1a, MT2a, MT2b, and MT3. Antibodies raised against MT1a and MT2a proteins expressed in *E. coli* were used to probe Western blots of these protein fractions and reacted with proteins of the sizes predicted for these MTs. In addition, expression of these proteins followed the same pattern observed for the corresponding mRNAs. MT1 protein was found predominantly in samples

```
MT1a  (N)      CGCGSSCKCGDSCS.C
MT1a  (C)      CSCGSNCSCGSNCN.C
MT2a  (N)      CGCGSGCKCGNGCGGC
MT2a  (C)      CK.GSDCKC.DPCT.C
MT3   (C)      CKCGSSCSCVN.CTCC
MT4a  (N)      CGCPSPCPGGNSCR.C
MT4a  (M)      CPCGEHCGC.NPCN.C
MT4a  (C)      CTCGEGCTC.ASCDT.

Consensus      C-CGS-C-C---C--C
```

Figure 14-3 Homology of cysteine-rich domains in *Arabidopsis* MTs. Amino acid sequences from the amino (N), middle (M), and carboxy (C) regions of representative MTs are aligned with cysteine residues shown in bold. The consensus sequence shows amino acid residues that are conserved in at least six of the domains.

obtained from root tissues whereas MT2 protein was more abundant in leaf tissue but could be induced by treatment of seedlings with copper. In these experiments on the purification and analysis of *Arabidopsis* MTs, precautions were taken to exclude oxygen during the protein isolation and purification procedures. It is likely that the difficulties in detecting these proteins may have arisen because of the instability of MTs and their tendency to oxidize under normal protein isolation conditions.

Although detection of plant MTs has been problematic, a number of alternative approaches have been used to demonstrate that these proteins are able to bind heavy metals. These have relied on expressing plant MT proteins in microbial hosts and either studying the properties of the purified protein or demonstrating an effect of MT expression on metal tolerance of the host organism. The *cup1* gene in *S. cerevisiae* encodes a MT that is required for copper tolerance; deletion of this gene results in a copper-sensitive phenotype. Expression of either *Arabidopsis* MT1a or MT2a under a constitutive high-level promoter was able to restore copper tolerance to this mutant at least as effectively as a mammalian MT (Zhou and Goldsbrough, 1994). The plant MTs were less effective at providing tolerance to cadmium but still provided some increased growth in the presence of this metal. Differences observed between the relative effects of MT1 and MT2 on metal tolerance of yeast might reflect the metal-binding properties of these MTs but could also be accounted for by variation in the level of expression of the introduced genes. Despite the clear effects on metal tolerance, the *Arabidopsis* MTs could not be detected in protein extracts prepared from the transformed yeast strains by immunoblotting, again indicating the difficulties in working with these proteins.

A similar approach was used to demonstrate that MT2a could affect Zn tolerance (Robinson et al., 1996). Deletion of the *Synechococcus* MT results in a Zn-sensitive phenotype. Replacement of the endogenous MT open reading frame with that encoding *Arabidopsis* MT2a was able to partially restore tolerance to Zn. Again, the difference in level of Zn tolerance between the normal and complemented strains may result from expression effects or indicate differences in metal binding between microbial and plant MTs.

There have been few studies aimed at identifying the metals that are coordinated with plant MTs in vivo. The affinity of a pea MT for different metals was assessed after expressing the protein as a GST fusion and measuring the amount of metal released at decreasing pH (Tommey et al., 1991). Under these conditions the pea MT had the greatest affinity for Cu followed by Cd and Zn. Whether these results reflect the affinity of Type 1 MTs for metals in vivo, or the metal affinities of other plant MTs, remains to be established. One explanation for the diverse MT gene family in plants is that different MT types are responsible for binding specific metals. While studies of MTs expressed in a heterologous system can indicate potential metal-binding properties, there may well be additional factors that influence this activity in plant tissues. For example, plant MTs are able to bind Cd in vitro even though it is clearly established that phytochelatins are required for Cd tolerance in plants. The question remains as to whether MTs have any role in the mechanism of tolerance to this metal in plants that express PCs.

Expression of MT Genes in Plants

The problems in detecting MT proteins in plant tissues have meant that essentially all studies on MT gene expression have relied on examining steady-state levels of MT RNAs. While it is important to understand this facet of gene expression, it fails to address the possible involvement of posttranscriptional processes in the function of MT genes. Nevertheless, the one study that was able to detect MT protein expression in *Arabidopsis* indicated no major differences between RNA and protein expression (Murphy et al., 1997). However, this was not an extensive examination of MT protein expression and may not have revealed some of the subtleties of MT gene regulation. With this in mind, we summarize the results of studies of MT RNA expression in the following paragraphs.

Some general observations can be made on the RNA expression of the specific MT types. Under normal growth conditions, Type 1 MT genes are expressed more in roots than in leaves. Types 2 and 3 share similar patterns of expression, with higher RNA levels in leaves than roots. Type 4 MTs, including the prototype Ec from wheat, are expressed in developing seeds. These generalities should be regarded with caution as they may represent the combined expression of more than one MT gene of a given type, and expression of an individual gene may be more specific.

A number of studies have investigated MT RNA expression using methods that allow tissue-specific localization of transcripts. In situ hybridization demonstrated that RNA of a bean Type 2 MT was localized predominantly in trichomes and at a much lower level in vascular tissues of leaves (Foley and Singh, 1994). Similar studies to determine the tissue-specific expression of other MT genes, in terms of both RNA and protein, are required. Reporter gene expression has been used to analyze the activity of the Type 1 $PsMT_A$ promoter in transgenic *Arabidopsis* (Fordham-Skelton et al., 1997). β-Glucuronidase (GUS) activity was higher in roots than in shoot tissues, in agreement with the overall pattern of RNA expression of Type 1 MT genes. The reporter gene was not expressed in the apical region of the root but was active throughout the remainder of the root. In shoots, GUS activity was observed in a variety of tissues, notably those that were undergoing senescence. Similar analysis of promoters from the *Arabidopsis* MT genes that are expressed in vegetative tissues is in progress (Bundithya and Goldsbrough, unpublished observations). The patterns of GUS expression driven by the promoters from both of the active MT1 genes are identical. In roots GUS activity is observed in the root tip and mature region of the root but not in the zones of elongation and differentiation. GUS activity driven by the MT1 promoters is restricted to vascular tissues in young leaves, declines in older leaves, but returns as leaves senesce. In contrast to the similar activities of the MT1 promoters, the MT2a and MT2b promoters have quite distinct patterns of expression. In roots the MT2a promoter is active only in root tips, whereas a low level of GUS expression is observed throughout the leaf and this increases in older leaves. The MT2b promoter is highly active in vascular tissues of roots, leaves, and floral organs, and is also expressed in the abscission zone at the base of the silique. These results indicate that expression of the MT gene family in plants is complex. A sound understanding of the functions of these genes will require not only analysis of individual genes but also more comprehensive examination of the entire gene family.

In addition to studies on the expression of MT genes in specific tissues, many reports have described the RNA expression of MT genes under a variety of environmental conditions. In some cases this reflects how these genes were cloned, by identification of genes whose RNA transcript level increased in response to specific treatments. Conditions that induce MT RNA expression in plants include sugar starvation (Chevalier et al., 1995; Hsieh et al., 1995), aluminum treatment (Snowden et al., 1995), iron deficiency (Okumura et al., 1991; Robinson et al., 1993), heat shock (Hsieh et al., 1995; Murphy and Taiz, 1995), wounding and virus infection (Choi et al., 1996) and copper treatment (Zhou and Goldsbrough, 1994; Hsieh et al., 1995; Snowden et al., 1995; Robinson et al., 1993). Expression of some MT RNAs also increases during embryogenesis (Kawashima et al., 1992; White and Rivin, 1995), fruit development (Ledger and Gardner, 1994; Cledennen and May, 1997), and leaf senescence (Buchanan-Wollaston, 1994; Coupe et al., 1995). The variety of conditions that have been shown to alter the abundance of MT RNAs is both surprising and intriguing. However, it is difficult to identify a single characteristic of these diverse conditions that can provide a unifying explanation for these results. It is worthwhile to emphasize here that MT gene expression in animals is influenced by a similarly diverse range of conditions (Kägi, 1991). Two aspects of plant MT gene expression are reviewed in more detail here: the responses to various metals and senescence.

Zhou and Goldsbrough (1994) demonstrated that RNA expression of MT genes in *Arabidopsis* could be induced by copper, and to a lesser degree by zinc and cadmium, but only for genes whose expression was relatively low in the absence of the inducing metal ion. It has since been shown that MT2a is the most highly responsive to copper and that MT protein expression is also induced under these conditions (Zhou and Goldsbrough, 1995; Murphy et al., 1997). Expression of MT2a RNA was also shown to be elevated in a copper-sensitive mutant of *Arabidopsis* that accumulated excess copper (van Vliet et al., 1995). Another indication that MT gene expression can be modulated in response to copper came from a comparative study of different *Arabidopsis* genotypes (Murphy and Taiz, 1995). Variation in copper tolerance among genotypes, measured by root growth inhibition, was highly correlated with the expression of MT2a RNA in copper-treated plants. Metal-induced expression of MT genes has also been reported in both maize and rice (Chevalier et al., 1995; Hsieh et al., 1995). A wheat MT gene, initially identified as being induced under aluminum stress, was subsequently shown to be induced by copper and cadmium as well as by calcium deficiency (Snowden et al., 1995). Iron deficiency has also been demonstrated to increase expression of MT genes in barley and pea (Okumura et al., 1991; Robinson et al., 1993). One explanation for this response is that activation of the iron reduction and transport system under conditions of iron deficiency results in uptake not only of iron but of a number of other ions, including copper, leading to elevated MT gene expression (Robinson et al., 1993). Although these examples demonstrate that MT genes can be regulated by metals such as copper, many other studies have shown no effect of these metals on MT RNA expression. A possible explanation for these results is that the studies were looking at a member of the MT gene family that is either nonresponsive to metals in general, or at least not in the tissues used for these experiments. In *Arabidopsis*, MT2a is the most copper-responsive MT gene, and this response is greatest in seedlings. These observations highlight the need to look at expression of the family of MT genes in these processes rather than assuming that a single gene is representative of the entire MT gene family.

Expression of MT genes has also been observed in senescing tissues. This was first reported for a *Brassica napus* Type 1 MT whose expression increased during leaf senescence (Buchanan-Wollaston, 1994) and similar results were obtained for a MT gene in *Sambucus nigra* (Coupe et al., 1995). Promoters of MT genes from *Arabidopsis* and pea have also been shown to have increased activity in senescing leaves (see discussion above). Buchanan-Wollaston (1994) proposed that expression of MT genes in senescing tissues may be required for chelation of metals released after degradation of proteins. Senescence is not merely the degeneration of a tissue but is instead a highly regulated process that allows recovery and recycling of resources from that tissue. Uncontrolled release of metal ions after hydrolysis of macromolecules could promote damage to cellular components, thereby disrupting the senescence program. Expression of MTs in senescing tissues would allow these metal ions to be chelated, reducing their reactivity and allowing the controlled mobilization of nutrients to continue before the tissue dies. The activity of *Arabidopsis* MT1 promoters in vascular tissues of senescing leaves may indicate that MTs are also involved in metal ion transport during this physiological process.

Functions of Metallothioneins

The amino acid sequences of plant MTs indicate that these proteins are likely involved in some aspects of metal metabolism in plants. When expressed in a variety of microbes, plant MTs function as metal-binding proteins that can mediate metal tolerance. The arrangements of cysteine residues in different MT types might influence the metals that are bound by MTs, and this could provide one explanation for the complexity of the MT gene family in plants. For example, Type 1 MTs might be involved primarily with metabolism of copper whereas Type 2 MTs are required for zinc. The expression of MT genes in response to a number of environmental and developmental conditions can also be interpreted as reflecting the involvement of these genes in metal homeostasis. However, direct evidence that MTs are required for a specific function, in metal metabolism, tolerance, or another process, is currently lacking and is likely to be forthcoming only when we can analyze the phenotypes of plants that lack individual or multiple MT genes. Strategies for manipulating the expression of MTs include antisense RNA or gene silencing through cosuppression. Identification of plants with insertions in MT genes that abolish expression is also feasible in *Arabidopsis* and some other species. There may be redundancy among MT genes, which will further complicate these efforts to define specific functions for MTs in plants. The potential functions for MTs include serving as an accessible storage pool for metal ions within the cell and the transport of metals within the cell from the plasmamembrane to other sites where metals are utilized, for example, incorporated into proteins. Although there are no data to indicate that MTs are targeted to specific organelles, this possibility highlights the need for more information about the localization of MTs within the cell and the components that interact with MTs.

CONCLUSION

A role for PCs in the detoxification of some heavy metals, particularly Cd, is clearly established. Nevertheless, the complexity of factors contributing to PC biosynthesis and function is becoming increasingly evident. To what extent manipulation of the expression of any single one of these factors can influence PC function and the capacity to detoxify heavy metals remains to be established. To date only a few of the genes controlling PC biosynthesis and function have been isolated. Clearly the isolation of the PC synthase gene will be a significant advance and is an obvious candidate for manipulating plants for increased capacity for phytoremediation.

Is there a role for MTs in phytoremediation? If MTs are demonstrated to function in the uptake, transport, compartmentation, and final storage of essential metals in normal metabolism, then MTs may perform the same functions with toxic metals that have chemical properties similar to those of copper and zinc. However, it is unlikely that MTs will function as the final ligand that protects the plant from the high concentrations of metals that are envisioned to accumulate in plants used for pytoremediation. Expression of a protein at the required level is likely to be difficult and, with the high cysteine content of MTs, would place a major demand on sulfur uptake and reduction. In addition, MTs appear designed to remain in the cytosol, whereas it would be preferable to partition toxic metals to an extracellular location or to the vacuole. Modification of MT structure or expression might enable these proteins to fulfill specific roles in plants used for phytoremediation. A number of studies have examined the effects of ectopic expression of MT genes from animals and yeast in plants, with some showing increased tolerance to cadmium. However, this strategy to increase metal tolerance still faces the problems of producing large amounts of a cysteine-rich protein. While MTs are likely involved in metal metabolism in plants, our limited knowledge about their function makes any prediction about how they may be utilized in phytoremediation highly speculative at this time.

REFERENCES

Alscher, R. G., 1989. Biosynthesis and antioxidant function of glutathione in plants. *Physiol. Planta* **77:**457–464.

Buchanan-Wollaston, V., 1994. Isolation of cDNA clones for genes that are expressed during leaf senescence in *Brassica napus*. Identification of a gene encoding a senescence-specific metallothionein-like protein. *Plant Physiol.* **105:**839–846.

Chen, J., and P. B. Goldsbrough, 1994. Increased activity of γ-glutamylcysteine synthetase in tomato cells selected for cadmium tolerance. *Plant Physiol.* **106**:233–239.

Chen, J. et al., 1997. Characterization of phytochelatin synthase from tomato. *Physiol. Plant* **101**:165–172.

Chevalier, C. et al., 1995. Molecular cloning and characterization of six cDNAs expressed during glucose starvation in excised maize (*Zea mays* L.) root tips. *Plant Mol. Biol.* **28**:473–485.

Choi, D. et al., 1996. Molecular cloning of a metallothionein-like gene from *Nicotiana glutinosa* L. and its induction by wounding and tobacco mosaic virus infection. *Plant Physiol.* **112**:353–359.

Clendennen, S. K., and G. D. May, 1997. Differential gene expression in ripening banana fruit. *Plant Physiol.* **115**:463–469.

Cobbett, C. S. et. al., 1998. The glutathione-deficient, cadmium-sensitive mutant, *cad2-1*, of *Arabidopsis thaliana* is deficient in γ-glutamylcysteine synthetase. *Plant J.* **16**:73–78.

Coupe, S. A. et al., 1995. Characterization of an mRNA encoding a metallothionein-like protein that accumulates during ethylene-promoted abscission of *Sambucus nigra* L. leaflets. *Planta* **197**:442–447.

Dameron, C. T. et al., 1989. Biosynthesis of cadmium sulfide quantum semiconductor crystallites. *Nature* **338**:596–597.

de Knecht, J. A. et al., 1994. Phytochelatins in cadmium-sensitive and cadmium-tolerant *Silene vulgaris*: Chain length distribution and sulfide incorporation. *Plant Physiol.* **104**:255–261.

Delhaize, E. et al., 1989. Poly(γ-glutamylcysteinyl)glycine synthesis in *Datura innoxia* and binding with cadmium. *Plant Physiol.* **89**:700–706.

Foley, R. C., and K. B. Singh, 1994. Isolation of a *Vicia faba* metallothionein-like gene. Expression in foliar trichomes. *Plant Mol. Biol.* **26**:435–444.

Fordham-Skelton, A. P. et al., 1997. GUS expression in *Arabidopsis* directed by 5′ regions of a pea metallothionein-like gene, *PsMT$_A$*. *Plant Mol. Biol.* **34**:659–668.

Gekeler, W. et al., 1989. Survey of the plant kingdom for the ability to bind heavy metals through phytochelatins. *Zeit. Naturfor. Sec. C Biosci.* **44**:361–369.

Glaeser, H. et al., 1991. Glutathione metabolism and heavy metal detoxification in *Schizosaccharomyces pombe*. *Current Genet.* **19**:207–213.

Grill, E. et al., 1985. Phytochelatins, the principal heavy-metal complexing peptides of higher plants. *Science* **230**:674–676.

Grill, E. et al., 1986. Homo-phytochelatins are heavy-metal-binding peptides of homo-glutathione containing Fabales. *FEBS Lett.* **205**:47–50.

Grill, E. et al., 1987. Phytochelatins, a class of heavy-metal-binding peptides from plants are functionally analogous to metallothioneins. *Proc. Natl. Acad. Sci.* (USA) **84**:439–443.

Grill, E. et al., 1988. Induction of heavy-metal binding phytochelatins by innoculation of cell cultures in standard media. *Plant Cell Rep.* **7**:375–378.

Grill, E. et al., 1989. Phytochelatins, the heavy-metal-binding peptides of plants, are synthesized from glutathione by a specific γ-glutamylcysteine dipeptidyl transpeptidase (phytochelatin synthase). *Proc. Natl. Acad. Sci.* (USA) **86**:6838–6842.

Grill, E. et al., 1991. Phytochelatins. *Meth. Enzymol.* **205**:333–341.

Gupta, S. C., and P. B. Goldsbrough, 1991. Phytochelatin accumulation and stress tolerance in tomato cells exposed to cadmium. *Plant Cell Rep.* **9**:466–469.

Harmens, H. et al., 1993. Increased zinc tolerance in *Silene vulgaris* (Moench) Garcke is not due to increased production of phytochelatins. *Plant Physiol.* **103:**1305–1309.

Hayashi, Y. et al., 1991. Two pathways in the biosynthesis of cadystins (γ-EC)$_n$G. *Biochem. Cell. Biol.* **69:**115–121.

Hell, R., and L. Bergmann, 1988. Glutathione synthetase in tobacco suspension cultures: Catalytic properties and localization. *Physiol. Planta* **72:**70–76.

Hell, R., and L. Bergmann, 1990. γ-Glutamylcysteine synthetase in higher plants: Catalytic properties and subcellular localization. *Planta* **180:**603–612.

Howden, R., and C. S. Cobbett, 1992. Cadmium-sensitive mutants of *Arabidopsis thaliana*. *Plant Physiol.* **100:**100–107.

Howden, R. et al., 1995a. Cadmium-sensitive, *cad1*, mutants of *Arabidopsis thaliana* are phytochelatin deficient. *Plant Physiol.* **107:**1059–1066.

Howden, R. et al., 1995b. A cadmium-sensitive, glutathione-deficient mutant of *Arabidopsis thaliana*. *Plant Physiol.* **107:**1067–1073.

Hsieh, H.-M. et al., 1995. A novel stress-inducible metallothionein-like gene from rice. *Plant Mol. Biol.* **28:**381–389.

Hudspeth, R. L. et al., 1996. Characterization and expression of metallothionein-like genes in cotton. *Plant Mol. Biol.* **31:**701–705.

Jackson, P. J. et al., 1987. Poly(γ-Glutamylcysteinyl)glycine: Its role in cadmium resistance in plant cells. *Proc. Natl. Acad. Sci.* (USA) **84:**6619–6623.

Juang, R.-H. et al., 1993. Two purine biosynthetic enzymes that are required for cadmium tolerance in *Schizosaccharomyces pombe* utilize cysteine sulfinate in vitro. *Arch. Biochem. Biophys.* **304:**392–401.

Kägi, J. H. R., 1991. Overview of metallothionein. *Meth. Enzymol.* **205:**613–626.

Kawashima, I. et al., 1992. Wheat E$_c$ metallothionein genes. *Eur. J. Biochem.* **209:**971–976.

Klapheck, S., 1988. Homoglutathione: Isolation, quantification and occurrence in legumes. *Physiol. Plant* **74:**727–732.

Klapheck, S. et al., 1988. Properties and localization of the homoglutathione synthetase from *Phaseolus coccineus*. *Physiol. Plant* **74:**733–739.

Klapheck, S. et al., 1994. Hydroxymethyl-phytochelatins [(γ-Glutamylcysteine)$_n$-serine] are metal induced peptides of the Poaceae. *Plant Physiol.* **104:**1325–1332.

Klapheck, S. et al., 1995. Synthesis of phytochelatins and homo-phytochelatins in *Pisum sativum* L. *Plant Physiol.* **107:**515–521.

Kneer, R., and M. H. Zenk, 1992. Phytochelatins protect plant enzymes from heavy metal poisoning. *Phytochemistry* **31:**2663–2667.

Kneer, R. et al., 1992. *Saccharomyces cerevisiae* and *Neurospora crassa* contain heavy metal sequestering phytochelatin. *Arch. Microbiol.* **157:**305–310.

Kondo, N. et al., 1983. Structure of cadystin, the unit peptide of cadmium-binding peptides induced in the fission yeast *Schizosaccharomyces pombe*. *Tetrahed. Lett.* **24:**925–928.

Lane, B. et al., 1987. The wheat-germ E$_c$ protein is a zinc-containing metallothionein. *Biochem. Cell Biol.* **65:**1001–1005.

Law, M. Y., and B. Halliwell, 1986. Purification and properties of glutathione synthetase from spinach (*Spinacia oleracea*) leaves. *Plant Sci.* **4:**185–191.

Ledger, S. E., and R. C. Gardner, 1994. Cloning and expression of five cDNAs for genes differentially expressed during fruit development of kiwifruit (*Actinidia deliciosa* var. deliciosa). *Plant Mol. Biol.* **25**:877–886.

Li, Z.-S. et al., 1997. A new pathway for vacuolar cadmium sequestration in *Saccharomyces cerevisiae*: YCF1-catalyzed transport of *bis*(glutathionato)cadmium. *Proc. Natl. Acad. Sci.* (USA) **94**:42–47.

Loeffler, S. et al., 1989. Termination of the phytochelatin synthase reaction through sequestration of heavy metals by the reaction product. *FEBS Lett.* **258**:42–46.

Lu, Y.-P. et al., 1997. *AtMRP1* of *Arabidopsis thaliana* encodes a glutathione *S*-conjugate pump: Isolation and functional definition of a plant ATP-binding cassette transporter gene. *Proc. Natl. Acad. Sci.* (USA) **94**:8243–8348.

Lu, Y.-P. et al., 1998. *AtMRP2*, an *Arabidopsis* ATP binding cassette transporter able to transport glutathione *S*-conjugates and chlorophyll catabolites: Functional comparisons with *AtMRP1*. *The Plant Cell* **10**:267–282.

Maitani, T. et al., 1996. The composition of metals bound to class III metallothionein (phytochelatin and its desglycyl peptide) induced by various metals in root cultures of *Rubia tinctorum*. *Plant Physiol.* **110**:1145–1150.

May, M. J., and C. J. Leaver, 1994. *Arabidopsis thaliana* γ-glutamylcysteine synthetase is structurally unrelated to mammalian, yeast, and *Escherichia coli* homologs. *Proc. Natl. Acad. Sci.* (USA) **91**:10059–10063.

May, M. J. et al., 1998. Glutathione homeostasis in plants: Implications for environmental sensing and plant development. *J. Exptl. Bot.* **49**:649–667.

Mehra, R. K. et al., 1988. Metal-specific synthesis of two metallothioneins and γ-glutamyl peptides in *Candida glabrata*. *Proc. Natl. Acad. Sci.* (USA) **85**:8815–8819.

Mehra, R. K. et al., 1994. Role of CdS quantum crystallites in cadmium resistance in *Candida glabrata*. *Biochem. Biophys. Res. Commun.* **200**:1193–1200.

Mehra, R. K. et al., 1995. Chain-length-dependent Pb(II)-coordination in phytochelatins. *Biochem. Biophys. Res. Commun.* **215**:730–736.

Mehra, R. K. et al., 1996a. Optical spectroscopic and reverse phase HPLC analyses of Hg(II) binding to phytochelatins. *Biochem. J.* **314**:73–82.

Mehra, R. K. et al., 1996b. Ag(I)-binding to phytochelatins. *J. Inorg. Biochem.* **61**:125–142.

Mendum, M. L. et al., 1990. Effect of glutathione on phytochelatin synthesis in tomato cells. *Plant Physiol.* **93**:484–488.

Meuwly, P. et al., 1995. Three families of thiol peptides are induced by cadmium in maize. *Plant J.* **7**:391–400.

Morelli, E., and E. Pratesi, 1997. Production of phytochelatins in the marine diatom *Phaeodactylum tricornutum* in response to copper and cadmium exposure. *Bull. Environ. Contam. Toxicol.* **59**:657–664.

Murphy, A., and L. Taiz, 1995. Comparison of metallothionein gene expression and nonprotein thiols in ten *Arabidopsis* ecotypes. *Plant Physiol.* **109**:945–954.

Murphy, A. et al., 1997. Purification and immunological identification of metallothioneins 1 and 2 from *Arabidopsis thaliana*. *Plant Physiol.* **113**:1293–1301.

Mutoh, N., and Y. Hayashi, 1988. Isolation of mutants of *Schizosaccharomyces pombe* unable to synthesize cadystin, small cadmium-binding peptides. *Biochem. Biophys. Res. Commun.* **151**:32–39.

Noctor, G., and C. H. Foyer, 1998. Ascorbate and glutathione: Keeping active oxygen under control. *Annu. Rev. Plant Physiol. Plant Mol. Biol.* **49:**249–279.

Okumura, N. et al., 1991. An iron deficiency-specific cDNA from barley roots having two homologous cysteine-rich MT domains. *Plant Mol. Biol.* **17:**531–533.

Ortiz, D. F. et al., 1992. Heavy-metal tolerance in the fission yeast requires an ATP-binding cassette-type vacuolar membrane transporter. *EMBO J.* **11:**3491–3499.

Ortiz, D. F. et al., 1995. Transport of metal-binding peptides by HMT1, a fission yeast ABC-type vacuolar membrane protein. *J. Biol. Chem.* **270:**4721–4728.

Rauser, W. E., 1990. Phytochelatins. *Annu. Rev. Biochem.* **59:**61–86.

Rauser, W. E., 1991. Cadmium-binding peptides from plants. *Meth. Enzymol.* **205:**319–333.

Rawlins, M. et al., 1995. Characterization of a cDNA encoding *Arabidopsis* glutathione synthetase. *FEBS Lett.* **376:**81–86.

Reese, R. N., and G. J. Wagner, 1987. Effects of buthionine sulfoximine on Cd-binding peptide levels in suspension-cultured tobacco cells treated with Cd, Zn, or Cu. *Plant Physiol.* **84:**574–577.

Reese, R. N., and D. R. Winge, 1988. Sulfide stabilization of the cadmium-γ-glutamyl peptide complex of *Schizosaccharomyces pombe*. *J. Biol. Chem.* **263:**12832–12835.

Reese, R. N. et al., 1992. Cadmium sulfide crystallites in Cd-(γ-EC)$_n$G peptide complexes from tomato. *Plant Physiol.* **98:**225–229.

Robinson, N. J. et al., 1993. Plant metallothioneins. *Biochem. J.* **295:**1–10.

Robinson, N. J. et al., 1996. Expression of the type 2 metallothionein-like gene *MT2* from *Arabidopsis thaliana* in Zn^{2+}-metallothionein deficient *Synechococcus* PCC 7942. Putative role for *MT2* in Zn^{2+}-metabolism. *Plant Mol. Biol.* **30:**1169–1179.

Salt, D. E., and W. E. Rauser, 1995. MgATP-dependent transport of phytochelatins across the tonoplast of oat roots. *Plant Physiol.* **107:**1293–1301.

Salt, D. E., and G. J. Wagner, 1993. Cadmium transport across tonoplast of vesicles from oat roots: Evidence for a $Cd^{2(+)}/H^{(+)}$ antiport activity. *J. Biol. Chem.* **68:**12297–12302.

Salt, D. E. et al., 1989. Copper phytochelatins of *Mimulus guttatus*. *Proc. Roy. Soc. Lond. B Biol. Sci.* **236:**79–89.

Scheller, H. B. et al., 1987. Phytochelatin synthesis and glutathione levels in response to heavy metals in tomato cells. *Plant Physiol.* **85:**1031–1035.

Snowden, K. C. et al., 1995. Aluminium-induced genes. Induction by toxic metals, low calcium, and wounding and pattern of expression in root tips. *Plant Physiol.* **107:**341–348.

Speiser, D. M. et al., 1992a. *Brassica juncea* produces a phytochelatin-cadmium-sulfide complex. *Plant Physiol.* **99:**817–821.

Speiser, D. M. et al., 1992b. Purine biosynthetic genes are required for cadmium tolerance in *Schizosaccharomyces pombe*. *Mol. Cell. Biol.* **12:**5301–5310.

Steffens, J. C. et al., 1986. Accumulation of non-protein metal-binding polypeptides (γ-glutamyl-cysteinyl)$_n$-glycine in selected cadmium-resistant tomato cells. *J. Biol. Chem.* **261:**13879–13882.

Steffens, J. C., and B. G. Williams, 1989. Increased activity of γ-glutamylcysteine synthetase in DMSO-permeabilized cadmium-resistant plant cells. In *Metal Ion Homeostasis: Molecular Biology and Biochemistry*, D. H. Hamer and D. R. Winge, eds., Alan R. Liss, New York, pp. 359–366.

Szczypka, M. S. et al., 1994. A yeast metal resistance protein similar to human cystic fibrosis transmembrane conductance regulator (CFTR) and multidrug resistance-associated protein. *J. Biol. Chem.* **269:**22853–22857.

Thumann, J. et al., 1991. Reactivation of metal-requiring apoenzymes by phytochelatin-metal complexes. *FEBS Lett.* **284:**66–69.

Tommey, Λ. M. et al., 1991. Expression of the pea gene *PsMT_A* in *E. coli. FEBS Lett.* **292:**48–52.

Ullmann, P. et al., 1996. Cloning of the *Arabidopsis thaliana* glutathione synthetase (GSH2) by functional complementation of a yeast *gsh2* mutant. *Eur. J. Biochem.* **236:**662–669.

van Vliet, C. et al., 1995. Copper-sensitive mutant of *Arabidopsis thaliana. Plant Physiol.* **109:**871–878.

Vogeli-Lange, R., and G. J. Wagner, 1990. Subcellular localization of cadmium and cadmium-binding peptides in tobacco leaves. Implication of a transport function for cadmium-binding peptides. *Plant Physiol.* **92:**1086–1093.

Wagner, G. J., 1984. Characterization of a cadmium-binding complex of cabbage leaves. *Plant Physiol.* **76:**797–-805.

Wang, C. L., and D. J. Oliver, 1996. Cloning of the cDNA and genomic clones for glutathione synthetase from *Arabidopsis thaliana* and complementation of a *gsh2* mutant in fission yeast. *Plant Mol. Biol.* **31:**1093–1104.

White, C. N., and C. J. Rivin, 1995. Characterization and expression of a cDNA encoding a seed-specific metallothionein in maize. *Plant Physiol.* **108:**831–832.

Zenk, M. H., 1996. Heavy metal detoxification in higher plants—a review. *Gene* **179:**21–30.

Zhou, J., and P. B. Goldsbrough, 1994. Functional homologs of fungal metallothionein genes from *Arabidopsis. Plant Cell* **6:**875–884.

Zhou, J., and P. B. Goldsbrough, 1995. Structure, organization and expression of the metallothionein gene family in *Arabidopsis. Mol. Gen. Genet.* **248:**318–328.

15

MOLECULAR MECHANISMS OF ION TRANSPORT IN PLANT CELLS

Mary Lou Guerinot

INTRODUCTION

Soils in many places around the world are now polluted with heavy metals such as Zn, Cd, Cu, Ni, or Co. In addition to posing a threat to human health, these metals can poison plants, livestock, and wildlife. The effective use of plants to clean up these metal-contaminated sites relies on the simple premise that plants already "know" how to extract metals from soils. Indeed, some plants are known to hyperaccumulate metals. By studying how plants transport metals, we hope to be able to design plants for particular phytoremediation applications.

Metal ions need to be transported from the soil solution into the root and then distributed throughout the plant, crossing both cellular and organellar membranes. Transport across the plant plasma membrane is driven by an electrochemical gradient of protons generated by plasma membrane H^+-ATPases. These primary transporters pump protons out of the cell, thereby creating pH and electrical potential differences across the plasma membrane. Secondary transport systems then utilize these gradients for many functions, including nutrient uptake. A number of genes involved in ion transport in plants have already been identified on the basis of function, via complementation of yeast mutants, or on the basis of sequence similarity. Not surprisingly, many of these genes belong to previously described transporter families including those encoding P-type ATPases (Axelsen and Palmgren, 1997) and Nramp proteins (Cellier et al., 1995). ZIP, a novel cation transporter family first identified in plants (Eide et al., 1996), seems to be ubiquitous; members of this family are also

Phytoremediation of Toxic Metals: Using Plants to Clean Up the Environment, edited by Ilya Raskin and Burt D. Ensley.
ISBN: 0-471-19254-6 ©2000 John Wiley & Sons, Inc.

found in trypanosomes, yeast, nematodes, and humans. With cloned genes in hand, it now becomes possible not only to carry out structure–function studies to address how various transporters move ions but also to investigate how transporters respond to changing ion levels. It is important to keep in mind that many ions, although essential, can also be toxic when present in excess [e.g., see Halliwell and Gutteridge (1992)]. Thus, plants may have tightly regulated uptake and efflux systems that enable them to control precisely their intracellular ion concentrations. Such systems may have to be circumvented in order for phytoremediation efforts with transgenic plants to succeed.

This chapter briefly reviews progress on the identification of plant genes involved in the transport of various heavy metals and what we know about the expression of these genes in plants. It is not yet prossible to review any studies on the use of such cloned plant genes in phytoremediation efforts. Hopefully, such studies lie ahead in the not-too-distant future. This chapter also discusses *Arabidopsis* mutants with interesting heavy-metal transport phenotypes. It will not be long before the entire genomic sequence of *Arabidopsis* becomes available (for an up-to-date look at the progress of the sequencing effort, see http://genome-www3.stanford.edu/atdb_welcome.html). Thus, as various mutants of *Arabidopsis* with metal accumulation defects are identified and mapped, we will soon be in a position to quickly know the molecular nature of the gene marked by the mutation.

COPPER

Genes Involved in Copper Transport

Functional expression cloning in yeast has been instrumental in identifying heavy-metal transport proteins from both plants and animals. This method involves construction of a cDNA library from the organism of interest in a yeast expression vector and screening this library for complementation of a yeast mutant defective for heavy-metal uptake. A putative copper transporter from *Arabidopsis*, COPT1, was isolated in this manner because its expression suppressed the growth defects of a yeast *ctr1-3* strain that lacks high-affinity copper uptake (Kampfenkel et al., 1995). Moreover, overexpression of *COPT1* in yeast conferred hypersensitivity to copper in the medium. Although copper uptake was not examined directly, the available data suggest that *COPT1* encodes a copper transport protein. In *Arabidopsis*, *COPT1* is expressed in flowers, stems, and leaves but is undetectable in roots; the lack of root expression may indicate that *COPT1* is not responsible for copper uptake from soil. The response of *COPT1* to copper deficiency has yet to be examined. Several ESTs in the database encode proteins with similarity to COPT1, suggesting that *COPT1* may be part of a small gene family in *Arabidopsis*.

Functional complementation of a yeast mutant does not necessarily identify proteins with amino acid similarity to the gene product encoded by the mutated yeast gene. However, COPT1 is 49% similar to Ctr1 and 56% similar to a newly described human copper transporter, hCTR1. h*CTR1*, like *COPT1*, was identified by its ability to rescue a yeast *ctr1* mutant (Zhou and Gitschier, 1997). COPT1, Ctr1p, and hCTR1 all have three potential transmembrane domains, and each contains an *N*-terminal putative metal-binding domain rich in methionine and serine residues. This metal-binding domain is predicted to lie on the extracellular surface and is similar to those found in several bacterial copper-binding proteins, including the *Enterococcus hirae* Cu ATPase (CopB) and the CopA and CopB proteins from *Pseudomonas syringae* (Silver and Phung, 1996). Two other genes in yeast [*CTR2* and *CTR3* (Kampfenkel et al., 1995; Knight et al., 1996)] and one other gene in humans [h*CTR2* (Zhou and Gitschier, 1997)] are similar to *COPT1*. Ctr2p and hCTR2 were originally identified as being similar to COPT1 and to hCTR1, respectively, in database searches (Kampfenkel et al., 1995). Neither *CTR2* nor h*CTR2* rescues a *ctr1-3* mutant, and neither has a recognizable metal-binding motif (Kampfenkel et al., 1995; Zhou and Gitschier, 1997). However, *CTR2* overexpression confers sensitivity to copper, and *ctr2* mutants are more resistant to copper, suggesting that Ctr2p functions as a low-affinity copper transporter in yeast. Ctr3p is a small, integral membrane protein identified by its ability to restore high-affinity copper uptake to a *ctr1-3*-deficient yeast mutant (Knight et al., 1996). Mutations in both Ctr1p and Ctr3p are required to completely eliminate high-affinity copper uptake. Interestingly, in many laboratory strains of yeast, *CTR3* is interrupted by a transposable element (Knight et al., 1996).

What ionic form of copper is actually a substrate for transport into cells via Ctr-like transporters? In yeast, high-affinity copper uptake requires plasma membrane reductases to reduce Cu(II) to Cu(I) (Hassett and Kosman, 1995). The reductases responsible for this are encoded by the *FRE* genes (Dancis et al., 1992; Georgatsou and Alexandraki, 1994; Martins et al., 1998). These reductases were originally identified because they reduce Fe(III) to Fe(II), a necessary step for iron transport in yeast (see discussion below). Whether plants need to reduce Cu(II) before transport is not yet clear. Genes encoding Fe(III) reductases have now been identified in *Arabidopsis* as well as mutants with altered Fe(III) reductase activities. This work is described below in the section on iron.

In addition to understanding how a heavy metal such as copper is taken into cells, we also need to understand how heavy metals are distributed intracellularly. Copper delivery within cells is mediated by soluble cytoplasmic proteins called "copper chaperones" [for a brief overview, see Valentine and Gralla (1997)]. These chaperones bind copper after it enters the cell and subsequently donate the bound copper to recipient protein(s) (Pufahl et al., 1997). Using such chaperones to deliver copper makes sense, given the potential toxicity of free copper ions in the cytoplasm. In *S. cerevisiae*, three such copper chaperones have been identified that transport copper from the plasma membrane transporter Ctr1p to three different cellular locations: Cox17p guides copper to the mitochondria for insertion into cytochrome *c* oxidase (Glerum et al., 1996), Lys7p targets copper to Cu–Zn superoxide dismutase (Culotta et al., 1997), and Atx1p directs copper to a post-Golgi compartment by way of Ccc2p, a P-type ATPase, for insertion into Fet3p, a multicopper oxidase essential for high affinity iron uptake (Lin et al., 1997).

So, how are these chaperones thought to function? Atx1p contains the highly conserved metal-binding motif MTCXXC that functions as the mercury binding site in the *Escherichia coli* MerP and MerA proteins and as the putative copper binding site for the copper ATPases from *E. hirae* (CopA), *S. cerevisiae* (Ccc2p), and humans (the Wilson and Menkes proteins). Pufahl et al. (1997) proposed that Cu(I) binds to this site in apo-Atx1p and the complex diffuses through the cytoplasm. After encountering its target, Cu-Atx1p associates with the metal binding domain of Ccc2p, which is identical to that found in Atx1p. The Cu(I) then undergoes a series of exchange reactions, moving from Atx1p to Ccc2p. Several plant ESTs, from *Arabidopsis* and rice, encode peptides having greater than 60% identity to Atx1p; each contain a presumptive metal-binding motif, MXCXXC. Although the role of these Atx1p-like proteins in metal homeostasis in plants has not yet been established, their similarity to Atx1p suggests a similar function and makes these genes interesting targets for further investigation. Recently, a functional homolog of *ATX1* was identified in Arabidopsis that is up-regulated 7-fold in leaves undergoing senescence (Himelblau et al., 1998). There are also several *Arabidopsis* sequences encoding peptides with similarity to Cox17p and Lys7p.

As mentioned above, Atx1p is thought to deliver its copper to Ccc2p, a Cu-ATPase that is the yeast homolog of Menkes and Wilson proteins in humans. It is worth reviewing how heavy-metal P-type ATPases such as Ccc2 are thought to function. P-type ATPases comprise a ubiquitous family of proteins involved in the active pumping of charged substrates across biological membranes (Møller et al., 1996). Their distinguishing feature is the formation of a phosphorylated intermediate during the reaction cycle (hence P-type) that involves shifting between two conformational states: E1, which represents the enzyme with a high affinity for ATP and the cation to be transported and E2, which has a low affinity for ATP and the cation. The Wilson protein has recently been shown to complement a *ccc2* mutant of yeast, restoring copper incorporation into the multicopper oxidase Fet3p (Hung et al., 1997). This remarkable conservation of function suggests that all eukaryotes probably have similar systems for intracellular copper transport. Both Menkes protein and Wilson protein are thought to pass copper from the cytoplasm into the secretory system. There is an *Arabidopsis* gene, recently sequenced as part of the genome effort, that has up to 80% similarity to Ccc2p and to the human Ccc2 homologues, Menkes protein and Wilson protein (ML Guerinot, unpublished data). Recently, the RAN1 gene was shown to encode a protein with similarity to copper transporting P-type ATPases, including yeast Ccc2 and the human Menkes/Wilson proteins (Hirayama et al., 1999). RAN1 can complement a yeast ccc2 mutant, demonstrating that it can function as a copper transporter. Most interestingly, RAN1 is required for ethylene signaling in *Arabidopsis*, suggesting that RAN1 delivers copper ions to create functional ethylene receptors. Using degenerate oligonucleotides based on residues conserved among metal-transporting P-type ATPases, an *Arabidopsis* cDNA has been identified encoding a protein, PAA1, which shows similarity to Cu-ATPases. PAA1 is 42.6% identical to the PacS Cu ATPase from *Synechococcus* and 37.9% identical to the Wilson Cu-ATPase (Tabata et al., 1997). It also has an *N*-terminal MTCXXC metal-binding motif. *PAA1* transcripts are not very abundant in either roots or shoots and are not copper-inducible.

Arabidopsis Mutants with Altered Copper Transport Properties

There are now a number of *Arabidopsis* mutants with altered sensitivities to copper. The mutant *cup1-1* is significantly more sensitive to Cu than wild type (van Vliet et al., 1995). This mutant is not deficient in phytochelatin biosynthesis or function, so its sensitivity cannot be attributed to an inability to sequester metals. As the mutant accumulates more Cu (and Cd) than wild type in its tissues, the sensitive phenotype may to be due to increased accumulation of Cu. Consistent with this, a copper-inducible, root-specific metallothionein gene, *MT2a*, is expressed in *cup1-1* roots under conditions in which it is not expressed in wild type (van Vliet et al., 1995). *cup1-1* could conceivably be a transport mutant in which Cu uptake is deregulated relative to wild type. It is also possible that *cup1-1* may have a defect in Fe metabolism. The *cup1-1* mutant is chlorotic, and this phenotype can be at least partly reversed by increasing the concentration of Fe in the medium. There is a documented relationship between iron and copper metabolism that has been seen in both yeast and plants (see text below). Murphy and Taiz (1995) have described 59 EMS, Cu-sensitive (*cus*) mutants of *Arabidopsis*. When grown with 30 μM CuCl$_2$, *cus* mutants develop marked toxicity symptoms whereas the growth of wild-type plants is actually stimulated at this concentration of copper. A dose response curve showed that one of the mutants, *cus1*, is only sensitive to copper in the 10–40-μM range, suggesting that the *cus1* mutation affects a mechanism that operates at low metal concentrations. Such a mechanism might be high-affinity copper transport. This pool of mutants is a rich source to be "mined" for genes involved in metal homeostatsis in plants.

IRON

Genes Involved in Iron Transport in Strategy I Plants

All plants except the grasses rely on a reductive mechanism to solubilize rhizosphere Fe(III), which is basically unavailable in aerobic environments at neutral pH (Guerinot and Yi, 1994). This strategy (termed *Strategy I*) is thought to be an obligatory step in iron uptake and is similar to the system used by *S. cerevisiae*. The initial reduction of Fe(III) is carried out by a plasma-membrane-bound Fe(III) chelate reductase. The resulting Fe(II) is then transported across the root epidermal cell membrane. Both the Fe(III) chelate reductase (Yi and Guerinot, 1996) and the Fe(II) transport activities (Fox et al., 1996) are enhanced under Fe deficiency. Similar results have also been seen with the green alga, *Chlamydomonas reinhardtii* (Eckhardt and Buckhout, 1998), suggesting that *Chlamydomonas* may prove to be another excellent model system in which to dissect iron uptake.

Genes encoding Fe(III) chelate reductases have been cloned from *S. cerevisiae* [*FRE1* (Dancis et al., 1992), *FRE2* (Georgatsou and Alexandraki, 1994)] and *S. pombe* [*Frp1* (Roman et al., 1993)]. Using degenerate PCR with primers designed against motifs common to the yeast Fe(III) reductase proteins, several genes (*froh*) that encode

Fe(III) reductases have been identified in *Arabidopsis* (Robinson et al., 1999). Each of these *froh* genes encodes a *b*-type cytochrome belonging to a larger family whose other members include the respiratory burst oxidase of mammalian neutrophils (gp91-phox).

In yeast, Fre1p and Fre2p also mediate Cu(II) reduction. Whereas there is good evidence that Fe is required in its reduced form for uptake by plants (Chaney et al., 1972; Yi and Guerinot, 1996), it is still not clear in plants whether Cu(II) needs to be reduced before transport and whether Fe(III) chelate reductases also reduce Cu(II) chelates. Because *Arabidopsis* has a copper transporter (COPT1) similar to the yeast Cu transporter Ctr1p that uses Cu(I) as a substrate, it may indeed turn out that plants also reduce Cu(II) before transport. Copper deficiency has been shown to induce Fe(III) chelate reductase activity in pea plants; deficiencies of other cations (K, Mg, Ca, Mn, and Zn) do not elicit a similar response (Cohen et al., 1997).

An *Arabidopsis* gene encoding a presumptive Fe(II) transporter, *IRT1* (iron-regulated transporter), was isolated because its expression in yeast could restore iron-limited growth to a yeast *fet3fet4* mutant defective in iron uptake (Eide et al., 1996). Consistent with its proposed role as a metal ion transporter, yeast cells expressing *IRT1* possess a novel iron uptake system that is specific for Fe(II) over Fe(III). Moreover, IRT1 is specific for iron over other potential substrates; Fe(II) uptake was not greatly inhibited by high concentrations of other physiologically relevant metal ions such as Cu(I), Cu(II), Mn(II), and Zn(II). Most interestingly, Cd has been shown to inhibit iron uptake by IRT1. This suggests that Cd may serve as a substrate for this transporter. This will need to be tested directly using radiolabeled Cd. In *Arabidopsis*, *IRT1* is expressed in roots and is induced by iron-deficient growth conditions. Based on these results, we proposed that IRT1 is an Fe(II) transporter that takes up iron from the soil. Interestingly, IRT1 bears no resemblance to either the high-affinity (Ftr1p) or the low-affinity (Fet4p) Fe transporters in yeast.

The significance of IRT1 in the field of metal uptake research is twofold. *IRT1* is the first Fe transporter gene to be isolated from plants and it provides a useful handle on the mechanism and regulation of Fe uptake in plants. Second, *IRT1* has led to the discovery of a new family of transporters involved in metal ion uptake. We have named this group the *ZIP* gene family (for *ZRT/IRT*-related proteins) for the first three members to be isolated and characterized: *ZRT1*, *ZRT2*, and *IRT1*. *ZRT1* and *ZRT2* encode Zn transporters in *S. cerevisiae* (Zhao and Eide, 1996a, 1996b). *ZIP* family genes are found in a diverse array of eukaryotic organisms. On the basis of our studies of *IRT1* and the two *ZRT* genes, we propose that the other genes in this family also function as metal transporters (see section below on zinc for more about *ZIP* genes).

Arabidopsis **Mutants with Altered Iron Transport Activities**

Arabidopsis mutants, *frd1* and *frd3*, that exhibit defects in Fe(III) reduction have been identified (Yi, 1995; Yi and Guerinot, 1996). *frd1* mutants do not show induction of Fe(III) chelate reductase activity under iron deficiency. Furthermore, *frd1* mutants have also lost the ability to reduce Cu(II) chelates. These mutants also do not translocate radiolabeled iron to the shoots when roots are presented with a tightly chelated form of Fe(III). This confirms that iron must be reduced before it can be transported. We now know that one of the *FRO* genes identified by PCR, *FRO2*, is allelic to *frd1* (Robinson et al., 1999). *frd3* mutants express Fe(III) reductase activity under both iron-deficient and iron-sufficient conditions. They also show constitutive expression of the presumptive iron transporter, *IRT1* (Eide et al., 1996). Taken together, these data suggest that *frd3* mutants may be regulatory mutants.

Genes Involved in Iron Transport in Strategy II Plants

Unlike Stategy I plants, Strategy II plants (the grasses) release phytosiderophores, low-molecular-weight Fe(III)-specific ligands, in response to iron deficiency. These molecules are nonproteinogenic amino acids synthesized from methionine via nicotianamine (NA) to give mugineic acids that efficiently chelate Fe(III) with their amino and carboxyl groups (Ma and Nomoto, 1996). A similar chelating strategy is used by a wide variety of bacteria and fungi (Guerinot, 1994). The Fe(III)–mugineic acid complexes are then internalized by specific transport systems that have yet to be fully characterized.

Efforts are under way to clone the genes involved in phytosiderophore synthesis and transport. The first step in phytosiderophore synthesis is the formation of NA. This reaction is catalyzed by NA synthase, which condenses three molecules of *S*-adenosylmethionine to form NA. All plants, not just grasses, synthesize NA, which is thought to function as a metal-chelating agent (Stephan et al., 1996); thus, this particular enzymatic reaction is not limited to the grasses. Genes encoding *S*-adenosylmethionine synthetase have been cloned from a number of plant species, including *Arabidopsis*. The gene encoding NA synthase has now been cloned from barley (Higuchi et al., 1999) and from tomato (Ling et al., 1999). After formation of NA, NA aminotransferase (NAAT) is then thought to transfer an amino group to produce an unstable intermediate that is rapidly reduced to form deoxymugineic acid. cDNAs from barley encoding NAAT have been identified; NAAT is strongly induced under iron deficiency (Takahashi et al., 1997). Using a subtractive hybridization approach, two other barley genes, *ids2* and *ids3*, which may function in synthesis of mugineic acid have been identified (Nakanishi et al., 1993; Okumura et al., 1994). Each encodes a protein with some similarity to 2-oxoglutarate dioxygenases, making them good candidates for the conversion of deoxymugineic acid and mugineic acid to epihydroxymugineic acid via hydroxylation.

To identify the Fe(III)-mugineic acid transporter, a yeast *ctr1* mutant that is unable to grow on iron deficient media was transformed with a barley cDNA expression library and clones that could use Fe(III)–mugineic acid as an iron source were isolated. One clone, designated *SFD1* [suppressor of ferrous (uptake) defect], can restore the ability of a *ctr1* mutant of yeast to grow on iron-deficient media when either Fe(III) mugineic acid or Fe(III) citrate are provided (Yamaguchi et al., 1997). SFD1 has no similarity to any protein of known function, and the exact mechanism by which the growth arrest is bypassed remains to be determined.

MANGANESE

Genes Involved in Manganese Transport

Members of the Nramp (natural-resistance associated macrophage protein) family have now been implicated in Mn transport in *S. cerevisiae* (Supek et al., 1996) and in cation transport in mammals (Fleming et al., 1997; Gunshin et al., 1997). Nramp proteins have 10 conserved transmembrane domains, a glycosylated loop (between TM6 and TM7) and a sequence signature (TMT[X]$_4$G[D/Q[X]$_4$GF in the TM8 to TM9 interval) that shares similarity with the permeation pore of the K^+ channel family (Cellier et al., 1996). Originally, Nramp1 was identified in mice because a mutation in this gene leads to susceptibility to intracellular pathogens such as *Mycobacterium*, *Salmonella*, and *Listeria*. If all members of the Nramp family function as metal transporters, this could nicely explain the original observations on Nramp1 [see Gunshin et al. (1997) and Supek et al. (1997) for models]. Briefly, Nramp is proposed to transport certain divalent cations, possibly Mn and/or Fe, from the extracellular milieu into the cytoplasm of a macrophage and after the generation of a phagosome removes these divalent cations from the organelle, thus depriving the invading microorganism of divalent cations needed for production of defense enzymes such as superoxide dismutase (Mn) and catalase (Fe). It is also possible that Fe^{2+} uptake by macrophages may allow production of toxic hydroxyl radicals via the Fenton reaction, killing pathogens in the phagosome as part of the defense mechanism.

Two ESTs from *Oriza sativa* and one from *Arabidopsis* that encode proteins with strong similarity to mammalian Nramp were used as hybridization probes to isolate three different cDNAs from rice (Belouchi et al., 1997). Northern blots indicate that *OsNramp1* is expressed primarily in roots whereas *OsNramp2* is expressed primarily in leaves. *OsNramp3* is expressed in both tissues. Recently, genome sequencing efforts have identified two genes encoding NRAMP-like proteins in *Arabidopsis*, one located on chromosome II and one on chromosome V. The two predicted proteins are 91% similar to each other, 83% similar to *OsNramp2* and 77% similar to Nramp from mouse (M. L. Guerinot, unpublished). As yet, there is no functional proof of what substrate(s) the plant Nramp proteins are transporting. We would also like to know whether any of the plant Nramp proteins respond to a deficiency of any particular metal. For example, dietary Fe deficiency upregulates the expression of *DCT1*, the rat isoform of human Nramp2, possibly via an iron-responsive element found in the 3′ UTR of *DCT1* (Gunshin et al., 1997). DCT1 has an unusually broad substrate range that includes Fe, Mn, Co, Cd, Cu, Ni, and Pb. It is not known whether other dietary deficiencies can also upregulate the expression of this transporter.

Compared to copper transport, we know very little at present about how manganese is transported intracellularly in any system. There is accumulating evidence from both yeast and *Arabidopsis* that Ca and Mn may be substrates for the same intracellular transporter (Lapinskas et al., 1995; Liang et al., 1997). One such transporter, *ECA1*, was originally isolated from an *Arabidopsis* cDNA library using a putative Ca^{2+} ATPase gene from tobacco as a hybridization probe (Liang et al., 1997). ECA1 was then further shown to rescue the Ca^{2+} pumping defect of a *pmr1* mutant of *S. cerevisiae*, PMR1 is a P-type, Ca^{2+} ATPase that localizes to the Golgi and has properties distinct from sarco/endoplasmic reticulum and plasma membrane calcium pumps (Sorin et al., 1997). ECA1 could also restore the growth of the *pmr1* mutant in Mn-containing medium, suggesting that this Ca^{2+} pump may also catalyze Mn transport. Mn has been shown to stimulate the formation of a phosphorylated intermediate of ECA1 in microsomes prepared from yeast expressing ECA1. Mn is likely to be sequestered in endomembrane compartments; ECA1 may be responsible for transport of both Ca and Mn into the lumen of the ER or the Golgi. Indeed, Pmr1p has been implicated in supplying both Mn and Ca to the Golgi in yeast (Lapinskas et al., 1995). An *Arabidopsis* line with a T-DNA insertion in *ACA3* (identical to *ECA1*) can now be used to examine the role of this pump in plants (J. Harper, personal communication). For example, one would predict that such a line would be more sensitive than wild type to high levels of Mn.

Arabidopsis Mutants with Altered Manganese Transport Activities

A mutant of *Arabidopsis*, *man1*, has been identified that accumulates Mn (7.5× than WT) as well Cu (4.6× than WT), Zn (2.8× than WT), and Mg (1.8× than WT) (Delhaize, 1996). This mutant also has constitutively high levels of Fe(III) chelate reductase in its roots. Interestingly, the *frd1* mutant which does not show induction of Fe(III) chelate reductase, also accumulates Mn and Zn but does not appear to accumulate any other cations (Yi and Guerinot, 1996). In addition to accumulating cations, expression of a presumptive Fe(II) transporter, *IRT1*, appears to be constitutive in the roots of *man1* plants (Guerinot, unpublished), suggesting that this mutant may have a defect in metal regulation. The inappropriate expression of *IRT1* may lead to the accumulation of a variety of cations.

ZINC

Genes Involved in Zinc Transport

Zinc is taken up from the soil solution as a divalent cation (Marschner, 1995). Once taken up, Zn is neither oxidized nor reduced; thus, the role of Zn in cells is based on its behavior as a divalent cation that has a strong tendency to form tetrahedral complexes [for a review, see Berg and Shi (1996)]. Despite the importance of Zn as an essential micronutrient for plant growth, relatively few studies have examined the mechanisms and regulation of Zn absorption by roots. Currently, there is little agreement on whether Zn enters via ion channels or via a divalent cation carrier and whether there is a link between uptake and metabolic energy transduction (Kochian, 1993).

Using a method similar to the one used to isolate *IRT1*, we isolated the *ZIP1, ZIP2*, and *ZIP3* genes of *Arabidopsis* by functional expression cloning in a *zrt1zrt2* mutant yeast strain; expression of these genes in yeast restored Zn-limited growth to this strain (Grotz et al., 1998). Biochemical analysis of metal uptake has demonstrated that these genes encode Zn transporters. Yeast expressing *ZIP1, ZIP2*, and *ZIP3* each have time-, temperature-, and concentration-dependent Zn uptake activity with apparent K_m values between 10 and 100 nM Zn(II). These values are similar to the levels of free Zn available in the rhizosphere (Norvell and Welch, 1993). Moreover, no Fe uptake activity has been detected with any of these proteins in uptake experiments using [55]Fe. We propose that each of these three genes plays a role in Zn transport in the plant. These represent the first Zn transporter genes to be cloned from any plant species.

All ZIP proteins are predicted to have eight transmembrane domains. Several of the transmembrane domains are amphipathic and are predicted to have a histidine residue on the hydrophilic side of the helix that is absolutely conserved among all family members. For example, Figure 15-1 shows a helical wheel representation of

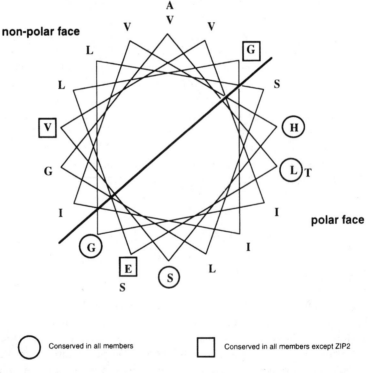

Figure 15-1   A helical wheel representation (as complied by the GCG program Helicalwheel) of residues 189–209 of IRT1, predicted by TOP PREDII to form a membrane spanning α helix. Residues that are identical in all seven family members are circled and those that are conserved in all except ZIP2 are boxed. Each residue is offset from the preceding one by 100°, the typical angle of rotation for an α helix.

transmembrane domain IV from *IRT1*. We are planning to mutagenize conserved residues that appear to cluster on the polar faces of the transmembrane helices; such residues could form an intramembranous heavy-metal binding site.

ZIP proteins range from 309 to 476 amino acids in length; this difference is due largely to the length between transmembrane domains III and IV. This region, which we have designated the "variable region," is particularly intriguing because in all but two members, this domain contains a histidine-rich motif that may serve as a metal-binding site. Similar domains have also been found in the Zn efflux transporters ZRC1 (Kamizono et al., 1989), COT1 (Conklin et al., 1992), ZnT-1 (Palmiter and Findley, 1995), and ZnT-2 (Palmiter et al., 1996). ZRC1, COT1, ZnT-1 and ZnT-2 all belong to a family of metal transport proteins designated the CDF (cation diffusion facilitator) family (Paulsen and Saier, 1997).

Central to the characterization of any new family of transport proteins is determining what provides the thermodynamic driving force to move the substrate in the desired direction. Evidence to date suggests that ZIP proteins are carriers rather than channels. This hypothesis is based primarily on the observation that low substrate concentrations saturate these transporters (apparent K_m values of 10–100 nM free metal ion); channel proteins generally require millimolar concentrations of substrates to saturate uptake [for a review, see Stein (1990)]. However, direct evidence for the carrier nature of these proteins is not yet available.

The driving force for metal uptake by ZIP proteins may be provided by direct energy input (e.g. ATP hydrolysis), coupling to the transport of another substrate (secondary active transport) or facilitated diffusion. ZIP proteins do not appear to utilize either primary or secondary active transport mechanisms. They do not contain ATPase domains, nor do they appear to require other subunits that could provide direct energy coupling. This hypothesis is supported by the ability of ZIP proteins to function when expressed in heterologous systems. *Arabidopsis* ZIP proteins are functional when expressed in yeast and, in the case of *IRT1*, *Xenopus* oocytes (D. Eide, unpublished). Thus, no other *Arabidopsis* proteins are required for their function. (However, this does not preclude the possibility that these proteins are interacting with heterologous subunits.) Secondary active transport of a metal ion would be coupled to either anion cotransport or proton/cation antiport. This also appears to be an unlikely mechanism for ZIP proteins; Fe(II) uptake by IRT1 is not inhibited by high concentrations (100 mM) of extracellular K^+, Ca^{2+}, or Na^+, nor does it require specific extracellular anions. Instead, we propose that these proteins mediate metal ion uptake by facilitated diffusion. Given the cationic nature of their substrates, the driving force of uptake would be provided by both the concentration gradient of the substrate as well as the membrane potential of the plasma membrane (Stein, 1990). The facilitated diffusion model of ZIP protein function is supported by our observation that iron uptake by yeast expressing *IRT1* is inhibited in cells that are starved for glucose or treated with an agent that selectively depolarizes the plasma membrane.

Recently, a gene, *ZAT1*, encoding a protein with homology to mammalian zinc efflux proteins was identified in *Arabidopsis*. Transgenic plants engineered to overexpress *ZAT1* exhibited enhanced zinc resistance and had higher levels of zinc in their roots, compared to wild type plants (van der Zaal et al., 1999).

SUMMARY

With many cloned genes already in hand, the obvious challenge now is to decipher the role of each transporter encoded by these genes. For most transporters, physiologically important information such as expression pattern and mechanism of regulation is still lacking. Various molecular approaches ultimately can tell not only in what tissue and cell types certain transporters are expressed but also where within a cell each is expressed. They can also indicate whether gene expression is directly influenced by changes in cation concentration. We are also now in a position to identify plant mutants carrying insertions in particular transporter genes (Krysan et al., 1996; McKinney et al., 1995); this will greatly help in assigning functions. Having cloned genes is also allowing us to undertake structure–function studies on the encoded proteins themselves. Many transporter activities have been well characterized at the electrophysiological level. Our ability to now combine such information with structural information about the proteins will hopefully lead to an understanding of the molecular mechanisms of transport. Finally, moving beyond how any one transporter functions, we need to keep in mind that ultimately we want to understand cation transport at the whole-plant level and to use such knowledge to create plants that bioaccumulate toxic cations such as cadmium and lead.

ACKNOWLEDGMENTS

The author is grateful to Rob McClung, Erin Connolly, Tama Fox, Natasha Grotz, and Elizabeth Rogers for constructive comments. Work in the author's laboratory is supported by grants from the National Science Foundation (IBN-9318093) and the Department of Energy (07-97ER20292).

REFERENCES

Axelsen, K. B., and M. G. Palmgren, 1997. Evolution of substrate specificities in the P-type ATPase superfamily. *J. Mol. Evol.* **46**:84–101.

Belouchi, A. et al., 1997. Cloning and characterization of the *OsNramp* family from *Oryza sativa*, a new family of membrane proteins possibly implicated in the transport of metal ions. *Plant Mol. Biol.* **33**:1085–1092.

Berg, J. M., and Y. Shi, 1996. The galvanization of biology: A growing appreciation for the roles of zinc. *Science* **271**:1081–1085.

Cellier, M. et al., 1996. Resistance to intracellular infections: Comparative genomic analysis of *Nramp*. *Trends Genet.* **12**:201–204.

Cellier, M. et al., 1995. Nramp defines a family of membrane proteins. *Proc. Natl. Acad. Sci. (USA)* **92**:10089–10093.

Chaney, R. L. et al., 1972. Obligatory reduction of ferric chelates in iron uptake by soybeans. *Plant Physiol.* **50**:208–213.

Cohen, C. K. et al., 1997. Induction of the root cell plasma membrane ferric reductase. *Plant Physiol.* **114:**1061–1069.

Conklin, D. S. et al., 1992. *COT1*, a gene involved in cobalt accumulation in *Saccharomyces cerevisiae*. *Mol. Cell. Biol.* **12:**3678–3688.

Culotta, V. C. et al., 1997. The copper chaperone for superoxide dismutase. *J. Biol. Chem.* **272:**23469–23472.

Dancis, A. et al., 1992. Ferric reductase of *Saccharomyces cerevisiae*: Molecular characterization, role in iron uptake, and transcriptional control by iron. *Proc. Natl. Acad. Sci.* (USA) **89:**3869–3873.

Delhaize, E., 1996. A metal-accumulator mutant of *Arabidopsis thaliana*. *Plant Physiol.* **111:**849–855.

Eckhardt, U., and T. J. Buckhout, 1998. Iron assimilation in *Chlamydomonas reinhardtii* involves ferric reduction and is similar to Strategy I higher plants. *J. Exper. Bot.* **9:**1219–1226.

Eide, D. et al., 1996. A novel iron-regulated metal transporter from plants identified by functional expression in yeast. *Proc. Natl. Acad. Sci.* (USA) **93:**5624–5628.

Fleming, M. D. et al., 1997. Microcytic anaemia mice have a mutation in *Nramp2*, a candidate iron transporter gene. *Nat. Genet.* **16:**383–386.

Fox, T. C. et al., 1996. Direct measurement of [59]labeled Fe^{2+} influx in roots of *Pisum sativum* using a chelator buffer system to control free Fe^{2+} in solution. *Plant Physiol.* **111:**93–100.

Georgatsou, E., and D. Alexandraki, 1994. Two distinctly regulated genes are required for ferric reduction, the first step of iron uptake in *Saccharomyces cerevisiae*. *Mol. Cell. Biol.* **14:**3065–3073.

Glerum, D. M. et al., 1996. Characterization of *COX17*, a yeast gene involved in copper metabolism and assembly of cytochrome oxidase. *J. Biol. Chem.* **271:**14504–14509.

Grotz, N. et al., 1998. Identification of a family of zinc transporter genes from *Arabidopsis* that respond to zinc dificiency. *Proc. Natl. Acad. Sci.* (USA) **95:**7220–7224.

Guerinot, M. L., 1994. Microbial iron transport. *Annu. Rev. Microbiol.* **48:**743–772.

Guerinot, M. L., and Y. Yi, 1994. Iron: Nutritious, noxious, and not readily available. *Plant Physiol.* **104:**815–820.

Gunshin, H. et al., 1997. Cloning and characterization of a mammalian proton-coupled metal-ion transporter. *Nature* **388:**482–488.

Halliwell, B., and J. M. C. Gutteridge, 1992. Biologically relevant metal ion-dependent hydroxyl radical generation. *FEBS Lett.* **307:**108–112.

Hassett, R., and D. J. Kosman, 1995. Evidence for Cu(II) reduction as a component of copper uptake by *Saccharomyces cerevisiae*. *J. Biol. Chem.* **270:**128–134.

Higuchi, K. et al., 1999. Cloning of nicotianamine synthase genes, novel genes involved in the biosynthesis of phytosiderophores. *Plant Physiol.* **119:**471–479.

Himelblau, E. et al., 1998. Identification of a functional homolog of the yeast copper homeostasis gene *ATX1* from *Arabidopsis*. *Plant Physiol.* **117:**1227–1234.

Hirayama, T. et al., 1999. Responsive-to-antagonist1, a Menkes/Wilson disease-related copper transporter, is required for ethylene signaling in *Arabidopsis*. *Cell* **97:**383–393.

Hung, J. H. et al., 1997. Biochemical characterization of the Wilson disease protein and functional expression in the yeast *Saccharomyces cerevisiae*. *J. Biol. Chem.* **272:**21461–21466.

Kamizono, A. et al., 1989. Identification of a gene conferring resistance to zinc and cadmium ions in the yeast *Saccharomyces cerevisiae*. *Mol. Gen. Genet.* **219**:161–167.

Kampfenkel, K. et al., 1995. Molecular characterization of a putative *Arabidopsis thaliana* copper transporter and its yeast homologue. *J. Biol. Chem.* **270**:28479–28486.

Knight, S. A. B. et al., 1996. A widespread transposable element masks expression of a yeast copper transport gene. *Genes Devel.* **10**:1917–1929.

Kochian, L., 1993. Zinc absorption from hydroponic solutions by plant roots. In *Zinc in Soils and Plants*, A. D. Robson, ed., Kluwer, Dordrecht and Boston, pp. 45–57.

Krysan, P. J. et al., 1996. Identification of transferred DNA insertions within *Arabidopsis* genes involved in signal transduction and ion transport. *Proc. Natl. Acad. Sci.* (USA) **93**:8145–8150.

Lapinskas, P. J. et al., 1995. Mutations in *PMR1* suppress oxidative damage in yeast cells lacking superoxide dismutase. *Mol. Cell Biol.* **15**:1382–1388.

Liang, F. et al., 1997. *ECA1* complements yeast mutants defective in Ca^{2+} pumps and encodes an endoplasmic reticulum-type Ca^{2+}-ATPase in *Arabidopsis thaliana*. *Proc. Natl. Acad. Sci.* (USA) **94**:8579–8584.

Lin, S.-J. et al., 1997. A role for the *Saccharomyces cerevisiae ATX1* gene in copper trafficking and iron transport. *J. Biol. Chem.* **272**:9215–9220.

Ling, H. Q. et al., 1999. Map-based cloning of *chloronerva*, a gene involved in iron uptake of higher plants encoding nicotianamine synthase. *Proc. Natl. Acad. Sci. U.S.A.* **96**:7098–7103.

Ma, J. F., and K. Nomoto, 1996. Effective regulation of iron acquisition in graminaceous plants. The role of mugineic acids as phytosiderophores. *Physiol. Plant.* **97**:609–617.

Marschner, H., 1995. *Mineral Nutrition of Higher Plants*, 2nd ed. Academic Press, Boston.

Martins, L. J. et al., 1998. Metalloregulation of *FRE1* and *FRE2* homologs in *Saccharomyces cerevisiae*. *J. Biol. Chem.* **273**:23716–23721.

McKinney, E. C. et al., 1995. Sequence-based identification of T-DNA insertion mutations in *Arabidopsis*: Actin mutants *act2-1* and *act4-1*. *Plant J.* **8**:613–622.

Møller, J. V. et al., 1996. Structural organization, ion transport and energy transduction of P-type ATPases. *Biochim. Biophys. Acta* **1286**:1–51.

Murphy, A., and L. Taiz, 1995. A new vertical mesh transfer technique for metal-tolerance studies in *Arabidopsis*. *Plant Physiol.* **108**:29–38.

Nakanishi, H. et al., 1993. Expression of a gene specific for iron deficiency (*Ids3*) in the roots of *Hordeum vulgare*. *Plant Cell Physiol.* **34**:401–410.

Norvell, W. A., and R. M. Welch, 1993. Growth and nutrient uptake by barley (*Hordeum vulgare* L. cv. Herta): Studies using an *N*-(2-hydroxyethyl)ethylenedinitrilotriacetic acid-buffered nutrient solution technique. I. Zinc ion requirements. *Plant Physiol.* **101**:619–625.

Okumura, N. et al., 1994. A dioxygenase gene (*Ids2*) expressed under iron deficiency conditions in the roots of *Hordeum vulgare*. *Plant Mol. Biol.* **25**:705–719.

Palmiter, R. D. et al., 1996. ZnT-2, a mammalian protein that confers resistance to zinc by facilitating vesicular sequestration. *EMBO J.* **15**:1784–1791.

Palmiter, R. D., and S. D. Findley, 1995. Cloning and functional characterization of a mammalian zinc transporter that confers resistance to zinc. *EMBO J.* **14**:639–649.

Paulsen, L. T., and M. H. Saier, 1997. A novel family of ubiquitous heavy metal ion transport proteins. *J. Membrane Biol.* **156**:99–103.

Pufahl, R. A. et al., 1997. Metal ion chaperone function of the soluble Cu(I) receptor Atx1. *Science* **278**:853–856.

Robinson, N. J. et al., 1999. A ferricchelate reductase for iron uptake from soils. *Nature* **397**:694–697.

Roman, D. G. et al., 1993. The fission yeast ferric reductase gene *frp1*⁺ is required for ferric iron uptake and encodes a protein that is homologous to the gp91-*phox* subunit of the human NADPH phagocyte oxidoreductase. *Mol. Cell. Biol.* **13**:4342–4350.

Silver, S., and L. T. Phung, 1996. Bacterial heavy metal resistance: New surprises. *Annu. Rev. Microbiol.* **50**:753–789.

Sorin, A. et al., 1997. PMR1, a Ca^{2+}-ATPase in yeast golgi, has properties distinct from sarco/endoplasmic reticulum and plasma membrane calcium pumps. *J. Biol. Chem.* **272**:9895–9901.

Stein, W. D., 1990. *Channels, Carriers, and Pumps. An Introduction to Membrane Transport.* Academic Press, San Diego.

Stephan, U. W. et al., 1996. The nicotianamine molecule is made-to-measure for complexation of metal micronutrients in plants. *BioMetals* **9**:84–90.

Supek, F. et al., 1996. A yeast manganese transporter related to the macrophage protein involved in conferring resistance to mycobacteria. *Proc. Natl. Acad. Sci.* (USA) **93**:5105–5110.

Supek, F. et al., 1997. Function of metal-ion homeostasis in the cell division cycle, mitochondrial protein processing, sensitivity to mycobacterial infection and brain function. *J. Exper. Biol.* **200**:321–330.

Tabata, K. et al., 1997. Cloning of a cDNA encoding a putative metal-transporting P-type ATPase from *Arabidopsis thaliana. Biochim. Biophys. Acta* **1326**:1–6.

Takahashi, M. et al., 1997. Cloning and sequencing of nicotianamine aminotransferase gene (*Naat*)—a key enzyme for the synthesis of mugineic acid-family phytosiderophores. *Proc. 9th Internatl. Symp. Iron Nutrition and Interaction in Plants.*

Valentine, J. S., and E. B. Gralla, 1997. Delivering copper inside yeast and human cells. *Science* **278**:817–818.

van der Zaal, E. J. et al., 1999. Overexpression of a zinc transporter gene from *Arabidopsis* can lead to enhanced zinc resistance and zinc accumulation. *Plant Physiol.* **119**:1–9.

van Vliet, C. et al., 1995. Copper-sensitive mutant of *Arabidopsis thaliana. Plant Physiol.* **109**:871–878.

Yamaguchi, H. et al., 1997. Isolation and characterization of a barley cDNA clone which restores growth defect of yeast ferrous uptake mutant, *ctr1. Proc. 9th Internatl. Symp. Iron Nutrition and Interaction in Plants*, Stuttgart, Germany.

Yi, Y., 1995. *Iron Uptake in Arabidopsis thaliana.* Ph.D. thesis, Dartmouth College, Hanover, NH.

Yi, Y., and M. L. Guerinot, 1996. Genetic evidence that induction of root Fe(III) chelate reductase activity is necessary for iron uptake under iron deficiency. *Plant J.* **10**:835–844.

Zhao, H., and D. Eide, 1996a. The yeast *ZRT1* gene encodes the zinc transporter of a high affinity uptake system induced by zinc limitation. *Proc. Natl. Acad. Sci.* (USA) **93**:2454–2458.

Zhao, H., and D. Eide, 1996b. The *ZRT2* gene encodes the low affinity zinc transporter in *Saccharomyces cerevisiae. J. Biol. Chem.* **271**:23203–23210.

Zhou, B., and J. Gitschier, 1997. *hCTR1*: A human gene for copper uptake identified by complementation in yeast. *Proc. Natl. Acad. Sci.* (USA) **94**:7481–7486.

INDEX